CHITIN

CHITIN

by
RICCARDO A. A. MUZZARELLI

Faculty of Medicine, University of Ancona,
Ancona, Italy 60100

PERGAMON PRESS

OXFORD · NEW YORK · TORONTO · SYDNEY · PARIS · FRANKFURT

U.K.	Pergamon Press Ltd., Headington Hill Hall, Oxford OX3 0BW, England
U.S.A.	Pergamon Press Inc., Maxwell House, Fairview Park, Elmsford, New York 10523, U.S.A.
CANADA	Pergamon of Canada Ltd., 75 The East Mall, Toronto, Ontario, Canada
AUSTRALIA	Pergamon Press (Aust.) Pty. Ltd., 19a Boundary Street, Rushcutters Bay, N.S.W. 2011, Australia
FRANCE	Pergamon Press SARL, 24 rue des Ecoles, 75240 Paris, Cedex 05, France
WEST GERMANY	Pergamon Press GmbH, 6242 Kronberg-Taunus, Pferdstrasse 1, Frankfurt-am-Main, West Germany

First edition 1977

REF
QD
321
.M93
1977
Cop 1

Library of Congress Cataloging in Publication Data

Muzzarelli, R A A 1937-
Chitin.
Bibliography: p.
Includes index.
1. Chitin. 2. Chitosan. I. Title.
QD321.M93 1977   547'.84   76-52421
ISBN 0-08-020367-1

In order to make this volume available as economically and rapidly as possible the author's typescript has been reproduced in its original form. This method unfortunately has its typographical limitations but it is hoped that they in no way distract the reader.

Printed in Great Britain by A. Wheaton & Co., Exeter

*This book is dedicated to
my son Corrado
(who is just learning to read).*

CONTENTS

FOREWORD

The publication of the present book by Professor Riccardo Muzzarelli marks a turning point in the history of chitin and in the place occupied by this biopolymer in the chemical world of mankind.

Thanks to the pioneering work of the last century, scientists learned, at first with apparently some apathy, that the arthropod shells and the fungal cell walls were built up of a particular substance to which the name of "chitin" was finally given. Chitin appeared at that time as nothing more than a simple fantasy of the chemistry of living beings, nothing more than an extra variety among the range of complex sugars so far discoverd.

It was in the fields of organic chemistry and crystallography that chitin raised a certain interest at the beginning of the present century. As a highly organized macromolecule, chitin stimulated the curiosity of the organic chemists of some famous laboratories such as Zechmeister's, Grassman's and Karrer's; physicists such as Meyer and, more recently, Rudall had become interested.

For their part, biologists new more about the exceptional resistance offered by chitin to the usual chemical and physical breakdown factors. At the same time, it was progressively realized that chitin was an extremely abundant material in nature. Zoologists became sure of the fact that chitin was not the sole apanage of fungi and Arthropoda, but was also the organic matrix in the cuticles and skeletons of a wide range of animal classes (with the exception of those belonging to the Echinoderm and Vertebrate lineages). These two facts led biologists to raise the question, what is the fate of "dead" chitin in nature. The existence of chitinoclastic bacteria was thus suspected, and a wide variety of such microorganisms were soon detected in different soil, mud and water samples. More unexpected was the finding of chitinolytic enzymes in the gut fluids of the edible snail. This was at first considered to be a remarkable isolated instance among animals, and was tentatively interpreted in different ways. A systematic exploration of the distribution of chitinolytic enzymes in the animal kingdom, later, led to the conclusion that chitinases and chitobiases could not only be frequently found in the tissues and glandular secretions of diverse animals, but could even be considered as primitive constituents of the hydrolytic equipment of the living cell. A closing chapter of the history of the knowledge of chitin metabolism began finally with the discovery of the enzymatic process of chitin biosynthesis.

From this stage of knowledge forward, it has come to light that the enzymatic processes controlling the synthesis and the breakdown of chitin play an important role in the fundamental mechanism of growth, differentiation, nutrition, and movement of a large number of species. It was then evident that some of the most striking peculiarities of the morphological, physio-logical and even ecological diversity of living forms could be explained, at

least partly, by the fate of a very limited number of genes which control the
enzymes involved in chitin synthesis and hydrolysis. It would thus seem
legitimate, now to speak of the evolution of chitin and chitinolysis as a
chapter of molecular biology.

Besides this considerable interest from a fundamental scientific point of
view, the natural history, chemistry and enzymatic control of chitin have
aroused much more interest in different fields of applied biological sciences.
The thorough knowledge of the properties of the chitinous cuticles and
envelopes is indeed essential for progressing in the study of contact insecticides, and for the understanding of the phenomenon of egg resistance and
hatching in parasitic worms like Nematods etcetera. Even in ecology and pedo-
logy, humus formation and autodepuration have something to do with chitin
degradation and synthesis. The list of examples is of course far from being
exhausted.

Today, a new place has developed for chitin in the interest of mankind: chitin
has entered the industrial age. Some modern technologies may now rely on the
use of chitin and of some of its derivatives, as shown by a large number of
patents, in waste treatment, food processing, agriculture and pharmaceuticals.
As a natural resource for numerous technological applications, chitin has to
be supplied in industrial quality and quantity. This problem has recently
been taken into account in the United States of America by the National Sea
Grant Program of the National Oceanographic and Atmospheric Administration.

The impressive evolution of the importance of chitin is going to find an
up-to-date expression with the organization of the First International
Conference on Chitin and Chitosan in April 1977, which will be chaired by
Professor Riccardo Muzzarelli. This Conference will be the first interdisciplinary meeting of all the scientists, researchers and industrialists of
the "chitin world".

As can be seen, this book by Riccardo Muzzarelli, Professor of Chemistry in
the Faculty of Medicine of the University of Ancona, Italy, arrives at just
the right time, when so many searchers and technicians are anxious to become
more fully acquainted with fundamental information on chitin. Professor
Muzzarelli is in an ideal position to make the best use of an exhaustive and
up-to-date documentation. Due to his leading position among the specialists of
chitin and chitosan, his constant interest in the biological and chemical
aspects in which these polysaccharides are implicated and to his relations
with the most advanced laboratories involved in chitin technology, no one else
was more competent to write this book.

We must show gratitude to him for his presentation of such a considerable
amount of data in such a lucid and attractive manner.

Liège, 1976   Charles Jeuniaux

ACKNOWLEDGEMENTS

My interest in chitin originated in 1966 from the observation of some of its chemical properties. In those years, it was rather odd to go to the beach and prepare..."what do you call that polymer from crabs", when so many synthetic resins were on the shelf. Thus, now, I wish to thank first of all my co-workers and graduate students for their enthusiasm; Rousselot & Kuhlmann in Paris and Alnaes Canning Co. in Kristiansund for their technical assistance. I am grateful to the National Sea Grant Program and to Marine Commodities International Inc. for supplying samples of chitin; also, to the Italian National Research Council, the International Atomic Energy Agency, the Ente Nazionale Idrocarburi and the North Atlantic Treaty Organization, for their financial support over many years.

As documented in the Bibliography section, the rate of growth in the number of research papers and patents concerning chitin is presently so great that I have conceived the idea of preparing a book to acquaint scientists newly interested in chitin, with fundamental and up-to-date information in various fields of research and application.

It is a pleasure to thank my friends and colleagues for kind advice and discussions, particularly all those who attended the Workshop on Chitin which was held in Brownsville, Texas, in 1975. I thank Prof. Ch. Jeuniaux for critically reading the manuscript and kindly writing the Foreword, and Dr. K. Beran, Prof. Y. Nozawa, Dr. L.C. Post, Dr. E. Cabib, Prof. S. Bartnicki-Garcia, Prof. G.G. Allan and Dr. A. Ferrero who supplied original illustrations. My thanks to Mr. P. Perceval who contributed something of his prodigious experience in the field of the industrial production of chitin; Drs. A.C. Grosscurt and A. Verloop for reading parts of the manuscript, Dr. V. Accorroni for preparing the drawings and Mr. R.A. DeAmicis for surveying the final text. I also acknowledge with thanks the contributions of those authors whose names appear under the Tables and Figures and the help received from all those who made available unpublished results.

A visit to several Laboratories and Libraries in the United States of America in September, 1976, in the context of a research activity supported by the North Atlantic Treaty Organization, has made possible the up-to-date completion of the text. This book includes some information made available for the first time in English; thanks are due to Dr. P. Santandrea who translated from Russian, Dr. A. Miglionico who translated from Japanese and, finally, to Mrs. M.G. Muzzarelli, who not only translated from German and Dutch, but also, to add to the great merits of the author's wife, collected and classified the bibliographic material.

Ancona, 1976   Riccardo A.A. Muzzarelli

INTRODUCTION

*..... je vais faire connaître la nature de
la substance qui forme le corps ou la base
charnue insoluble du champignon, et que je
désignerai sous le nom de fungine.*

Henri Braconnot (1811).

Chitin was described for the first time by Braconnot (1811) who was professor
of Natural History, director of the Botanical Garden and member of the Academy
of Sciences of Nancy, France. In the pursuit of his researches on mushrooms,
he treated *Agaricus volvaceus* and other mushrooms with diluted warm alkali and
isolated chitin, possibly slightly contaminated with proteins. From the dry
distillation of this product, called fungine, he obtained a liquid that, upon
distillation with potassium hydroxide, yielded ammonia. He described the lack
of reaction of chitin with diluted alkali solutions, the compound formed with
tannin, the degradation produced by sulfuric acid, the release of acetic acid
and several other reactions of chitin. He stated that "fungine seems to contain
more nitrogen than wood" and concluded that it is "a quite distinct substance
among those identified in plants".

Odier (1823) in an article on insects wrote: "It is most remarkable to find
out in the insect structure the same substance that forms the structure of the
plants". He called this substance chitin from the greek term χιτών (tunic,
envelope); even though he failed to detect nitrogen in chitin, he established
for the first time a relationship between the insect cuticle and plant tissue.
The discussion of the differences between cellulose and chitin, which was
continued for more than one century, was started by Payen (1843). The controversy about cellulose and chitin has permeated the whole development of the
information about chitin; for instance cuticles of cellulose were reported and
chitin occurrence in budding yeasts and tubercle bacilli was excluded in
reports of recent date. For this reason the present book will mention cellulose
only when pertinent and necessary.

The chitinous exoskeleton of the silkworm *Bombyx mori* was isolated in toto by
Lassaigne (1843) by the action of warm potassium hydroxide; he prepared a large
selection of insects exoskeletons by treating the whole insect with potassium
hydroxide and potassium hypochlorite, to demonstrate the presence of chitin in
flying organs and internal parts. Upon charring chitin with potassium, he
demonstrated the presence of nitrogen in chitin on the basis of the potassium
cyanide formed.

Ledderhose (1878) clearly indicated that chitin is composed of glucosamine and
acetic acid and wrote the hydrolysis equation. Gilson (1894) confirmed the
presence of glucosamine.

1

Chitosan was discovered by Rouget (1859). He found that chitin which has been
boiled in a very concentrated potassium hydroxide solution becomes soluble in
organic acids. This modified chitin, as he called it, was colored violet by
diluted solutions of iodine and acid, whereas chitin was stained brown.
Modified chitin was studied again by Hoppe-Seyler (1894) who named it chitosan.
Von Furth & Russo (1906) showed that chitosan forms crystallizable salts with
acids and is precipitated from acid solutions by alkaloids precipitants. Lowy
(1909) studied the elementary composition and the crystalline form of chitosan
sulfate which is one of the most insoluble of the chitosan salts.

The first reviews on the occurrence and distribution of chitin were those by
Van Wisselingh (1898), Wester (1910), Von Wettstein (1921), Levene & Lopez-
-Suarez (1925) and Von Franciis (1930).

The advancement of the knowledge about chitin has been very uneven, because
many misleading data have been reported, ranging from erratic tests to
structures including anhydroglucosidic bonds directed at 180°. Real difficulty
was also encountered because of the variety of sources of chitin; few chemists
and physicists have been involved in the study of chitin until recently and
this may help explain why the information about chitin is still so scattered
and poorly integrated.

During the first half of the present century, research on chitin was mostly
directed toward the study of its occurrence in living organisms, its degradation by bacteria, its uses in resin technology and its chemistry, as can be
seen in the "Bibliography" section, which is arranged in chronological and
alphabetical order and includes full titles. On the basis of the nature of the
chemical and bacterial decompositon products, the identity of chitin obtained
from fungi and from crab shells was reported by Rammelberg (1931) and confirmed
in the following years by elemental analysis and chemical behavior. The splitting of chitin was recognized as an important activity of the flora of the soil.
Several hundred dissimilar cultures from a variety of sources were reported
to have in common the ability to attack chitin and some of them were introduced
as new species in recognized genera.

Chitinoclastic bacteria were reported by Zobell (1937) as widely distributed
in marine sediments, animals and sea water and some were isolated from marine
sand, mud, water, crabs and sea-weeds. Under anaerobic conditions the microorganisms that reduce sulfates destroy chitin with evolution of ammonia; some
of these bacteria were fully described.

With x-ray analysis as the most reliable method for the determination of chitin
and cellulose in cell walls, the occurrence of these polymers in fungi was investigated. The older, generally used microchemical methods for chitin and
lignin were found of little reliability compared to x-ray analysis. In none of
the organisms investigated was found a juxtaposition of chitin and cellulose,
at variance with earlier reports. The sharp separation of Phycomycetes into
chitin and cellulose fungi was confirmed. With the lower Ascomycetes the presence or absence of chitin is uniform within single families. From those
experimental results, Frey (1950) concluded that the occurrence of cellulose
or chitin in fungi does have a phylogenetic significance.

The chemical constitution of chitin was studied by several authors: the idea
that chitin is a polymer of acetylated glucosamine has still been an object of
controversy as chitin was believed for a long time to contain an unspecified

nitrogen compound in addition to glucosamine, possibly due to protein
contamination. Many authors sustained the view that acetic acid evolved upon
hydrolysis was not part of the polymer, but was a secondary decomposition
product together with other acids. In 1931, upon acetolysis of lobster chitin,
a disaccharide of glucosamine was obtained as a crystalline octaacetate
derivative and it was proposed as a repeating unit for the study of chitin, as
cellobiose was already for cellulose research. The same year, other authors
re-acetylated chitosan and degraded the chitin thus obtained with snail
chitinase; however, they still had doubts about the chemical form of nitrogen
in chitin. After hydrolyzing chitin in different ways, it was understood that
chitin is a polysaccharide of glucosamine (Purchase & Braun, 1946) and a
patent was granted to Matsushima (1948) for the production of glucosamine from
crab shells.

In 1950, chitosan was clearly described as a polymer of glucosamine and the
structure of chitin was studied by periodate oxidation; results supported the
structure tailored on the cellulose model.

The first books on chitin were published only in the fifties: they are
the well-known books on the integument of the arthropods by Richards (1951)
and on the biochemistry of aminosugars by Kent & Whitehouse (1955). While
chitin has supported life since its origins on earth, it is a matter of fact
that most of the information presently available on chitin has been obtained
after 1950. A review on chitin by Tracey (1955) presented detection procedures
and quantitative analytical methods. Jeuniaux (1963) published a book on
chitin and chitinolysis and Bernfeld (1963) published on the biogenesis of
natural compounds, including chitin. The same year, the book by Reese (1963)
considered chitin in the context of the discussion of the advances in the
enzymic hydrolysis of cellulose. Brimacombe & Webber (1964) wrote a comprehensive monograph on chitin and BeMiller (1965) also contributed a short monograph mainly dealing with its production. Rudall (1967) discussed the chitin-
protein complexes: this topic was also the object of books by Friedman (1970),
Hohnke & Scheer (1970) and Hunt (1970) devoted to the zoological importance of
chitin, and by Jeuniaux (1971) about chitinous structures and the meaning of
chitin in the biochemical evolution. Chitin in plant biochemistry was also
dealt with by Goodwin & Mercer (1972) while chitin in bacterial cell surface
was treated by Glaser (1973). A bibliography of selected publications on
chitin and its derivatives was published by Pariser & Boch (1972).

The book by Muzzarelli (1973) on the chelating properties of alginic acid,
substituted celluloses, chitin, chitosan and other polymers, brought together,
for the first time, interdisciplinary information about chitin, while the book
on biopolymers by Walton & Blackwell (1973) included parts on structural and
biological aspects of chitin. Advances in the chemistry of chitin were reported
in the book by Brimacombe (1973) and the one by Miller (1973) and in a review
by Whistler (1973).

Preston (1974) and Rockstein (1974) published two books partially concerning
chitin in plant cell walls and in insects, respectively. Brown (1975) mentioned
chitin in his book on structural materials in animals, while Jeuniaux (1975)
described the cuticular chitin complexes. Aminosaccharide chemistry was treated
by Williams (1975); Whistler & BeMiller (1976) published a book on general
methods in glycosaminoglycan chemistry. The biology of the arthropod cuticle
is the subject of the book by Neville (1975) and the insect integument is
treated in a book edited by Hepburn (1976) which commemorates the book by

Richards (1951), 25 years after its publication.

In this context and in view of the rapidly increasing interest in chitin
chemistry and applications, the present book is offered as a contribution that
may provide a chemical background to all those who are carrying out research
or are investigating applications. I hope that the chemical approach incorporating the contributions of biochemists, physicists, industrialists, botanists,
entomologists, physiologists and scientists in other fields, will confer to
the book a really interdisciplinary character to make the book useful in many
branches of advanced research. As a teacher, I also hope that the book will
mean a starting point for graduate students, wishing to undertake research in
the many fields involving chitin.

CHAPTER 1

ENZYMIC SYNTHESIS OF CHITIN AND CHITOSAN

OCCURRENCE OF CHITIN

Chitin is very widely distributed, especially in animals, and exists also in less evolved taxonomic groups, such as Protozoa. Early established genes appear to control the biosynthesis of chitin and this biosynthetic ability has been retained by numerous diblastic animals and by most of the triblastic Protostomia; it was lost, however, at the beginning of the deuterostomian evolutionary lineage with the remarkable exception of Tunicata, the peritrophic membrane of which does not contain chitin (Jeuniaux, 1963, 1965).

Vertebrata, Urochordata, Pterobranchia, Enteropneusta and Echinoderma have utilized other polysaccharides, namely cellulose, chondroitin sulfate, collagen and keratins for building their supporting structures.

In plants, chitinous cell walls are only found in those forms, such as fungi and molds, which, like animals, find considerable nitrogen in their food. On the contrary, photosynthetic plants utilize nitrogen-free sugars almost exclusively for their supporting structures; chitin is however believed to constitute the cell membrane of some lower green plants like Chlorophycea. Information about occurrence and chemical features of chitinous structures is assembled in Table 1.1.

Chitinous structures are mainly of ectodermal origin in multicellular animals and form the characteristic exoskeleton of most of the invertebrates in contrast with collagenous structures which almost entirely are of mesodermal origin. Chitin exceptionally constitutes more than half of the total organic matter in chitinous structures. Higher concentrations up to 85 % are found in Arthropoda, which are particularly able to synthesize chitin. There is no apparent relationship between the proportion of chitin and the degree of calcification, hardness or flexibility of the structure, as shells of Gastropoda and Lamellibranchia contain only small amounts of chitin.

Chitin is associated with other polysaccharides in the fungal cell walls, while in animal forms chitin is associated with proteins. In hard structures, the proteins are tanned by phenolic derivatives. Rudall (1955) has emphasized the frequent independence of collagen secretion and chitin synthesis.

The distribution of the three crystallographic forms of chitin does not appear to be related to taxonomy. The various chitinous supporting structures exhibit striking differences in morphology, chemical composition and physical characteristics.

Chitin is synthesized according to the scheme in Table 1.2, where several enzymes are shown to take part in the biosynthesis together with chitin synthetase, whose official name is uridine diphosphate-2-acetamido-2-deoxy-D-glucose: :chitin-4-β-acetamidodeoxyglucosyl transferase, E.C. 2.4.1.16. The present

5

TABLE 1.1. Distribution and chemical features of the diverse chitinous structures in living organisms.

Organisms	Examples	Structures	Minerals	Chitin % org.	Other organic constituents
FUNGI					
Ascomyceta, *Aspergillus flavus, fumigatus, niger, Penicillium notatum, Neurospora crassa.*		cell walls and structural membranes of mycelia stalks and spores		traces to 45	glucans, mannan or other polysaccharides
Basidiomyceta, *Agaricus campestris, Armillaria mellea, Boletus edulis, Schyzophyllum commune.*					
Phycomyceta, *Blastocladiella emersonii, Mucor rouxii, Mortierella vinacea, Phycomyces Blakesleanus, Rhizopus rhizopodiformis, Apodachlya.*					
Imperfecti, *Candida albicans, Alternaria kikuchiana.*					
ALGAE					
Chlorophyceae, *Ulva lactuca, Valonia ventricosa.*				+	cellulose
PROTOZOA					
Rhizopoda, *Pelomyxa*		cyst wall		+	–
Plagiopyxidae		shell	silica	+	–
Allogromia		shell	iron	+	proteins, lipids
Vasicola ciliata		cyst wall		+	proteins
CNIDARIA					
Hydrozoa					
Hydroidea		perisarc		3-30 α	proteins (tanned)
Milleporina, *Millepora alcicornis*		coenosteum	CaCO₃	+ α	–
Siphonophora		pneumatophore		+ α	–
Anthozoa, *Metridium, Pocillopora*		"skeleton"	CaCO₃	+	proteins
Scyphozoa, *Aurelia*		podocyst		+	proteins
ASCHELMINTES					
Rotifera		egg inner membr.		14	proteins
Nematoda, *Ascaris lombricoides*		egg middle membr.		16	proteins
Acanthocephala		egg capsule		+	–
Priapulida, *Priapulus caudatus*		cuticle		+	proteins (tanned)

./.

Taxon	Structure	Mineral	Chitin %	Type	Associated substance
ENDOPROCTA, *Pedicellina cernua*	cuticle		+		tanned proteins
BRYOZOA, *Cristatella mucedo*	ectocyst	(CaCO$_3$)	2-6		proteins
PHORONIDA	tubes		13		proteins
BRACHIOPODA					
Articulata	stalk cuticle		4		-
Inarticulata	stalk cuticle		+	γ	collagen
	shell	CaCO$_3$	29	β	-
ECHIURIDA	hooked chaetae		+		-
ANNELIDA					
Polychaeta, *Aphrodite, Amphinome*	chaetae		20-38	β	quinone-tanned prot.
Eunicidae	jaws		0.3		proteins
Oligochaeta, *Lombricus terrestris*	chaetae		+	β	
all	peritrophic membr.		+		proteins
MOLLUSCA					
Polyplacophora, *Chiton*	shell plates	CaCO$_3$	12		proteins
	radula	iron	+		proteins
Gastropoda, *Helix pomatia*	mother of pearl	CaCO$_3$	3-7		conchiolin
	radula	iron sil.	20	α	tanned proteins
	jaws		+		tanned proteins
Cephalopoda, *Sepia officinalis, Nautilus pompilius, Loligo paelei, Octopus vulgaris, Ommatostrephes sagittatus*	calcified shell	CaCO$_3$	3-26	β	conchiolin
	pen		18	β	conchagen
	jaws and radula		20	α	tanned proteins
	stomach cuticle			γ	
Lamellibranchia, *Nucula nitida, Ostrea virginica, Venus Mercenaria mercenaria*	shells	CaCO$_3$	up to 8		conchiolin
	gastric shield		17		-
ARTHROPODA					
Crustacea, *Palinurus vulgaris, Homarus vulgaris, Cancer magister, pagurus, Carcinus maenas, Astacus fluviatilis, Euphausia superba*	calcified cuticle CaCO$_3$		58-85	α	arthropodin+sclerotin
	intersegmental m. CaCO$_3$		48-80	α	arthropodin
Insecta, *Drosophila melanogaster, Locusta migratoria, Periplaneta americana, Bombyx mori, Calliphora sp., Sarcophaga bullata, Tenebrio molitor, Schistocerca gregaria*	hardened cuticle		20-60	α	arthropodin+sclerotin
	unhardened cuticle		20-60	α	arthropodin+sclerotin
Arachnida and Chilopoda, *Galeodes, Scolopendra*					
all *Musca domestica*					
POGONOPHORA, *Oligobrachia ivanovi*	peritrophic membr.		4-22		protein+mucin
	tubes		33	β	proteins

TABLE 1.2. Pathway of chitin synthesis.

Trehalose in blood
↓
Glucose
↓   hexokinase
Glucose 6-phosphate
↓   glucose phosphate isomerase, E.C. 5.3.1.9
Fructose 6-phosphate

                                  glutamine-fructose 6-phosphate aminotransferase,
                                                          E.C. 2.6.1.16
            ← glutamine
            → glutamic acid
↓
Glucosamine 6-phosphate
                                  phosphoglucosamine transacetylase also called
            ← acetyl-Co-A   glucosamine 6-phosphate N-acetyltransferase
            → Co-A
↓
N-acetylglucosamine 6-phosphate
↓   phosphoacetylglucosamine mutase
N-acetylglucosamine 1-phosphate
                                  uridinediphosphate-N-acetylglucosamine
            ← UTP   pyrophosphorylase, E.C.2.7.7.23
            → pyrophosphate
↓
Uridinediphosphate N-acetylglucosamine
                                  chitin synthetase   E.C. 2.4.1.16
                                  chitin synthetase activating factor (proteinase B)
                                                          E.C.3.4.22.9
                                  (and chitin deacetylase)
            → UDP
↓
Native chitin (partially deacetylated chitin)

knowledge about the biosynthesis of chitin is mainly based on studies by Lund
& Kent (1961), Meenakshi & Sheer (1961), Candy & Kilby (1962), Jaworski, Wang
& Marco (1963), Gilmour (1965), Carey (1965), Lipke, Grainger & Siakotos
(1965), Condoulis & Locke (1966), Porter & Jaworski (1966), Speck, Urich &
Hertz-Huebner (1972), Gwinn & Stevenson (1973), Araki & Ito (1974), Selitren-
nikoṿ, Allin & Sonneborn (1976), Ulane & Cabib (1976) and many others. The
enzyme has been isolated from *Perspectania eridania*.

Arthropods.

The intersegmental membranes of the arthropods are flexible chitinous
structures. The cuticular sclerites also have the same characteristics just
after molting; the chitinous procuticle is laid down by the epidermal cell
layer and it shows internal laminations in the light microscope. It is

perforated by the dermal ducts and the pore canals. The bulk of the procuticle
is a chitin + protein complex, the protein fraction consisting of "arthropodins" which are cross-linked by tanning reactions.

The innermost layer of the calcified cuticle of crustacea is not calcified;
it is currently referred to as endocuticle because it lies between the
calcified procuticle and the epidermis and has characteristics similar to
those of the intersegmental membrane.

The peritrophic membrane of the arthropods and some other invertebrates which
contains the food in the mesenteron and the feces in the postintestine, is a
very thin and pliable membrane . The wing hinges and the elastic thoracic
ligaments of the insect flight system are other flexible chitinous structures
which contain the rubber-like protein called resilin.

Quinone tanning of the proteins makes the sclerites of the insect and arachnids
differ from intersegmental membranes. The hardening of the exocuticle by
tanning is often accompanied by a general darkening due to melanin formation.
It seems that sclerotization and melanization are independent from each other
and are catalyzed by two different phenol oxidases.

In molluscan shells, chitin occurs at low percentages and is often associated
with a fibrous protein, conchiolin, in which sclerotization by quinone tanning
seems to be entirely lacking. In the Decapoda, the shell is an internal
skeletal structure calcified in the Sepioidea, but non-calcified in the
Theuthioidea (pen of the squid). This internal structure is reduced in the
Octopoda. According to Stegemann (1963) the protein fraction in these non-
calcified shells can be converted to gelatin by steam: the remaining chitin
is linked to a small protein moiety.

The sclerites of crustaceans can be interpreted as intersegmental membranes
whose procuticle is almost entirely calcified; calcium carbonate and, to a
lesser extent, calcium phosphate, are the salts involved. Calcium carbonate
occurs as micro- or macro-crystals of calcite or, in rare cases, vaterite.
The hardening of calcified cuticles is, however, initiated by protein
sclerotization prior to deposition of calcium salts. The amount of protein in
the calcified cuticle is much lower than in the flexible procuticle or in the
sclerotized exocuticle of insects, as a result of the calcification. It can be
imagined that mineral deposits replace most of the protein in the hardening
process (Richards, 1951). Hardness is rarely provided by minerals other than
calcium salts. The shells of some Rhizopoda and the teeth of the radula of
some mollusks are hardened by iron oxide or silica deposits.

Meenakshi & Scheer (1961) listed glucose, maltose, maltotriose, maltotetraose
and glucose 6-phosphate as major blood carbohydrates in the crab. The carbon
from glucose is incorporated into chitin during the early post-molt period,
but not during intermolt. Glucosamine was reported by Parvathy (1970) to occur
only during the premolt stage due to resorption of chitin in preparation for
the shedding of the cuticle. Uridine diphosphate and uridine diphosphate-N-
acetylglucosamine were absent in the blood. Further confirmation of the above
findings was given by Parvathy (1971).

Yano (1974) injected glucose 1-^{14}C into the blood of the shore crab *Gaetice
depressus*: during the period of enducuticle formation, the radioactivity was
rapidly incorporated into chitin within hours of the injection. Throughout the

the molting cycle, the highest value of incorporation was observed the day after molting, suggesting that the most active synthesis occurs at the beginning of endocuticle formation and slows down thereafter. Speck & Urich (1969) carried out similar experiments on the crayfish *Orconectes limosus* and found that just after molting a major part of the glucose was incorporated into chitin, while intermolt animals metabolize glucose to aminoacids.

Coles (1966) demonstrated the in vitro incorporation of labeled glucose and glucosamine into insoluble polysaccharide by the isolated epidermal cell which secretes rubber-like cuticle in the desert locust *Schistocerca gregaria*. The incorporation of $|^3H|$-N-acetylglucosamine at the different stages of the larval-adult molting cycle, however, differs from results obtained with glucose. Before ecdysis, this precursor is incorporated into chitin at a rate 5 to 7 times greater than that of glucose. After ecdysis, at first this rate corresponds to that of glucose, but this high level is maintained during the first seven days of the imaginal life. The different incorporation rates of glucose and N-acetylglucosamine can presumably be explained in the following way: before ecdysis acetylglucosamine is used in synthesizing chitin; the aminosugar is set free in greater amounts by the action of chitinase in the exuvial fluid (Waterhouse, Hackman & McKellar, 1961) on the larval cuticle (Surholt, 1975). For a comparatively low rate of synthesis at this stage, its amount seems to suffice. Moreover, after ecdysis N-acetylglucosamine is no more available and the synthesis depends on the glucose supply. In this process the amination of glucose 6-phosphate is apparently a primary site of the regulation of chitin ex novo synthesis (Surholt, 1975).

Results on the in vivo incorporation of labeled glucose and N-acetylglucosamine obtained on various crustaceans (Hohnke, 1971; Hornung & Stevenson 1971; Stevenson, 1972; Speck & Urich, 1972; Gwinn & Stevenson, 1973) are not directly comparable with the studies on *Locusta*. In the crustaceans the integument presumably has additional functions like storage of reserve compounds, and undergoes cyclic changes. Nevertheless the incorporation of the two chitin precursors during the crustacean molting cycle is similar to the process in the locust. The glucose incorporation also reaches a maximum after molting, while N-acetylglucosamine is greatly incorporated before molting and for a long time afterwards (Stevenson, 1972).

The epidermal cells have the most important function as they are capable of synthesizing all precursors of chitin, from glucose to uridine diphosphate-N--acetylglucosamine (Surholt, 1975). Candy & Kilby (1962) demonstrated that they contain all the necessary enzymes. Chitin synthesis from labeled glucose in vitro by isolated integument preparations is a further proof. The results obtained after injection of labeled N-acetylglucosamine demostrate that the epidermal cells are, in addition, able to utilize sugars already aminated. N-Acetylglucosamine is adsorbed remarkably fast by the epidermal cells and, during the first minutes, converted almost quantitatively into uridine diphosphate-N-acetylglucosamine, while in the hemolimph labeled acetyl-glucosamine circulates almost exclusively. The synthesis of chitin precursors takes place most readily in locusts in the stage of highest chitin synthesis, 24 hours after the larval-adult ecdysis; in comparison, other stages (24 hours and 10 days after ecdysis) show a much smaller content or only traces of chitin precursors.

In *Persectania* the activity of chitin synthetase is highest during the late final larval instar and prepupal instar, correlating with maximal cuticle

deposition (Porter & Jaworski, 1965). The uridine coenzyme assists in the
formation of the glycosidic link. In the southern armyworm *Prodenia eridania*,
labeled uridine diphosphate-N-acetylglucosamine was incorporated into an
insoluble product from which it could be subsequently released, and the product
was identified as chitin by hydrolysis with chitinase (Jaworski, Wang & Marco,
1963; Porter & Jaworski, 1965). Electron microscope autoradiography shows that
labeled precursors are incorporated into chitin outside the cell and that the
incorporation of glucosamine is rapid (Weis-Fogh, 1970).

Many experiments have been made to study the synthesis of chitin in vitro by
cultured tissues. In the arthropods, even slight damage of the epidermal cells
leads to a cessation of chitin synthesis. Chitin synthetase is, in contrast to
the other enzymes of the above depicted pathway, integrated in a very specific
way in the cell structure and possibly unstable (Surholt, 1975).

β-Ecdysone is required for the induction of cuticle deposition by cultured
tissues and for the localization of $|^{3}\text{H}|$-glucosamine in chitinous areas of the
cuticle. The activation of epidermal cells by ecdysone and the appearance of
the new chitin-bearing cuticle on cockroach leg regenerates has been studied
by time-lapse microphotography and the sequence of the events has been
described by Marks (1973): these include the hormonal induction of morphogenesis and the biosynthesis of the primary structural materials, protein and
chitin. In an adequate medium, radiolabel from glucose and N-acetylglucosamine
was incorporated into the cuticle and when the explants were treated with
chitinase, the labeled material was released into the medium. This indicates
that the complete biochemical pathway for chitin synthesis is operative in
vitro (Marks & Sowa, 1976).

Fungi.

Peberdy & Moore (1975) reported studies on *Mortierella vinacea* whose 2000 *g*
"cell wall" fraction shows high activity of chitin synthetase which corresponds
with the findings in the Mucorales *Mucor rouxii* (McMurrough, Flores-Carreon &
Bartnicki-Garcia, 1971), *Phycomyces blakesleeanus* (Jan, 1974) and *Cunninghamella elegans* (Moore & Peberdy, 1975). In other organisms studied (Glaser & Brown,
1957; Jaworski, Wang & Carpenter, 1965; Porter & Jaworski, 1966; Plessmann-
-Camargo, Dietrich, Sonneborn & Strominger, 1967) the enzyme was said to be
located mainly in the mitochondrial and microsomal fractions.

On the basis of their findings with *Mucor rouxii*, McMurrough, Flores-Carreon &
Bartnicki-Garcia (1971) proposed that the 100,000 *g* "microsomal fraction" is
the site of chitin synthetase formation before its migration to the cell wall.
The difficulty of removing chitin synthetase from the cell wall fractions of
Mortierella vinacea and *Mucor rouxii* suggests that the enzyme is bound to the
wall. However, the electron micrographs of wall fractions of *Phycomyces blakesleeanus* show high contamination of the walls with attached membranous
material. A membranous location of chitin synthetase in the yeast form of *Mucor
rouxii* was indicated by Ruiz-Herrera & Bartnicki-Garcia (1974).

Porter & Jaworski (1966) suggested that freezing and thawing of chitin
synthetase fractions may release inhibitor or hydrolytic enzymes which act upon
the substrate of the chitin formed. The presence of activators and inhibitors
of chitin synthetase have been demonstrated in yeast cells (Cabib & Keller,
1971; Cabib & Ulane, 1973). The presence of a heat-stable, pH-dependent

inhibitor of chitin synthetase, similar to that found in *Mortierella vinacea*, has been reported in *Mucor rouxii* (McMurrough & Bartnicki-Garcia, 1973) and it has been purified as a protein of molecular weight 8500 from *Saccharomyces cerevisiae* (Ulane & Cabib, 1974). Activators of chitin synthetase have not so far been isolated from filamentous fungi, although activation has been demonstrated in *Mucor rouxii* and *Aspergillus nidulans*.

The formation of diacetylchitobiose as an intermediate in chitin synthesis has been reported in *Blastocladiella emersonii* and diacetylchitobiose was formed in place of chitin in the presence of low uridine diphosphate-N-acetylglucosamine concentrations. In *Mortierella vinacea*, Peberdy & Moore (1975) did not detect diacetylchitobiose; in *Mucor rouxii*, diacetylchitobiose was detected only with high concentrations of uridine diphosphate-N-acetylglucosamine. (Diacetylchitobiose is the name given by chemists to the dimer of N-acetylglucosamine; this compound is also called chitobiose as it is the substrate of chitobiase).

In chitin synthesis, N-acetylglucosamine and soluble chitodextrins have been described as activators. Activation was also found to be induced by N-acetylchitodextrins in *Mucor rouxii* and by mannose, glucose, cellobiose and glycerol (Keller & Cabib, 1971) in *Saccharomyces cerevisiae*. N-Acetylglucosamine is thought to bear the role of an allosteric effector (Glaser & Brown, 1957). Activation of *Mortierella vinacea* chitin synthetase by N-acetylglucosamine was low compared to other organisms (Peberdy & Moore, 1975). Soluble chitodextrins were found to behave as inhibitors of chitin synthesis leading to large reductions in net product formation. In growing hyphae the tips are the sites of wall synthesis (Bartnicki-Garcia & Lippman, 1969; Grove & Bracker, 1970; Gooday, 1974).

Magnesium ions and N-acetylglucosamine greatly stimulate the enzyme activity. The enzyme activity increases with the increase of magnesium concentration in the reaction mixture, reaches maximum when it is between 15 and 25 mM and starts to decrease when the magnesium concentration is above 25 mM. The enzyme activity increases with the increase of N-acetylglucosamine concentration in the reaction mixture and reaches a plateau when N-acetylglucosamine exceeds 0.1 M (Jan, 1974).

The amount of chitin synthesized in the reaction mixture increases linearly with time for at least the first two hours. It also varies linearly with the concentration of the enzyme in the reaction mixture, in the range of 0.5 to 5 mg$\times$ml^{-1} protein in the enzyme preparation. The optimal temperature is about 28° with half-maximal temperature at about 17° and 42°. The pH optimum is about 6.5, as shown in Fig. 1.1.

Among numerous metal ions investigated by Peberdy & Moore (1975), only magnesium showed a marked stimulation in activity over the control, as shown in Table 1.3. Cobalt ions showed a slight increase in activity, but zinc, barium, calcium and ferrous ions caused the greatest inhibition of chitin formation. Variation in magnesium chloride concentration between 0.6 and 100 mM showed an increase in activity up to 20 mM, after which there was a progressive inhibition of the enzyme.

Moore & Peberdy (1975) disrupted mycelial material in buffer in a carbon dioxide-cooled homogenizer for 30 sec and centrifuged to produce 4 fractions; cell wall (2000 *g*), mitochondrial (10,000 *g*), microsomal (100,000 *g*) and a

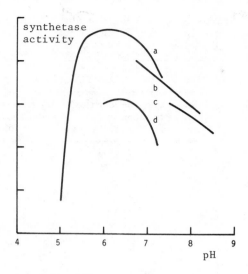

Fig. 1.1. Dependence of chitin synthetase activity on pH
and buffer systems. The assay was done under standard
conditions with the exception of the pH and the buffer
used. a = imidazole acetate; b = HEPES; c = tris-HCl;
d = phosphate. Jan, 1974.

soluble cytoplasmatic fraction (supernatant) using 50 mM KH_2PO_4 + KOH buffer,
pH 6.5 containing 1 mM $MgCl_2$ to suspend the pellets. The enzyme fractions were
stored at 20°.

The procedure for measuring chitin synthetase activity was based on the method
of Glaser & Brown (1957). The assay mixture (100 µl) contained 50 µl enzyme,
20 mM $MgCl_2$, 25 mM KH_2PO_4 + KOH buffer and 50 nmol uridine diphosphate-N-
acetylglucosamine labeled at the uridine carbon. The mixture was incubated
for 15 min at 26° and the reaction was stopped by adding 10 µl glacial acetic
acid. Chitin formation was determined by descending chromatography. 50 µl
aliquots of the reaction mixture were spotted onto strips of paper and the
chromatograms were developed for 6 hr using ethanol 95 % + acetic acid (7:3,
v/v). The chromatograms were scanned to locate the reaction products: chitin
was found at the origin.

Variation of the uridine diphosphate-N-acetylglucosamine concentration over a
range of 0.2 - 4.0 mM was carried out in the presence of unaltered N-acetylglucosamine concentration (20 mM). Extrapolation of the double reciprocal plot
of the rate of product formation against substrate concentration gave an
apparent K_m for uridine phosphate-N-acetylglucosamine of 1.23 mM. The apparent
K_m for N-acetylglucosamine was obtained by varying its concentration between
0 and 20 mM whilst maintaining uridine diphosphate-N-acetylglucosamine
concentration at 1 mM. The Lineweaver-Burk plot, when extrapolated, gave an
apparent K_m value for N-acetylglucosamine of 2.08 mM (Moore & Peberdy, 1975).

Several nucleotide sugars with structures analogous to that of the substrate

Table 1.3. Effect of cations on chitin synthetase activity. The reaction mixture contained 0.4 mM magnesium chloride and other cations as chlorides at a final concentration of 10 mM. Peberdy & Moore, 1975.

	Enzyme activity	
Cation	N-acetylglucosamine incorporated $nmol \times min^{-1} \times mg^{-1}$ prot.	Percentage of control level
none	0.90	100.0
Fe^{3+}	0.26	28.5
Mn^{2+}	0.80	88.0
Mg^{2+}	1.88	206.6
Ca^{2+}	0.36	39.6
Na^{+}	0.89	97.8
K^{+}	0.78	85.7
Co^{2+}	1.02	112.1
Cu^{2+}	0.48	52.7
Zn^{2+}	0.15	16.5
NH_4^{+}	0.85	93.4
Ba^{2+}	0.35	38.5

were also tested for their effects on the enzyme activity. The following compounds were found to inhibit about 20 % of the enzyme activity: uridine diphosphate glucose, uridine diphosphate mannose, uridine diphosphate xylose, uridine diphosphate glucuronic acid and uridine diphosphate galacturonic acid. Uridine diphosphate, one of the end products of the chitin synthetase catalyzed reaction, inhibited 75 % of the enzyme activity at a concentration of 0.5 mM. Adenosine triphosphate at a concentration of 0.25 mM raised the enzyme activity by 33 % . However, at concentrations higher than 1 mM, adenosine triphosphate is inhibitory. 3':5'-Cyclic adenosine monophosphate inhibited 20 % of the enzyme activity at the concentration of 2 mM but at lower concentrations it was ineffective.

Yeasts.

Beran, Holan & Baldrian (1972) described the occurrence of crystalline chitin with crystal size about 6 nm in the wall complex of the yeast *Saccharomyces cerevisiae*. The complex is localized in the electron-transparent encircling region and the primary septum. It was assumed that in addition to chitin, another polysaccharide may be present in the primary septum, and it was shown to be mannan by several authors. The literature on the chitin occurrence during bud formation in yeasts is abundant.

Although an extensive body of knowledge has accumulated about the descriptive aspects of morphogenesis, the relevant molecular mechanisms are little known, in spite of very numerous biochemical studies which have elucidated a great

number of metabolic pathways, including those which lead to the biosynthesis
of structural components. Nevertheless the products have been usually obtained
as amorphous chemical compounds and not as components of an organized structure.

The study of the chitin synthesis in the budding cycle and septum formation in
yeast has been considered with interest because it offers the possibility of
constructing a model in which the chitin synthesis under study is considered
in a specific portion of the cell and during a specific time in the cell cycle.

During vegetative growth of *Saccharomyces cerevisiae*, as well as other yeasts,
cell division occurs by budding. The bud that appears on the surface of the
cell progressively enlarges, until the nucleus is divided into two nuclei.
Finally, a septum is formed between the two cells and the plasma membranes are
pinched off on both sides as shown in Fig. 1.2.

The texture of these septa is similar to that of the cell walls. The cells
separate in an asymmetric fashion. The bud scar remains on the surface of the
mother cell, as can be seen in Fig. 1.3, while the daughter cell only shows a
flattened region, the birth scar.

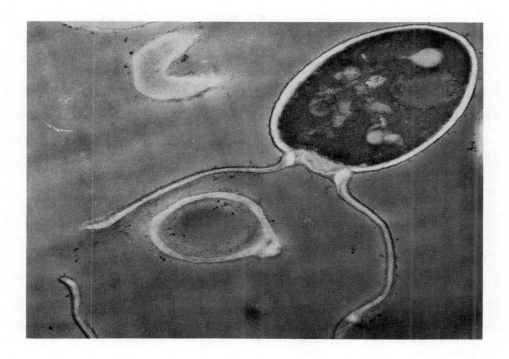

Fig. 1.2. Empty mother cell with associated daughter cell,
showing primary septum and encircling region. Magnification
× 11,520. Courtesy of K. Beran.

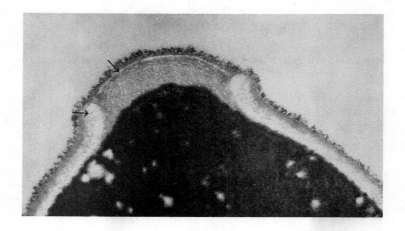

Fig. 1.3. A section of a bud scar in *Saccharomyces cere-*
visiae, showing the chitin primary septum (arrows)
embedded in the scar. Magnification × 33,600. Courtesy
of E. Cabib.

The presence of small amounts of chitin in the yeast cell wall, with the major
carbohydrates, glucans and mannan, was reported by Houwink & Kreger (1953).
Bacon, Davidson, Jones & Taylor (1966) detected a high concentration of
hexosamine, probably chitin, in isolated and partially degraded bud scars from
Saccharomyces cerevisiae. This observation has been reinvestigated by Cabib,
Ulane & Bowers (1975) with the use of specific enzymes. The scar material
contains about 50 % glucan and 15 % chitin. After treatment with chitinase the
ridge of the bud scar is eliminated and the remaining wall appears to be
flatter. Conversely, incubation with glucanase uncovers a chitin disk, which
still retains the circular ridge, as shown in Fig. 1.4. It may be concluded
that a chitin disk with an annular ridge is incorporated between glucan layers
in the bud scars. This is the location of the primary septum, which would be
composed of chitin, whereas glucan would be the component of the secondary
septum.

The yeast primary septum has the well-defined geometrical shape shown in Fig.
1.4. It is a disk with a thicker annular zone. It consists of chitin which is
localized at a precise site in the cell, the junction between mother cell and
bud, and it is formed during a specific period of the cell cycle. The primary
septum appears to meet all the foundamental conditions required in a suitable
system for the study of morphogenesis at the molecular level. In order to
understand how septum formation is triggered at a certain time and site of the
cell, it is essential to know the enzymatic components necessary for chitin
synthesis and to build a model.

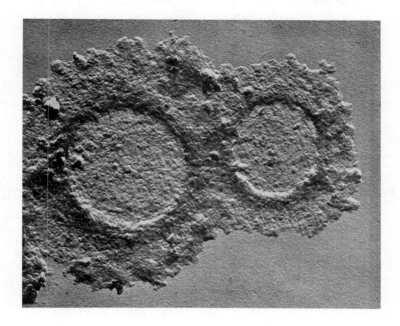

Fig. 1.4. Two chitin primary septa in *Saccharomyces cere-*
visiae after extraction of the other components of the
bud scars by chemical and enzymatic methods. Magnification
× 30,000; metal shadowing. Courtesy of E. Cabib.

ENZYMATIC MODEL OF CHITIN SYNTHESIS.

Properties of chitin synthetase.

The biosynthesis of chitin represents the first case in which substantial
evidence was presented for the formation of a polysaccharide from a sugar
nucleotide. The original report by Glaser & Brown (1957) was concerned with
the enzyme from *Neurospora crassa*. Also the systems of the fungi *Allomyces*
macrogynus (Porter & Jaworski, 1966), *Mucor rouxii* (McMurrough & Bartnicki-
Garcia, 1971; McMurrough, Flores-Carreon & Bartnicki-Garcia, 1971) and *Bla-*
stocladiella emersonii (Plessmann-Camargo, Dietrich, Sonneborn & Strominger,
1967; Selitrennikov, Allin & Sonneborn, 1976) were described in detail and
found to have similar properties. All these enzymes are particulate, and all
are activated by free N-acetylglucosamine. In several cases, activation by
chitodextrin and stimulation by divalent cations were observed. The reaction
product was generally chitin and, in some cases, also chitobiose. *Blastocla-*
diella emersonii zoospore germination is accompanied by de novo construction

of a chitinous cell wall. A large increase in total hexosamine occurs during germination to build up chitin. While the enzymatic apparatus to provide substrates for the chitin synthetase reaction is present in the zoospore, it does not in fact contain sufficient substrates: the flux of metabolites in the pathway to uridine diphosphate-N-acetylglucosamine must undergo an abrupt increase which would involve relief from inhibition of activity of the first enzyme in the pathway. Lowering of the uridine diphosphate-N-acetylglucosamine concentration, perhaps induced by the chitin synthetase reaction, may be an explanation; however, it should be recalled that uridine nucleotides, especially UTP and UDP, counteract the inhibitory effects of uridine diphosphate-N-acetylglucosamine. Thus, an elevation of UTP concentration could be equivalent, on relieving inhibition of the first enzyme, to a decreasing of uridine diphosphate-N-acetylglucosamine concentration.

Keller & Cabib (1971) obtained a particulate preparation of chitin synthetase from both *Saccharomyces carlsbergensis* and *Saccharomyces cerevisiae* by lysis of yeast protoplasts. Gooday & Rousset-Hall (1975) prepared chitin synthetase from *Coprinus cinereus* and reported that these preparations needed only a divalent cation and uridine diphosphate-N-acetylglucosamine for chitin synthesis and showed no primary requirement. The sole glycosyl product was characterized as macromolecular chitin. They incubated the particulate enzyme with other nucleotide sugars: uridine diphosphate-$|$U-^{14}C$|$ glucose, uridine diphosphate--$|$U-^{14}C$|$ galactose and guanosine diphosphate-$|$U-^{14}C$|$ glucose with and without the addition of uridine diphosphate-N-acetylglucosamine. In no case was a chromatographically immobile product (i.e. chitin) formed and no glucan or galactan synthetase activity was found under these conditions.

Chitin synthetase, as described by Rousset-Hall & Gooday (1975), McMurrough & Bartnicki-Garcia (1971) and by Plessmann-Camargo, Dietrich, Sonneborn & Strominger (1967) does not show hyperbolic Michaelis-Menten kinetics, and the Lineweaver-Burk plot double reciprocal plots are not linear. These plots for other fungal preparations appear to be linear only at higher substrate concentrations, usually in the presence of N-acetylglucosamine. It was suggested that N-acetylglucosamine is a positive allosteric effector, binding to the enzyme to cause a conformational change so that it more readily binds uridine diphosphate-N-acetylglucosamine.

The mechanism of action of N-acetylglucosamine and other activators, and their significance in vivo, are not clear. Concentrations of free N-acetylglucosamine at the level found effective in vitro are extremely unlikely to occur in the cell. McMurrough & Bartnicki-Garcia (1971) have detected a very small incorporation of labeled N-acetylglucosamine into chitin in the presence of unlabeled uridine diphosphate-N-acetylglucosamine, and Cabib, Ulane & Bowers (1975) have confirmed these results with the yeast enzyme. It is conceivable that N-acetylglucosamine serves as a primer, but this appears unlikely because chitodextrins are ineffective in the yeast system. Another possibility is that N-acetylglucosamine is first transferred from uridine diphosphate-N-acetyl-glucosamine to an unknown acceptor, X, forming an intermediate. In a further step, the aminosugar would be transferred to a growing chitin chain. N-acetyl-glucosamine would prevent the hydrolysis of the intermediate by a mass action effect; in vivo the intermediate may be protected by hydrophobic environment, and the stabilizing effect of the free aminosugar would be necessary.

Zymogen + activating factor + inhibitor system.

Heat-stable inhibitors might act either by preventing activation of the
zymogen or by inhibiting active chitin synthetase. The availability of zymogen
and activating factor in separate fractions permitted investigation of these
alternatives. It was found that the inhibitor was devoid of action on
previously activated chitin synthetase, but blocked completely the activation
step (Cabib & Farkas, 1971). The relationship between the different components
of the chitin synthetase system according to Cabib, Farkas, Ulane & Bowers
(1973) is:

$$\text{Chitin synthetase zymogen}$$

$$\text{Activating factor} \longrightarrow \Big\downarrow \mathbf{X} \longleftarrow \begin{array}{l}\text{Activating factor} \\ \text{inhibitor}\end{array}$$

$$\text{Active chitin synthetase}$$

$$n \text{ UDP-N-acetylglucosamine} + \text{particle} \longrightarrow n \text{ UDP} + (\text{N-acetylglucosamine})_n\text{-particle}$$

Chitin synthetase zymogen.

Chitin synthetase behaves like a zymogen, which can be activated in vitro to
give a ten-fold increase in activity by trypsin as well as by a chitin
synthetase activating factor from yeast. Cabib & Farkas (1971) have proposed
that this factor is responsible also for in vivo activation of chitin
synthetase: according to their model this factor is brought into contact with
the plasmalemma-associated pre-chitin synthetase in the area of bud constriction, thus initiating the formation of the septum.

The function of yeast bud scar ring has been suggested as being to stabilize
mechanically the constriction between the mother and the emerging daughter
cell (Cabib & Bowers, 1971; Hayashibe & Katohada, 1973). Evidence from the
literature in the last few years suggests that chitin is synthesized from the
very beginning of the budding process. These pieces of evidence include the
timing of chitin synthesis in synchronous cultures, the early appearance of
electron-transparent areas (Marchant & Smith, 1968) which constitute a part of
the evolving bud scar rim and which are probably composed of chitin (Cabib &
Bowers, 1971), and of wall structures stainable by primulin and brightener CFW
(Seichertova, Beran, Holan & Pokorny, 1973).

A new regulatory model was suggested by Hasilik (1974) which attempts to bring
the morphology and chronology of chitin synthesis into accord with these
observations. In the cell wall area in which a bud is to be formed, a tran-
sient increase in the activity of chitin synthetase per unit area occurs.
Proteinase B is suggested to catalyze a sequential activation and inactivation
of chitin synthetase. It is reasonable to assume that the maximum activity of
proteinase B will occur in the center of the area in which the bud is to be
formed. Correspondently, the activation-inactivation sequence will be rapidly
completed in the center, while proceeding much slower at the perifery, allowing
formation of a ring of chitin. The delayed termination of chitin synthesis in
this area would be a prerequisite for its participation in the septum formation.
This latter event has been discussed by Cabib & Farkas (1971).

When appropriate precautions are taken to prevent action by the activating

factor, chitin synthetase can be isolated in an almost completely inactive
form (Cabib, 1972; Cabib & Ulane, 1973). The zymogen appears to be very stable.
No loss of latent activity was detected when the particles were stored for one
year at -70° in a suspension containing 33 % glycerol. Even after treatment
with cholate, most of the zymogen could be converted into active enzyme with
either activating factor or trypsin.

Knowledge about the structure of the zymogen and its precise relationship to
the active enzyme must await its solubilization and purification. The zymogen
content of the cell appeared to be practically constant, regardless of the
growth phase or of the medium used.

Duran, Bowers & Cabib (1975) carried out a pretreatment of yeast protoplasts
with concanavalin A to obtain intact membranes. The bulk of the chitin
synthetase was found in the zymogen form. Glutaraldehyde inactivated chitin
synthetase when it was added to a lysate, but not when applied to intact
protoplasts. Chitin synthetase is so oriented in the membrane that it is also
accessible from the inside of the cell. These results confirm that the chitin
synthetase zymogen is associated with the plasma membrane, a basic assumption
for the explanation of localized activation of the enzyme and initiation of
septum formation.

Activating factor.

The properties of the activating factor are those expected from an enzyme. It
catalyzes a time-dependent and concentration-dependent activation of zymogen
and is inactivated by heating at 100°. The nature of its action was assumed to
be proteolytic, when it was found that trypsin could mimic its effect. The
availability of an inhibitor of the activating factor help to prove this point.
It was found that the neutral proteolytic activity present in preparations of
activating factor could be completely blocked by adding purified inhibitor
(Cabib & Ulane, 1973). Activation of zymogens by proteolytic action is a well
known process in biological systems, as found in gastric and pancreatic
enzymes, in the proinsulin-insulin transformation and in blood clotting.
The activating proteinase has been purified by affinity chromatography by
Ulane & Cabib (1976) and found to be active within a wide range of pH values,
with an optimum between 6.5 and 7.0. The enzyme has been identified as
proteinase B (see Tab. 1.2) and its specificity, as compared to trypsin, has
been shown.

Inhibitor.

The inhibitor behaves like a heat-stable protein; this property has been used
to advantage by extracting the inhibitor from yeast in hot water to a good
degree of purity. The inhibitor of the yeast protease which activates chitin
synthetase zymogen has been purified from *Saccharomyces cerevisiae* by Ulane &
Cabib (1974). The molecular weight of the inhibitor, as determined by gel
filtration, is estimated to be about 8,500 with no evidence of the presence of
subunits. No cysteine, methionine, arginine and tryptophan has been found; the
terminal group is threonine. The inhibitors from *Saccharomyces cerevisiae* and
Saccharomyces carlsbergensis are of equivalent molecular size, but differ in
electrical charge.

PRODUCTION OF CHITIN SYNTHETASE.

According to Ruiz-Herrera, Sing, Van der Woude & Bartnicki-Garcia (1975), the production of chitin synthetase can be made by the following procedure; spores of *Mucor rouxii* (1.5×10^8) are inoculated into 0.5 l of liquid yeast extract + peptone + glucose medium and incubated in a shaker at 28° for 12 hr. A gas stream (30 % CO_2 + 70 % N_2) is bubbled through the medium during the entire incubation period.

Cell-free extract preparation.

Yeast cells from 1.5 l of medium were harvested on sintered-glass filters, washed twice with 200 ml of ice-cold phosphate buffer, pH 6.0, containing 10 mM $MgCl_2$, resuspended in 20 ml of buffer, mixed with 20 ml of glass beads and shaken for 30 sec in a homogenizer. Cell-free extracts were centrifuged at 1000 *g* for 5 min to remove whole cells and cell walls; the supernatant was recentrifuged at 54,000 *g* for 45 min. The soluble supernatant and the pellet were separated and the mixed membrane fraction was resuspended in 4 ml of buffer and treated with 500 µg of acid protease from *Rhizopus chinensis* at 22°. This treatment is necessary to activate chitin synthetase zymogen (Ruiz-Herrera & Bartnicki-Garcia, 1974); it has no detectable effect on the subsequent "solubilization". After 30 min, 15 ml of the ice-cold buffer at pH 6.0 were added and the samples were centrifuged at 54,000 *g* for 45 min and washed once with buffer by centrifugation. The pellet was resuspended in buffer and incubated at 0° for 30 min with 2 mM uridine diphosphate-N-acetylglucosamine (substrate), 20 mM N-acetylglucosamine (activator) and 0.2 mM adenosine triphosphate. After this treatment, samples were centrifuged at 80,000 *g* for 1 hr to give transparent supernatant of "solubilized" enzyme.

About 15 to 20 % of the total chitin synthetase activity (but only 2 to 3 % of the total protein) remained buoyant. When this soluble supernatant was brought at room temperature, turbidity appeared in about 20 min, followed by precipitation of a fibrous material.

Isolation of unbound chitin synthetase.

The soluble fraction (54,000 *g* supernatant) of the crude cell-free extract was subjected to gel filtration in a Sepharose column. Active fractions were pooled and treated with 80 µg×ml^{-1} of bovine pancreas ribonuclease for 30 min at 30°. There was no loss of chitin synthetase activity by this treatment designed to eliminate ribosome contamination. The turbid suspension was centrifuged for 20 min at 10,000 *g* and the clear supernatant was fractionated.

For a chitin synthetase assay, reaction mixtures contained: labeled uridine diphosphate-N-acetylglucosamine, N-acetylglucosamine, adenosine triphosphate, magnesium chloride, potassium sodium hydrogen phosphate at pH 6 and acid protease. After one hr, glacial acetic acid was added to stop the reaction. The radioactivity incorporated into chitin was measured by filtration of the entire incubation mixture through glass filters for subsequent liquid scintillation counting. Ruiz-Herrera, Sing, Van Der Woude & Bartnicki-Garcia (1975), on the basis of their findings related to the electron microscopic observations of the product obtained by using the above procedures, concluded

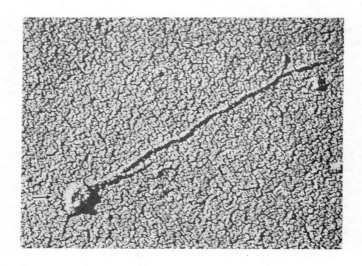

Fig. 1.5. Intimate association of chitin synthetase granule
with chitin microfibril ends. Two microfibrils appear
wrapped around each other. Magnification × 81,600. Ruiz-
-Herrera, Sing, Van Der Woude & Bartnicki-Garcia, 1975.

that chitin microfibrils are synthesized in vitro by granules of chitin
synthetase, as illustrated in Fig. 1.5, measuring about 35 ÷ 100 nm in
diameter.

The granules occur freely suspended in the 54,000 g supernatant (unbound
enzyme) or can be recovered from the mixed membrane fraction by substrate
induced "solubilization". The walls of the yeast and mycelial forms of *Mucor
rouxii* have significant levels of chitin synthetase activity and may conceiva-
bly contain similar enzyme granules. In the related fungus *Gilbertella persi-
caria*, Bracker & Halderson (1971) found globular elements associated with the
microfibrils.

The results obtained by the above mentioned authors demonstrate that the
synthesis of chitin chains and their assembly into microfibrils can be attained
with an enzyme preparation free of membranes; prove that the glycosyl transfer
from a nucleoside diphosphate sugar operates in the formation of cell wall
microfibrils; show that cell wall microfibrils can be formed in the absence of
a living cell or its membranes; support the belief that the polymer synthesis
and fibril assembly in the living cell takes place in the wall (Preston &
Goodman, 1968); suggest that other wall components, such as matrix materials,
are not indispensable for microfibril elaboration; support the end -synthesis
theory of microfibril elaboration via a terminal enzyme granule as advocated
by Preston (1974) and open the route to the in vitro assembly of a whole microfibrillar cell wall by a large chitin synthetase granule, i.e. a multienzyme

aggregate, which would collectively form the molecular chains and assemble
them into a microfibril. The elaboration of chitin microfibrils and the fine
structure of isolated chitinsynthetase particles, called chitosomes, from *Mucor rouxii* have been studied by Bracker, Ruiz-Herrera & Bartnicki-Garcia (to
appear). Chitosomes, in Fig. 1.6, are spheroidal organelles mostly 45 ÷ 65 nm
in diameter; they undergo a seemingly irreversible series of transformations
when substrate and activator are added. A coiled microfibril appears inside
the chitosome, the shell is opened and an extended microfibril arises from the
fibroid particle.

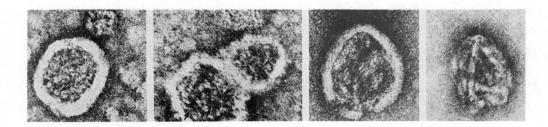

Fig. 1.6. Transformations of chitosomes. From left to right:
showing internal microgranules; incubated, with microgranules and fibrous elements; with fibroid material; coiled
and bent fibrillar material after shedding of the shell.
Magnification × 344,000. Courtesy of S. Bartnicki-Garcia.

CHITIN ASSOCIATIONS.

The chitinous structures, with the possible exception of those constructed by
molds and fungi, are built up of a glycoprotein framework, in which chitin is
covalently linked to proteins. At this level of organization, the chitinous
structures exhibit high tensile strength but are essentially pliable and
flexible, allowing movements and limited expansion (Jeuniaux, 1975; Hepburn,
1976). After sclerotization of the proteins, the resulting chitin-tanned
protein complex gains considerable stability and confers to the structure
hardness, rigidity and resistance to enzymatic hydrolysis.

Sclerotization of insect cuticles can follow two routes: oxidation of N-acetyl-
dopamine in either the aromatic ring leading to quinone tanning, or in the β-
position of the aliphatic side chain, leading to an uncolored type of cross-
link. In Crustacea, the presence of di- and tri-tyrosine as well as some
brominated derivatives has been demonstrated by Welinder, Roepstorff & Andersen
(1976) in the calcified exocuticle; this indicates an alternative way for
stabilizing organic matrix, besides quinone tanning. Quinones should be formed
by the action of a diphenol oxidase upon lower diphenols, and should form
cross-links between peptide chains by reacting with available amino groups.

The chitin-protein complexes can be completed by the deposition of other substances, such as waxes and lipoproteins, giving the structures the property of impermeability. Such chitinous structures have been extensively exploited by animals in the development of a number of different functions. The biosynthesis of chitin-protein complexes has been subjected, during the course of animal evolution, to a number of "morphological radiations".

Chitinous structures have been primarily exploited as protective envelopes, forming the theca and cyst walls of some Rhizopoda and Ciliata, the periderm or perisarc of Hydrozoa and Endoprocta, the tubes of Phoronida and Pogonophora, the shells of inarticulate Brachiopoda and of Mollusca. Chitinous structures are also used as envelopes of eggs or of latent forms of life such as cysts. By providing an adequate support for muscle insertions, chitinous structures contribute to the formation of locomotor appendages or chaetae and of prehensile and masticatory organs such as jaws and radula. In Brachiopoda, a chitinous cuticle is used to insure the fixation of the organism to the substrate. Some buoyancy organs are also built of chitin, as in Siphonophora and Cephalopoda. The chitinous peritrophic membrane of Arthropoda and Annelida plays a protective role with respect to the intestinal mucosa and a role in the formation of the feces.

The chitin-protein framework seems to provide a convenient support for calcification and silicification; as a result of this process of mineralization of their chitinous structures, the organisms appear to realize an economy of protein material, as do many crustaceans, or an economy of both chitin and protein, as in the case of molluscs. On the other hand, calcified exoskeletons also provides a rigid structure insuring the stability of colonies such as in Hydrozoa and Bryozoa.

At the top of the evolutionary lineage of protostomian invertebrates, the chitin-protein complexes in their most stable tanned form allow the formation of rigid planes at the expenses of a minimal weight of material. Insects have developed extended portions of their teguments, which, owing to their rigidity and to their low weight have successfully been transformed into functional wings. The wing hinges have the particular properties of a rubber-like cuticle due to the linkage of chitin to resilin. The realization of the flight system by insects is the consequence of the particular features of the different types of glycoprotein complexes that chitin can form by linkage with arthropodin, sclerotin or resilin.

Chitin bound carotenoids.

Carotenoids occur in many different species of insects, usually as conjugates with proteins and with chitin. The colors of most carotenoids are yellow, red or orange and these are the colors they give to the tissues. Carotenoids produce the green colors of many insects. They include lutein, β-carotene, astaxanthin and their derivatives.

The pink or red color of crustacean shells is also due to carotenoids: the crab *Taliepus nuttallii* possesses a red-purplish epicuticle, for instance; crabs, as many animal species, convert ingested yellow plant carotenoids into oxygenated, and thus more polar, orange or red ketoderivatives and in some instances, conjugate the latter to give chromoproteins or calcareous esters, like in hydrochoral skeletons. Crabs find an ample source of carotene in brown seaweeds but the carotene function in their metabolism is unknown.

While calcareous exoskeletons of marine crustaceans are relatively rich in chitin, but poor in protein content (88 % *vs* 12 % respectively in the organic portion of *Carcinus maenas*), insects and arachnids have relative proportions of hardened protein to chitin of the reverse order.

Very little pigment can be recovered from chitinous material from crustacea by exposure to the usual carotenoid solvents and/or protein denaturants, unless the shells have been decalcified. However, aqueous citric acid, disodium salt of ethylenediaminetetraacetic acid and hydrochloric acid dissolve calcium carbonate but leave all pigments firmly associated with the tough and pliable chitinous pieces; carotenoids, then, are not bound to the calcareous part. It has also been discounted that carotenoids are bound to the scleroproteins as occurs for instance in some feathers. Boiling chips of red carapace in strong alkali, before or after decalcification, does not release the carotenoids: the absence of protein in the samples treated in this way indicates that carotenoids are bound to chitin itself.

Decalcified chips of cleaned carapace yield their pigments to ethanol or acetone; warm acetic acid diluted 1:1 extracts the carotenoids while attacking the carbonate material; other ways of removing carotenoids are cold formic acid on carapace and mixtures of ammonium sulfate + sulfuric acid on chitosan. In view of this experimental evidence, it can be said that carotenoids are combined with chitin amino groups by carbonylamino or Schiff's base linkages, according to Fox (1973):

$$R_2C{=}O \; + \; RCH_2NH_2 \; \rightarrow \; R_2C{<}^{OH}_{NHCH_2R} \; \rightarrow \; R_2C{=}NCH_2R \; + \; H_2O$$

Table 1.4. Actions of reagents on chitin-bound carotenoids.

Reagent	Chitin	Pigment
Aqueous sodium hydroxide	none	none
Dil. hydrochloric acid, hot	dissolves carbonate	none
Conc. hydrochloric acid, hot	dissolves carbonate and chitin	releases as red solution
Conc. nitric acid, hot	dissolves	destroys
Dil. sulfuric acid with ammonium sulfate	none	releases
Pyridine, hot	none	leaches and dissolves
Ethanol or acetone, warm	none	dissolves readily, if decalcified
Potassium borohydride	none	none

The action of several reagents on chitin-bound carotenoids are summarized in Tab. 1.4.

In the *Taliepus* material there are one 4-keto and three 4,4'-diketo-β-carotene derivatives firmly bonded to the exoskeletal chitin.

During the chitosan production operations, the carotenoids escape removal as the aqueous sodium hydroxide solutions at high temperature are not effective in this respect, and therefore chitosan is often yellowish or grey.

Muzzarelli & Rocchetti (1974) have published data on the enhanced capacity of chitosan for transition metal ions, after treatment with sulfate + sulfuric acid solutions; chitosan in fact, is not only a basic polymer that can collect oxyanions, but is also a chelating polymer (Muzzarelli, 1973). The treatment mentioned introduces a higher degree of crystallinity in chitosan, while removing carotenoids: these so far uncorrelated facts are perhaps reciprocally dependent. In other words, the removal of carotenoids by sulfuric acid and sulfate makes available more free amino groups and enhances the capacity of chitosan for transition metal ions; in addition to this, the removal of carotenoids allows a better orientation of the polymeric chains, with sharper and more numerous x-ray diffraction lines.

Chitin and glucans.

The composition and the architecture of the fungal wall is now known in a variety of fungi. A basic pattern of organization of the hyphal walls of *Agaricus bisporus* and *Schizophyllum commune* includes an inner layer composed of chitin microfibrils in a β-glucan matrix also containing protein, and an outer layer of KOH-soluble α-glucan; a layer of β-glucan mucilage may also be present on the outer surface. Treatment with β-glucanase is essential before chitinase can effectively attack the KOH-extracted walls; thus, β-glucan and protein may form the matrix of a fibrillar chitinous layer which forms the inner compact surface of the wall, but is looser and uneven on the outer side.

Quantitative differences among fungi have been revealed by chemical analysis: for instance *Agaricus bisporus* (isolated walls) contains 7 to 8 times more chitin and protein than *Schizophillum commune*. O'Brien & Ralph (1966) examined the alkali-insoluble portion of walls of Basidiomycetes and Ascomycetes whose chitin content varies from 26 to 65 % and glucan content from 22 to 67 %. A model of the *Agaricus bisporus* wall based on the observations of Michalenko, Hohl & Rast (1976) is presented in Fig. 1.6: a water-soluble, poorly organized glucan mucilage lies on the wall. The interface of KOH-soluble glucan and β-glucan + chitin is irregular and there is a compositional and structural gradient across it due to the loosely-textured side of the β-glucan. The concept of completely discrete layers of single components (α-glucan, protein, β-glucan and chitin) has been rejected.

The chemical composition of *Agaricus bisporus* cell walls was determined by Temeriusz (1975) after hydrolysis at 110° for 2 hr in 0.1 N hydrochloric acid: the hydrolyzate, obtained with a 65 % yield, consisted of 22 % aminoacids, 72 % chitin, 3 % cellulose, glucosamine and mineral salts.

The mycelial cell walls of Ascomycetes are composed largely of a complex mixture of polysaccharides including neutral glucans together with chitin. On complete hydrolysis, glucose and glucosamine are the main products, accompanied by galactose, mannose, arabinose and galactosamine. A 50 % portion of the cell wall polysaccharides remains insoluble even in alkali; it is composed of chitin and glucans. When the insoluble material is submitted to an acetylation procedure identical to that used for the preparation of cellulose acetate, it is possible to solubilize a large portion of the glucan as acetate.

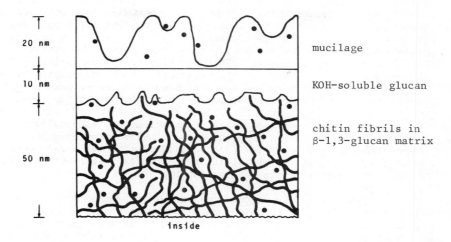

Fig. 1.7. Model of the hyphal wall of *Agaricus bisporus*.
Cystine-containing protein (●) occurs throughout the
wall. Michalenko, Hohl & Rast, 1976.

The action of nitrous acid on the insoluble fraction brings about striking
changes in solubility with respect to both aqueous sodium hydroxide and dimethylsulfoxide. Nitrous oxide causes oxidative deamination, as discussed in
Chapter 3.

Although it is not known with certainty whether covalent linkages occur between
anhydroglucose and anhydroglucosamine residues in cell walls, the isolation of
chloroform-soluble glucan acetates containing only small amounts of nitrogen
indicates that the alkali-resistant substance is not a heteropolymer, but it
is likely composed of chitin and glucan which may or may not be bound to one
another (Stagg & Feather, 1973).

The predominant linkages in the glucan in the Ascomycetes preparations are
1-3 and 1-4. Nuclear magnetic resonance scans on the methylated fractions
obtained after nitrous acid treatment indicate 80 % α and 20 % β glycosidic
linkages as evidenced by the chemical shift values for the C^1 proton signals
against known standards.

The presence of β-(1→3)-glucans in fungi has been reported by Bartnicki-Garcia
(1968) In the case of pathogenic fungi like *Paracoccidioides brasiliensis*,
Blastomyces dermatitidis and *Histoplasma capsulatum* and *farciminosum*, the main
difference is the amount, rather than the nature, of the constituent. In all
cases studied, the β-(1→3)-glucan seemed to be an essentially linear polysaccharide (Kanetsuna & Carbonell, 1970 and 1971; San-Blas & Carbonell, 1974).

The possibility of the existence of a compact mesh between the chitin and the
β-(1→3)-glucan is indicated by the more drastic hydrolysis brought about by
β-(1→3)-glucanase and chitinase together. One given polymer is exposed to the
action of the corresponding enzyme upon partial removal of the accompanying

polymer, exposing in this way hidden fibrils. According to San-Blas & Carbonell (1974), the different kinds of fibers do not mix in the cell walls but rather appear as conglomerates where only one type of fiber is found. The cell wall seems to be formed by a juxtaposition of zones in which either chitin or β-(1→3)-glucan or accompanying α-polysaccharides are present. This is in part sustained by the preferential synthesis of glucosamine on the apical region of the hyphae and in the buds of the yeast cell (Cabib & Bowers, 1971) and the localization of chitin in the bud scar rims of *Saccharomyces carlsbergensis* and *cerevisiae* (Seichertova, Beran, Holan & Pokorny, 1975).

The *Polyporus tumulosus* cell walls contain four polysaccharides, including chitin, α-(1→3)-glucan, β-(1→3)-glucan and xylomannan. Wessels, Kreger, Marchant, Regensburg & De Vries (1972) indicated that the α-(1→3)-glucans are located on the outer surface of the hyphal walls as microcrystalline layers, with a characteristic morphological structure.

In the structures mentioned, chitin is usually predominant. Cellulin granules found uniquely in the Leptomitales, were isolated by Lee & Aronson (1975) from *Apodachlya* sp. The granules, prepared free of cell walls and cytoplasmatic contaminants, contain 60 % chitin and 39 % β-(1→3)- and β-(1→6)-glucans.

Recent findings on the ultrastructure of the cell wall of a dermatophyte, *Epi-*

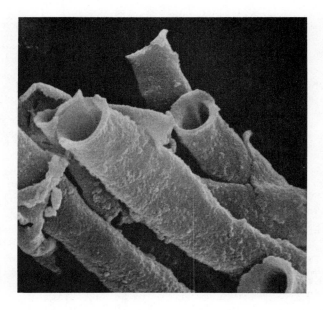

Fig. 1.8. Purified cell walls of *Epidermophyton floccosum* showing fibrous texture and absence of cytoplasmatic content. Magnification × 8,000. Courtesy of Y. Nozawa.

dermophyton floccosum, throw light on the nature of the association of chitin with polysaccharides and proteins. Mycelia broken by blendor disruption, after prolonged fractionation and sonication, showed that the cytoplasmatic contents inside the mycelium are removed. After acetone drying it was possible to observe the inside wall of the mycelium: in general, as shown in Fig. 1.8, rough external and smooth internal surface structures are observed. Short and thick fibrils in bundles are observed on the outer surface, whilst long and thin fibrils are seen on the inner surface. These fibrils were largely digested by chitinase and their x-ray diffraction spectrum corresponded to the chitin spectrum.

Chitin in these filamentous fungi is accompanied by other polysaccharides; the alkali/acid resistant fraction is 95 % N-acetylglucosamine. The chitin core is masked by glucan and galactomannan fibrils which together constitute the cell wall material in this fungus (Nozawa, Kitajima & Ito, 1973).

The involvement of protein in the cell walls has not been clarified, yet it can be considered that protein molecules may serve to link chitin to carbohydrates. The aminoacid analysis of *Epidermophyton floccosum* shows significant amounts of threonine, serine, aspartic acid and glutamic acid which may contribute to linkages between carbohydrates and protein moieties in the cell walls. Also noteworthy is the high content of alanine and glycine, since both aminoacids are involved in the cross-linking of peptidoglycans in bacterial cell walls.

Chitin associations with proteins.

Experimental evidence of covalent bonds between chitin and proteins has been obtained by many authors. N-Acetylglucosamine and chitin can react with α-aminoacids, especially tyrosine, peptides and cuticular proteins (Hackman, 1955), to give stable complexes, dissociable, however, upon pH changes. The organic material of insect cuticles, decalcified crab cuticles and squid skeletal pen dispersed in lithium thiocyanide, can be reprecipitated with acetone to give a series of chitin + protein fractions, presumably in the form of a glycoprotein complex (Hackman, 1960; Hackman & Goldberg, 1958; Foster & Hackman, 1957). The proportion of protein in these complexes varies according to the isolation method.

It seems that chitin, whether in its α- or β- form is covalently linked to arthropodins or sclerotins to form more or less stable glycoproteins probably through aspartyl and histidyl residues. Owing to the stability of these complexes in hot alkali and their instability in hot acids, the linkage could probably be as in N-acetylglucosamine, that is, between a carboxyl to the amino group of glucosamine. Other types of bond may also occur. Stegemann (1963) reported that, in the internal shell of cephalopods, the greater part of the protein moiety (conchagen) can be removed by hot water, while the remaining chitin is bound to a protein containing large amounts of aspartic acid.

Arthropodins are characterized by low proportion of glycine, absence of sulfur aminoacids, and high proportion of tyrosine. Fraenkel & Rudall (1947) and Richards (1947 and 1951) have shown that the molecular spacings of this protein in the extended configuration agree with those of chitin; the identity of these lattice spacings would permit a mixed crystallization and chemical bonding. Arthropodins are soluble in water at high temperature; a suitable solvent for their extraction has been proposed by Trim (1941) and their analyses have been performed by many authors (Jeuniaux, 1971). Sclerotins are water-insoluble.

Enzymic studies and acidic hydrolysis of cuticles have also led to the conclusion that a stable linkage exists, proteins being linked to chitin through a non-aromatic aminoacid (Lipke & Geoghegan, 1971; Lipke, Grainger & Siakotos, 1965). Fragments of mucoprotein with glycosyl, N-acetylglucosaminyl and peptidyl residues were isolated.

Chitin, as it occurs naturally in association with proteins, might be expected to have properties slightly different from those of chitin when isolated as a powder: for instance, the enzymic digestion of chitin preparations is far from complete, while the chitin in the endocuticle at the time of molting is digested by enzymes.

Herzog, Grossmann & Lieflaender (1975) have pointed out that lithium rhodanide, urea and anhydrous formic acid have little effect on the stability of the chitin-protein complex, while pronase and papain remove most of the protein; anhydrous formamide at 140° and 1 N NaOH at 100° remove all of it.

On the basis of this summarized information and the more detailed information on chitin stereochemistry and enzymic degradation to be found in the following Chapters, it can be said that the isolation of chitin, for both research and industrial production purposes, is a rather delicate operation that should take into account many facts if reproducible results and grade definition are desired.

CHITOSAN PRODUCTION BY CHITIN DEACETYLASE.

The first observation about the occurrence of chitosan in plant cell walls and, more broadly, in living systems was made by Kreger (1954) who reported its occurrence in the mycelia and sporangiophores of *Phycomyces blakesleeanus*.

Chitosan was later demonstrated to be the most abundant component of the cell walls of filamentous and yeast-like forms of *Mucor rouxii* grown respectively under air and under carbon dioxide. The course of breakage and purification of cell walls was followed by light microscope: in preparations treated with iodine solutions the cell walls became pink to violet in color, contrasting with either the bright yellow color of cytoplasmic debris or the reddish-brown staining of intact cells. This differential staining is based on the reaction of I_3^- with chitosan, which, as reported in Tab. 1.5, is the major component of the cell walls of *Mucor rouxii*. Only a minor fraction of the aminosugar was found to be soluble in alkali, while it was readily soluble in acetic acid; the Elson-Morgan reaction was used to determine the 2-aminosugar on the acid hydrolyzate. The x-ray diffraction spectra were recorded on the cell walls of *Mucor rouxii*, and electron microscope observations were made after digestion with 1 N hydrochloric acid at 100° for 30 min to remove most of the protein and non-nitrogenous carbohydrate as well as all the chitosan to observe chitin microfibrils.

Chitosan, as a polycation, is probably neutralized by the concurrent presence of large amounts of phosphate and some polyuronides in the cell walls of *Mucor*

Table 1.5. Composition of cell walls of *Mucor rouxii*. Bartnicki-Garcia & Nickerson, 1962.

Component	Filaments %	Yeasts %
Readily extracted lipids	2.0	0.8
Bound lipids	5.8	4.9
Chitosan	32.7	27.9
Chitin	9.4	8.4
Unidentified sugars	2.4	3.1
Fucose	3.8	3.2
Mannose	1.6	8.9
Galactose	1.6	1.1
Other carbohydrates	1.7	0.9
Protein	6.3	10.3
Purines and pyrimidines	0.6	1.3
Phosphate	23.3	22.1
Magnesium	1.0	n.d.
Calcium	1.0	n.d.

Table 1.6. Chemical differentiation of the cell wall in the life cycle of *Mucor rouxii*. Bartnicki-Garcia, 1968.

Wall component	Yeasts	Hyphae	Sporangiophores	Spores
Chitin	8.4	9.4	18.0	2.1
Chitosan	27.9	32.7	20.6	9.5
Mannose	8.9	1.6	0.9	4.8
Fucose	3.2	3.8	2.1	0.0
Galactose	1.1	1.6	0.8	0.0
Glucuronic acid	12.2	11.8	25.0	1.9
Glucose	0.0	0.0	0.1	42.6
Protein	10.3	6.3	9.2	16.1
Lipid	5.7	7.8	4.8	9.8
Phosphate	22.1	23.3	0.8	2.6
Melanin	0.0	0.0	0.0	10.3

rouxii. A peculiar feature is the absence of glucose polymers (Bartnicki-Garcia & Nickerson, 1962). During the life cycle of *Mucor rouxii*, the chemical differentiation of the cell walls takes place, as reported in Tab. 1.6, where it is evident that the highest proportion of chitosan is present in the hyphae stage (Bartnicki-Garcia, 1968).

In view of the well established pathway of chitin biosynthesis as summarized in Tab. 1.2, Araki & Ito (1974 and 1975) verified that chitosan is produced through enzymatic deacetylation of chitin in eleven microorganisms.

Preparation of chitin deacetylase.

Mucor rouxii was grown under shaking at 30° for two days in a culture medium containing 0.3 % yeast extract, 1 % polypeptone and 2 % glucose. The pH was adjusted to 4.5. The mycelia (damp weight 73 g), obtained from a 10-1 culture by filtration through a sintered-glass funnel, were disrupted by grinding with glass powder in a mortar at 2° for one hr. After dilution with 20 mM Tris-Cl buffer, pH 7.2 (final volume 400 ml per 25 g mycelia), the homogenate was centrifuged at 2500 g for 20 min. The resulting supernatant was used as the crude enzyme (Araki & Ito, 1975).

Chitin deacetylase was partially purified as follows: a portion (200 ml) of the 20,000 g supernatant fraction was brought to 63 % saturation with ammonium sulfate at 4° by the addition of solid ammonium sulfate and centrifuged at 20,000 g for 15 min. The supernatant was brought to 85 % saturation by adding further solid ammonium sulfate. The precipitate collected by centrifugation was dissolved in and dialyzed against 5 mM sodium acetate buffer, pH 5.3, and then applied to a 20-ml carboxymethylcellulose column in the same buffer and eluted with a linear gradient of sodium acetate, pH 5.5., from 5 to 100 mM at a flow rate of 60 ml×hr^{-1}. The enzyme eluted with about 50 mM acetate. The pooled active fractions were concentrated to 1 ml by filtration through a

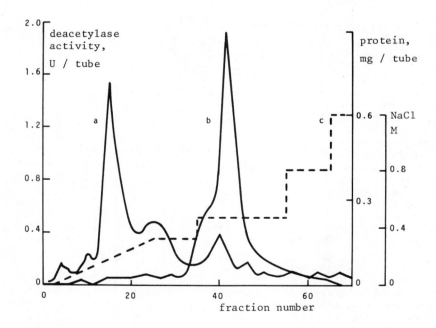

Fig. 1.9. Elution pattern of chitin deacetylase from a 1 × 3.5 cm DEAE-cellulose column. a = protein; b = deacetylase activity; c = sodium chloride. Araki & Ito, 1975.

collodion bag, dialyzed against 20 mM Tris-Cl, pH 7.2 and then put on a 3-ml diethylaminoethylcellulose column in 20 mM Tris-Cl, pH 7.2. As shown in Fig. 1.9, the column was eluted at a flow rate of 12 ml×hr^{-1} successively with a linear gradient (40 ml) of sodium chloride from 0 to 0.3 M, 15 ml 0.3 M NaCl and 30 ml 0.45 M NaCl, each in 20 mM Tris-Cl, pH 7.2, and the active fractions were pooled to be used as the purified enzyme.

O-Hydroxyethyl-|acetyl-^{3}H|-chitin, whose preparation is reported on page 123, was used together with unlabeled O-hydroxyethylchitin as substrate for the assay of chitindeacetylase. Unlabeled O-hydroxyethylchitin was prepared from O-hydroxyethylchitosan, using unlabeled acetic anhydride. Colloidal chitin, whose preparation is reported on page 57, was also used a substrate.

Properties of chitin deacetylase.

Under standard assay conditions, a linear relationship was obtained between amounts of enzyme (or time of incubation) and liberation of radioactivity until 8 nmol acetate had been released. The deacetylase reaction has a pH optimum at about 5.5, and the activity at this pH varies markedly with the buffer employed. This effect of the buffer on deacetylase activity appears to be ascribable to inhibition by the anionic component of the buffer. Thus, some kinds of carboxylic acid, particularly acetic acid, inhibited the deacetylase reaction in Tes buffer at pH 5.5 , while formic acid or isobutyric acid had little effect. The inhibition by acetic acid appears to be competitive and a K_i value of 13.3 mM was determied for this acid (see Fig. 1.10). The K_m value for glycolchitin was 0.87 g×l^{-1} or 2.6 mM with respect to monosaccharide residues. The reaction was also inhibited by glycolchitosan; the extents of inhibition were 30, 70 and 90 % at concentrations of glycolchitosan of 1, 2, and 9 µg per 50 µl, respectively. In contrast with peptidoglycan deacetylase, which has been shown to require cobaltous or manganous ions for full activity, chitin deacetylase does not require these metal ions. As indicated in Tab. 1.7, zinc slightly stimulates the enzyme.

Chitin deacetylase catalyzes the deacetylation of glycolchitin and N-acetyl-chitooligoses but not of cell wall peptidoglycan. Araki & Ito (1975) suggest that the 3-O-lactic ether moiety of the N-acetylmuramic acid residues in peptidoglycan probably interfere with the catalytic cleavage of the acetamido groups present not only in the N-acetylmuramic acid residues, but also in the N-acetylglucosamine residues. By the analogy of this case, a rather small extent of deacetylation about 30 %, obtained with glycolchitin as substrate, may be accounted for in terms of interference by the O-hydroxyethyl groups linked to the glucosamine residues at C^3 and C^6. Since in the reaction with this enzyme the tetramer of N-acetylglucosamine gave compounds presumed as the tetrasaccharides with one or three deacylated amino groups, this enzyme seems to be able to hydrolyze in vivo most of the acetamido groups of the natural substrate chitin. On the other hand, this enzyme is inactive toward the dimeric and monomeric N-acetylglucosamine derivatives including the chitin intermediate uridine diphosphate-N-acetylglucosamine, N-acetylglucosamine 1-phosphate and N-acetylglucosamine 6-phosphate (Tab. 1.8).

The occurrence of deacetylase appears to be biologically significant in two respects. First, the removal of the N-acetyl groups may have a great effect on the sensitivity of the polysaccharide to some specific enzyme such as lysozyme. Thus, chitin deacetylase as well as peptidoglycan deacetylase may function in

Table 1.7. Effect of divalent cations and other materials on deacetylase activity. The assay was carried out with 0.04 µg protein of the purified enzyme. The activity is referred to a control value. Araki & Ito, 1975.

Added substance	Concn., mM	Relative activity
none		100
2-Mercaptoetahnol	1	128
	10	120
EDTA	1	89
	10	14
zinc chloride	2	119
	10	94
cobalt chloride	1	73
	10	48
manganous chloride	1	68
sodium chloride	40	70

Table 1.8. Deacetylation of chitin and N-acetylchitooligoses. The substrate and the purified enzyme (1.9 µg protein) were in 0.5 ml of 50 mM Tes-NaOH buffer at pH 5.5; after incubation for 48 hr at 30°, free amino groups were assayed by the dinitrophenylation method. Araki & Ito, 1975.

Substrate	Substrate as glucosamine, nmol	Dinitrophenyl-glucosamine formed,	
		nmol	%
Glycolchitin	2000	244.0	12.2
Colloidal chitin	820	35.3	4.3
Powdered chitin	2000	30.3	1.5
N-acetylglucosamine	2000	0.0	0.0
Dimer	1640	0.9	0.05
Trimer	1130	16.8	1.5
Tetramer	1660	89.3	5.4
Pentamer	1620	230.0	14.2

converting cell wall structures to a form resistant to enzymatic degradation. The other possible significance of the deacetylase reaction relates to the polycationic nature of the product. In the cell walls of *Mucor rouxii* and *Neurospora crassa*, the polysaccharides which contain N-deacylated aminosugar residues are most likely present in combination with anionic materials such as polyphosphate and polyuronide.

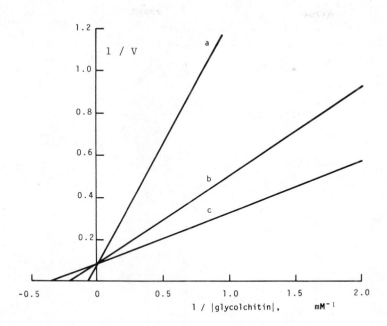

Fig. 1.10. Dependence of reaction rate on the concentration
of glycolchitin and acetic acid. Reciprocal plot of reaction
rate *vs* glycolchitin concentration. a = in 40 mM acetic
acid; b = in 10 mM acetic acid; c = in the absence of acetic
acid. Araki & Ito, 1975.

CHITIN BIOSYNTHESIS INHIBITION.

Polyoxins.

Polyoxins A to M are mixtures of peptidyl-pyrimidine nucleoside antibiotics
that are produced by a variety of *Streptomyces cacoi* and are being used as
fungicides in Japan. Polyoxins are competitive inhibitors of fungal chitin
synthetase.

In the context of research carried out mainly by Jan (1974), Otha, Kakiki &
Misato (1970), Hori, Kakiki, Suzuki & Misato (1971 and 1974) and by Hori, Eguchi, Kakiki & Misato (1974) on the action of polyoxins listed in Tab. 1.9 on
chitinsynthetase, it was found that the antibiotic polyoxin D inhibits the
chitinsynthetase activity in the cell-free system of *Piricularia orizae* and
Neurospora crassa. The inhibition is competitive respect to uridine diphosphate-N-acetylglucosamine and specific for the chitin synthetase, as the
results in Fig. 1.11 indicate. Studies on the structure of polyoxins indicate

Table 1.9. Polyoxins.

Polyoxin	R_1	R_2	R_3
A	$-CH_2OH$	$CH_3CH=C$ (COOH, N−)	$-OH$
B	$-CH_2OH$	$-OH$	$-OH$
D	$-COOH$	$-OH$	$-OH$
E	$-COOH$	$-OH$	$-H$
F	$-COOH$	$CH_3CH=C$ (COOH, N−)	$-OH$
G	$-CH_2OH$	$-OH$	$-H$
H	$-CH_3$	$CH_3CH=C$ (COOH, N−)	$-OH$
J	$-CH_3$	$-OH$	$-OH$
K	$-H$	$CH_3CH=C$ (COOH, N−)	$-OH$
L	$-H$	$-OH$	$-OH$
M	$-H$	$-OH$	$-H$
C		R $-OH$	
I		$CH_3CH=C$ (COOH, N−)	

that the mode of action of their inhibition of the biosynthesis of cell wall chitin is probably connected to their structural similarity to uridine di-phosphate-N-acetylglucosamine. In vivo, polyoxin D inhibits the growth and causes swelling of the mycelia of many fungi.

From the pK_i *vs* pH profiles for polyoxin L and M whose pK_a values for the amino groups are 7.0 and 7.4, a drop in the inhibition effect was noted at pH 7.2 and 7.4 respectively. Therefore, Hori, Kakiki & Misato (1974) concluded that this is due to the disappearence of the positive charge on one of the amino groups of the polyoxin. Since the position of the amino group is important for the binding of the polyoxins to the enzyme, the negatively charged group of the enzyme should have a specific position. An imidazole group has also been proposed as an essential part of the active site, which would interact with the pyrophosphate moiety of uridine diphosphate-N-acetyl-glucosamine in the catalytic process of the enzyme.

Trichoderma viride mycelia growing in the presence of polyoxin D become

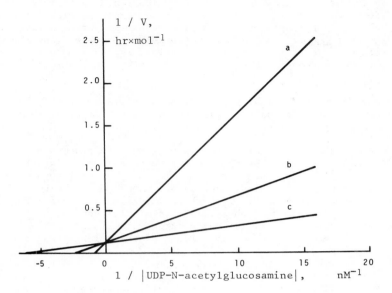

Fig. 1.11. Reciprocal plot of rate of N-acetylglucosamine
incorporation *vs* UDP-N-acetylglucosamine concentration.
a = in 0.05 mM polyoxin D; b = in 0.005 mM polyoxin D;
c = in the absence of polyoxin D. The apparent K_m for
UDP-N-acetylglucosamine is 0.6 mM. Jan, 1974.

irregular and lose their rigidity, showing bulges along the hypha. Under the
electron microscope, the features of the cell wall and cytoplasmatic content
appear normal; nevertheless, incubation with ^{14}C-glucose permits one to see
that polyoxin D (100 µg×ml^{-1}) partially inhibits the biosynthesis of chitin up
to 83 % (Benitez, Villa & Garcia-Acha, 1976).

Insecticides.

One of the most widely known insecticides in the past, DDT, was scarcely
studied in its connections to chitin: Richards & Cutkomp (1946) showed that
chitin surface binds DDT and noticed a correlation between the possession of
a chitinous cuticle and sensitivity to DDT. They hypothesized that DDT is
selectively concentrated in the cuticle. Lord (1948) determined that chitin,
as a colloidal suspension, adsorbs DDT and compared the data with those on
wool, cellulose and other substances.

In fact, as the cuticle becomes more sclerotized, it becomes more lipophilic
(Fraenkel & Rudall, 1940). The cuticle would then hold lipid soluble compounds.
The high lipid content observed in some resistant insects, but not in others,

would depend on the availability of lipids for deposition in the cuticle and would be a manifestation of sclerotization. The lipid further increases the insecticide-holding power of the cuticle. Bradleigh & Law (1971) demonstrated that the cuticular composition of *Heliotis virescens* larvae differs according to the degree of penetration of DDT in the cuticle. Protein and lipid contents are greater in the cuticle of resistant larvae, which is more sclerotized. Dietary levels of ascorbic acid influence resistance and sclerotization in the larvae.

As everybody knows, DDT has been prohibited after years of use; it is a striking fact that during such a long period, so little research has been conducted on the interactions of DDT with chitin.

More extended and sound research is presently being performed on a new class of insecticides which are able to interfere with the chitin synthesis. The most thorough studies have been on the following: 1-(4-chlorophenyl)-3- -(2,6-difluorobenzoyl)urea also known as diflubenzuron or PH-6040, available under the trademark "Dimilin"; 1-(4-chlorophenyl)-3-(2,6-dichlorobenzoyl)urea or PH-6038 and 1-(3,4-dichlorophenyl)-3-(2,6-dichlorobenzoyl)urea or DU-19111. After the discovery of DU-19111, many hundreds of related benzoylphenylureas were synthesized and screened on the ground of their larvicidal activity.

DU-19111 has been administered to various insects, from the Colorado potato beetles to the houseflies. One important early observation was that death was invariably connected with the process of molting. Larvae of all instars survived feeding on plants sprayed with DU-19111 until their next molt, when they suddenly perished.

The compounds are therefore larvicides, their most conspicuous feature being the abortion of molting. In the surviving larvae, however, disturbances in the endocuticular structure were evident not only during the molting process, but also during the subsequent postecdysial process of endocuticle formation.

Soft larval endocuticle consists mainly of chitin and protein, probably integrated as glucoprotein. The chitin matrix is constructed according to an elaborate plan involving a circadian rhythm; the disturbancies are elicited in this structure by the substituted ureas probably by a defect in the formation of this fibrillar tissue. Radiochemical techniques have been applied to study this. Incorporation of ^{14}C-glucose into chitin has been demonstrated in several insects by many authors, some of the most recent work being by Tate & Wimer (1974), and has been ascertained that glucose ^{14}C is incorporated into chitin during the early post-molt period, but not during intermolt (Meenakshi & Scheer, 1961) by a chitin synthetase in subcellular particles. The incorporation of glucose into the endocuticular chitin of fifth-instar larvae of *Pieris brassicae* was completely blocked by DU-19111 (Van Daalen, Meltzer, Mulder & Wellinga, 1972; Post & Vincent, 1973.

In Fig. 1.12 autoradiograms of ^{3}H-glucose incorporation in *Pieris brassicae* endocuticle are reported. In the normal larva the pattern of cuticular radioactivity is totally similar to that found by Condoulis & Locke (1966) in *Calpodes ethlius*. As a result of its rapid conversion to chitin and its correspondingly rapid clearance from the active metabolic system, the labeled glucose evokes a narrow zone of radioactivity which has moved some way up into the cuticular matrix. By contrast, the DU-19111-treated larva lacks this layer and practically no radioactivity is present in the endocuticle. The rapid

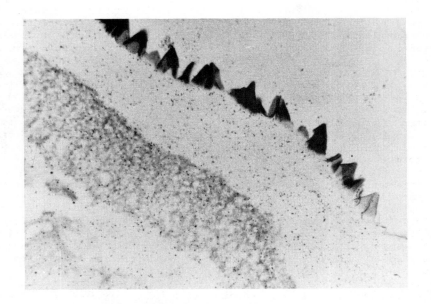

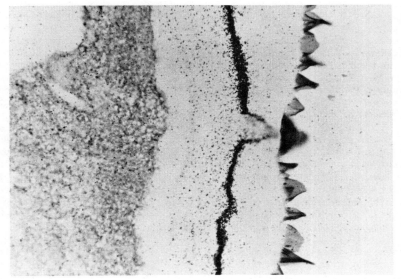

Fig. 1.12. Incorporation patterns of radioactivity in endocuticle after injection of 4.5 μCi of D-^{3}H-glucose into 5th-instar larvae of *Pieris brassicae*. a = larva fed on foliage treated with a 30 ppm suspension of DU-19111; b = normal larva. Magnification × 600. Post, De Jong & Vincent, 1974.

incorporation of glucose into endocuticle in normal larvae illustrates the
efficiency of the conversion of glucose into chitin, and therefore the failure
of the incorporation into chitin in treated larvae reflects the failure of the
construction of the chitin chain.

In order to ascertain what stage of the biosynthetical chain from glucose to
chitin was actually blocked, labeled glucose was injected into DU-19111-
treated larvae and normal larvae and their content of uridine diphosphate-N-
acetylglucosamine was analyzed. it was expected that the latter compound, as
the ultimate precursor of chitin would accumulate if the final step, namely
polycondensation, were impaired; on the other hand, should a previous stage be
inhibited by the insecticide, a reduced level of labeled uridine diphosphate-
-N-acetylglucosamine would be found.

After injection of D-^{14}C-glucose, the larvae were homogenized to yield deproteinized extracts with which quantitative paper chromatography was carried
out. Typical histograms of radioactivity are reported in Fig. 1.13, where a
remarkable feature is the N-acetylglucosamine peak for the sample originated
by DU-19111-treated larvae, different from the normal sample. Since N-acetylglucosamine as such takes no part in chitin biosynthesis, this compound is not
normally found in insects except during apolysis, when the old cuticle is
broken down. The relatively large amounts of this compound reflects an
aberration in the chitin biosynthesis, elicited by the insecticide. The
uridine diphosphate-N-acetylglucosamine level is not significantly affected by
DU-19111, and therefore the polycondensing enzyme continues to accept the
precursor, but in the process of connecting the N-acetylglucosamine moiety to
the chitin chain, it drops the monomer.This hypothesis is relevant to DU-19111
alone, because with diflubenzuron different chromatographic patterns were
obtained (Post, De Jong & Vincent, 1974). With diflubenzuron, Deul, De Jong &
Kortenbach (1976) did not observe an increase in N-acetylglucosamine and were
unable to distinguish uridine diphosphate-N-acetylglucosamine from glucosamine
6-phosphate; they pointed out, however, that the accumulation of uridine
diphosphate-N-acetylglucosamine is likely to be balanced by the inhibition of
the amination of fructose 6-phosphate.

As a macroscopic consequence of this disruption of the chitin synthesis,
the larvae could be seen moving within their intact exuviae, but they were
unable to split the latter and to wriggle out. After some time, the larvae
usually lost some fluid, gradually blackened and finally died while still
attached to the leaves in the typical molting position. At limiting concentrations of the insecticide, some larvae actually succeeded in splitting the
exuviae and in partially emerging, but they were unable to undergo ecdysis and
so died, again with loss of moisture and blackening.

In a larva that has ingested PH-6038, the endocuticle thickness remains
constant, whereas within 48 hr after ecdysis the endocuticle of a normal larva
more than doubles its thickness. Histological preparations show that the
epicuticle and the ectocuticle of the treated larvae are normal as well as the
intestinal and the tracheal linings. Dying larvae always display an important
lack of adipose tissue.

Data in Tab. 1.10 demonstrate that diflubenzuron inhibits the incorporation of
glucose in chitin specifically, as other tissue fractions show normal values
(Deul, De Jong & Kortenbach, 1976).

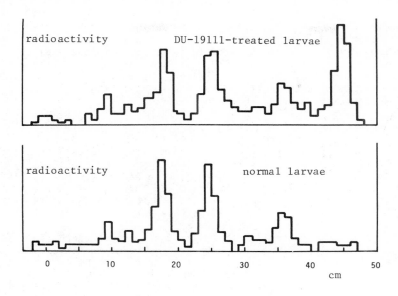

Fig. 1.13. Histograms of radioactivity of larvae extracts:
DU-19111-treated larvae and normal larvae. A significant
difference consists in the peak at 45 cm, assigned to
N-acetylglucosamine. Post, De Jong & Vincent, 1974.

Table 1.10. Effect of diflubenzuron on chitin synthesis studied by following
the incorporation of radioactive glucose in tissue fractions. Values are
expressed as μg glucose per larva with standard deviations. Deul, De Jong &
Kortenbach, 1976.

Tissue fraction	Control	After diflubenzuron treatment
Hemolymph + gut contents	0.38 ± 0.05	0.44 ± 0.04
KOH hydrolysate	0.56 ± 0.06	0.62 ± 0.07
glycogen	0.13 ± 0.04	0.15 ± 0.02
Feces	0.31 ± 0.06	0.31 ± 0.07
Chitin	0.54 ± 0.04	0.003 ± 0.003
Dry weight of cuticles	2.63 ± 0.19	0.73 ± 0.18

In *Spodeptera littoralis* larvae, mortality in feeding experiments occurred in prepupal or pupal stage. Diflubenzuron incorporated into wheat bran baits at $2.5 \div 10,000$ $\mu g \times g^{-1}$ killed $70 \div 90$ % of the insects (Ascher & Nemny, 1976).

Diflubenzuron causes ovicidal effects on *Musca domestica* by topical application to, or after oral uptake by, gravid females. In all cases the larvae in the eggs develop normally and lie apparently fully grown in the eggs at the moment that normal eclosion would occur. Sometimes the larva can split the egg wall but it is unable to leave the egg. In larvae, diflubenzuron disturbs the formation of chitin in the cuticle and this is most likely followed by an incapacity to use the muscles (Grosscurt, 1976).

The laboratory experiments have been extended under outdoors conditions and in field evaluations. Air milled wettable powders were used. Because of their retarded action, especially with slowly decomposing species, attention was paid to the time of application in order to keep damage economically acceptable and early spraying when eggs are about to hatch was suggested. The damage factor may become crucial when, apart from the larvae, the adults also feed on the crops, like the Colorado potato beetle (Mulder & Swennen, 1973; Turnipseed, Heinrichs, Da Silva & Todd, 1974). Some data are reported in Tab. 1.11.

These compounds seem suited for the control of insects where it is sufficient to prevent the development of the adult or to prevent the development of further generations as is usually the case in forestry. Diflubenzuron gave excellent season-long control of volvetbean caterpillar *Anticarsia gemmatalis*. Adequate control of the soybean looper *Pseudoplusia includens* and reduced numbers of mexican bean beetle *Epilachna varivestis* larvae could be afforded in field tests by foliage application of diflubenzuron, carried out in Georgia, South Carolina and Brazil; aerial spraying of pines infested with *Thaumetopoea pityocampa* eggs decreased the proportion of destroyed needles to 16 %, as compared with 46 % in the controls (Ribrioux & Dolbeau, 1975). Diflubenzuron was released from silicate and castorwax capsules when immersed in water. The amount released was sufficient to cause significant mortality of *Aedes aegypti* larvae after $1 \div 9$ days of immersion (Sjogren, 1976).

One major problem with every potential insecticide is the possibility of development of resistance to it by the target pests. Various levels of cross-resistance toward diflubenzuron by the insecticide-resistant strains of the housefly were revealed by the application of this compound at two dosage levels (Cerf & Georghiou, 1974). One susceptible strain was compared against seven other strains after topical application to white prepupae of a solution of diflubenzuron in tetrahydrofuran. The relatively high level of cross-resistance toward diflubenzuron developed by the housefly invites a judicious use of such a new chemical against presently susceptible populations under conditions which minimize the degree of selection pressure.

The results concerning cross-resistance to these inhibitors of chitin synthesis are important from the point of view of the ingestion and feeding habits of some insects. In fact, on the grounds that these insecticides are stomach poisons, it was believed that certain insects would escape death, such as sucking insects and spider mites, borers and miners; so, many useful and harmless insects would be safeguarded. On the contrary, it was seen that entry via the stomach is not an essential requirement for toxicity, as the insecticide can be assumed also after topical application. Incidentally, Lord (1948) and Bradleigh & Law (1971) demonstrated that the action of DDT was also due to

Table 1.11. Mortality percent of Colorado potato beetle larvae treated with DDT or with inhibitors of chitin synthesis. Mulder & Swennen, 1973.

Compound	ppm	Initial 3 d	Initial 7 d	Reinfection after 1 week 7 d	Reinfection after 2 weeks 7 d
DDT	100	100	100	70	0
25 %-m.o.	30	97	100	10	
	10	27	73	0	
PH-6038	100	20	100	64	27
10 %-1i	30	0	99	70	
	10	0	53	0	
PH-6040	100	13	98	43	7
5 %-1i	30	0	73	23	
	10	0	18	0	

sorption by chitin and resistant larvae were those able to develop a higher degree of sclerotization of the cuticle by varying the protein and lipid content.

While warm-blooded animals seem unaffected by the insecticidal levels of the substituted ureas, as the lethal dose for rats is 3160 mg×kg^{-1} when orally administered, it must be seen that a long term use of insecticides of this class might affect many organisms necessary to a correct ecological equilibrium such as fungi, marine plankton, etcetera. Miura & Takahashi (1973 and 1974) published results of simulated field studies. The water flea population was markedly reduced after diflubenzuron treatment, but slowly recovered later. Typical symptoms were lessened filter feeding and movements; reproduction was also suspended for a few days. Copepods populations responded to the treatment in the same way, but the magnitude of reduction was small and the recovery period was short. The seed shrimp population showed no harmful effect.

Water fleas, copepods and shrimps are the most common and abundant organisms found in the mosquito breeding habitats examined by Miura & Takahashi (1974). The side effects of diflubenzuron against crustaceans when it is used at the rate required for *Aedes nigromaculis* control, are such that tadpole and clam shrimp populations could be reduced, but would probably recover. Less severe side effects are reported when diflubenzuron is used for *Culex tarsalis* control.

The report by Miura & Takahashi (1974) shows that, when diflubenzuron is applied by hand sprayer to small shallow intermittent ponds, tadpole and clam shrimp populations are eliminated. While water flea populations were markedly reduced, they recovered within a week.

During a research on the chitin synthesis inhibition in the cockroach by diflubenzuron and polyoxin D, Sowa & Marks (1975) showed that leg tissues synthesize chitin and incorporate exogenous glucosamine and acetylglucosamine in the biosynthesis of chitin. Chitin synthesis in cultivated cockroach blood was studied by Ritter & Bray (1968).

Measurement of chitin production is a sensitive and quantitative assay for
molting hormone activity and for inhibitors of chitin synthesis; inhibitors
were detected at a concentration of 10^{-10} M by affinity of fluorescent
chitinase for chitin and the thiazine red dichroic method for the determination
of birefringence. Diflubenzuron was found to be concentrated by the tissue and
a general affinity of insect tissue for diflubenzuron was established. This
high affinity would be responsible for the high efficiency of this insecticide.

The findings of Sowa & Marks (1975) that polyoxin D inhibits insect chitin
synthetase (I_{50} = 7.5×10^{-7} M) as effectively as polyoxin B inhibits fungal
chitin synthetase (>I_{50} = 1.0×10^{-7} M), (Hori, Eguchi & Kakiki, 1974) together
with reports on the inhibition of the chitin synthesis in preparations of the
toadstool *Coprinus cinereus* (Gooday, 1972) and of *Schizophyllum commune* (De
Vries & Wessels, 1975) by polyoxin D, indicate a similarity between insect and
fungal chitin synthetases as well.

The degradation of diflubenzuron and the nature of its metabolites in
terrestrial and hydro-soils were studied with the aid of preparations radio-
labeled at three different positions in the molecule. It was found that,
because of its low solubility in water, the half-life of diflubenzuron depends
on the particle size. Bijlo (1975) reported that the microbial degradation of
diflubenzuron in soils produces 4-chlorophenylurea and 2,6-difluorobenzoic
acid, the half-life being 2 weeks for 5 µm particles and 3 ÷ 5 months for 5 -
10 µm particles. These primary degradation products are metabolized further
rapidly; possible uptake of the degradation products by crops is currently
being studied: diflubenzuron itself is not taken up by crops from the soil.

The fate of diflubenzuron is being studied with the aid of model ecosystems
including plants, soils, algae, snails, fishes and aquatic microflora and
fauna. Preliminary results indicate that there is no extensive bioaccumulation
of diflubenzuron either by direct absorption or by ingestion through the food
chain. These investigations are necessary before the use of chitin synthesis
inhibitors becomes generalized (Muzzarelli, 1976). Apparently, soybeans, apple
and cotton orchards can be protected with these insecticides.

Insecticides based on chitinases are described in Chapter 4.

CHAPTER 2

STEREOCHEMISTRY AND
PHYSICAL CHARACTERIZATION

The understanding of the physico-chemical characteristics of chitin and chitosan requires a model. A model for the macromolecule of chitin was elaborated a long time ago, and has been recently refined; it can be imagined that a similar model for chitosan will be produced soon.

The model of chitin refers to a polymer where anhydroglucose units having acetylated amino groups on C^2 are arranged in a regular way: however, it should be clearly stated, before examining this information, that the model is not representative of chitin in general, but is only representative of parts of a chitin specimen which shows a higher degree of order. Moreover, as will be described in Chapter 3, certain chitins from various sources are more or less acetylated and may include saccharide units other than aminated anhydroglucose.

Whilst the discussion of the x-ray diffraction techniques used to obtain information on structure is beyond the scope of this book, it may be recalled that x-ray spectra of the α-, β- and γ-chitins have been published by Rudall (1963), and the best crystalline picture of α-chitin is as shown by Blackwell, Parker & Rudall (1967). Highly crystalline α-chitin is illustrated by Rudall (1976). Many other authors have published on this subject, giving not only theoretical results but also information for chitin identification and for the elucidation of chitin combinations with other compounds. X-ray diffraction spectra of commercial purified crab chitin and chitosans are illustrated in Fig. 2.1.

CHITIN POLYMORPHISM.

Chitobiose, $C_{16}H_{28}O_{11}N_2$, is the dimer of chitin. Two crystalline forms of chitobiose have been prepared by Mo & Jensen (1976) who determined the structures. The orthorhombic form is a monohydrate of the α-anomer, space group $P2_12_12_1$, a = 1.1017, b = 1.3066, c = 1.3896 nm and Z = 4, data measured with MoK_α radiation. Results indicate anomeric disorder with approximately 90 % α- and 10 % β-anomer in the crystal. The monoclinic form belongs to space groups $P2_1$, a = 1.1569, b = 0.8920, c = 1.1086 nm, β = 99.00°, Z = 2, data measured with CuK_α radiation; this is a trihydrate of β-chitobiose. There are distinct dissimilarities between the conformation of the two crystal structures of chitobiose, in particular at the glycosidic bonds. Thus, the intramolecular hydrogen bond $O^{3'}-H\cdots O^5$, commonly observed in (1→4) β-linked disaccharides is found in the monoclinic but not in the orthorhombic crystal.

This information about the dimer of chitin already shows that chitin is not such a simple polymer to describe. Even though much work is still to be done, we possess a valid picture of the conformation of chitin.

45

chitin   chitosan

Fig. 2.1. X-ray diffraction spectra of commercial chitin
and chitosan. Courtesy of A. Ferrero.

Chitin occurs in three polymorphic forms which differ in the arrangement of
the molecular chains within the crystal cell. Typical x-ray spectra showing
these differences in molluscan samples are illustrated by Rudall (1963) and in
arthropodan samples by Rudall & Kenchington (1973). α-Chitin is the tightly
compacted, most crystalline polymorphic form where the chains are arranged in
an anti-parallel fashion; β-chitin is the form where the chains are parallel
and γ-chitin is the form where two chains are "up" to every one "down".

By far the most abundant polymorphic form is α-chitin which is found in the
arthropod cuticles and in certain fungi. β-Chitin exists in a crystalline
hydrate (Dweltz, 1961; Blackwell, 1969) which accounts for its lower stability
since water can penetrate between the chains of the lattice. Also γ-chitin can
be transformed into α-chitin by treating it with lithium thiocyanate (Rudall &
Kenchington, 1973). γ-Chitin has been found in the cocoons of the beetles
Ptinus tectus and *Rhynchaenus fagi*.

The three forms of chitin have been found in different parts of the same
organism, namely in the squid *Loligo* whose beak contains α-chitin, whose pen
contains β-chitin and whose stomach linings contain γ-chitin; this fact
indicates that the three forms are relevant to the different functions and not
to animal grouping.

Rudall (1963) expressed the opinion that β-chitin chains in 6 N hydrochloric acid fold back upon themselves thus becoming α-chitin. The α-form is more stable since this form is assumed by chitin samples, for instance the β-chitin of Polychetae, when reprecipitated from acids. All three structures are stable to boiling 5 % sodium hydroxide. Rudall (1955) has hypothesized that chitin exists in the three forms because of its diversity of functions. In plants it serves as an alternative to cellulose; in animals as an alternative to collagen. Where extreme hardness is called for, the α-chitin is usually found, frequently sclerotized and encrusted with mineral deposits; β- and γ-chitins seem to be associated with collagen-type proteins providing toughness, flexibility and mobility, and may have physiological functions other than that of support, such as control of electrolytes and transport of a polyanionic nature.

Conformation of α-chitin.

The conformation of α-chitin has been the object of many investigations. Carlstrom (1957) has proposed the buckled chain structure arranged in an orthorhombic cell with dimensions a = 0.476, b = 1.030, c = 1.885 nm with the space group $P2_12_12_1$. Ramakrishnan & Prasad (1972) applied the least square rigid-body refinement technique for the elucidation of the structure of chitin. The refinement has been done using x-ray diffraction intensity data taken in groups to account for the overlap of reflections.

The chitin chain can be treated as a poly(N-acetyl-D-glucosamine) helix. The spacings in the x-ray diffraction photographs of this polymer give the values of the unit translation as 0.515 nm and the number of residues per turn of the helix as 2. The unit cell contains two chains running in opposite directions, each constituted of two screw-related units of N-acetyl-D-glucosamine. To obtain the dimensions of the monomer unit, data from various crystal structure reports on different compounds containing D-glucose and related compounds were used. A planar conformation was assumed for the aminoacetyl group, as this is very similar to the peptide group.

Two such monomer units linked at the glycosidic oxygen give the chitobiose unit of the chitin chain which actually repeats along the b (fiber) axis. Any conformation of two such linked units can be indicated by two dihedral angles ϕ and ψ. The reference conformation where these angles are 0° corresponds to the one where the atoms O^1, C^1 and O^4 of the first unit and atoms $O^{4\prime}$, $C^{4\prime}$ and $O^{1\prime}$ of the second unit are in a plane and the covalent bond directions C^1-H^1 and $C^{4\prime}-H^{4\prime}$ are respectively nearly cis to the bonds $O^{4\prime}-C^{4\prime}$ and C^1-O^1. The definition of the initial conformation was the same as assumed by Sundararajan & Rao (1969). The angle at the glycosidic bond has been taken as 116°, the same as obtained in the crystal structure of cellobiose compounds by Chu & Jeffrey (1968) and by Ham & Williams (1970).

For the complete specification of the orientation of the chain in the unit cell four parameters were defined: q_1 = rotational parameter which represents the rotation of the chain around the fiber axis; q_2 = translational parameter which represents the relative shift of the two anti-parallel chains in the unit cell along the fiber axis; q_3 = rotational parameter which represents the rotation of the hydroxyl groups attached to C^6, about the bond C^5-C^6; q_4 = rotational parameter which represents the rotation of the aminoacetyl group

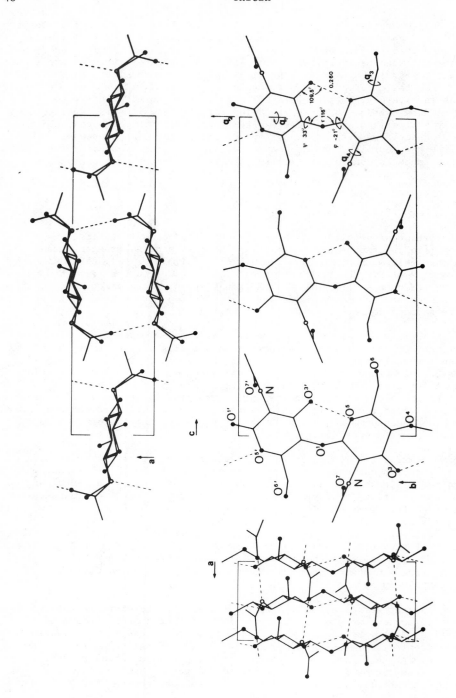

Fig. 2.2. Projections of the proposed model for α-chitin.

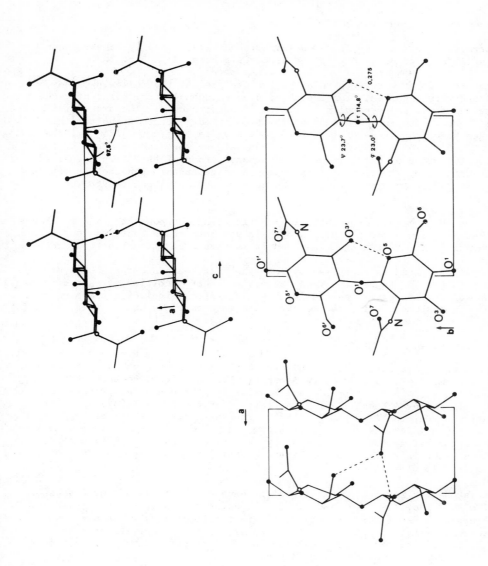

Fig. 2.3. Projections of the proposed model for β-chitin.

attached to C^2, about the bond C^2-N^2.

The structure obtained by Prasad & Ramakrishnan (1972) and by Ramakrishnan & Prasad (1972) according to this research, is presented in Fig. 2.2.

In the α-chitin structure there are therefore present the following bonds, which are very important from the chemical and technological point of view, insofar as they confer peculiar macroscopic physico-chemical characteristics: O-H···O bond, between $O^{3'}$ and O^5 of the two linked glucosamine units; this hydrogen bond is a feature of a rigid body and its length is 0.260 nm, the angle $C^{3'}-O^{3'}···O^5$ is 109.5°; N-H···O bond between the N and H atoms of the aminoacetyl group of one unit and the carbonyl group of the chain translated by one unit along the a axis. This hydrogen is dependent upon the parameters q_1 and q_4 and is independent of q_2 and q_3. The hydrogen bond length is 0.267 nm.

Examining the complete structure of α-chitin using the contact criteria it was remarked that there are neither interchain nor intrachain contacts less than the normal limit-allowed distances and it is thus free of steric hindrance.

The model described is supported by both the x-ray data and the good hydrogen bonding values and therefore it can be assumed that α-chitin corresponding to the above model occurs in several biological systems, including the mandibular tendon of lobster on which the measurements were carried out.

Conformation of β-chitin.

The unit cell of β-chitin is monoclinic, with dimensions a = 0.485, b = 1.038, c = 0.926 nm, β = 97.5° and space group $P2_1$. The unit cell contains two sugar residues related to the two-fold screw axis. The structure, as proposed by Blackwell, Parker & Rudall (1967) was refined by Blackwell (1969) and, more recently, by Gardner & Blackwell (1975). The sugar residues are arranged with β-(1→4)-glycosidic linkages on a 2_1 screw axis. This chain is the same as that used for the refinement of cellulose by Gardner & Blackwell (1974) except that the O^2-H group is replaced by the trans planar -NHCOCH$_3$ group which is a characteristic of chitin.

The disaccharide residue has a glycosidic bond angle of 114.8° and glycosidic torsion angles of φ = 23.0° and χ = 23.7°, using the convention followed by Sundararajan & Rao (1969). An $O^{3'}-H···O^5$ intramolecular hydrogen bond is present with length 0.275 nm. In this conformation, the chitin chain is completeley rigid, except for the side chains that can undergo the rotations described under the parameters q_3 and q_4.

Three models have been worked out by these authors. The first one was stereochemically acceptable but did not provide for hydrogen bonding involving the -CH$_2$OH group. The second one, containing an intrasheet $O^{6'}_a-H···O^7$ hydrogen bond was found to be acceptable. A third structure, containing an intersheet $O^6-H···O^{7'}$ hydrogen bond as previously proposed, was found to have a short O^6 C^8 contact, thus the third model containing the intersheet bonding was rejected, on the ground that it contains unbonded short distances. The choice between the other two models was not simple because the differences were not very singificant and there is a 50-50 chance that either structure is correct.

On the basis of infrared data indicating that the $-CH_2OH$ group should be hydrogen bonded, the structure containing the $O^{6'}-H\cdots O^7$ intrasheet hydrogen bond is proposed as the structure of β-chitin. The $N-H\cdots O^7$ hydrogen bond is 0.267 nm and the intrasheet bond is 0.289 nm. The structure contains no other intermolecular contacts of less than 0.32 nm and it is presented in Fig. 2.3.

The orientation of the chain about the helix axis, defined as packing parameter Φ, whose increase corresponds to anticlockwise rotation of the chain about the c axis, was refined to a value of Φ = 110.4°. In the proposed structure the O^3-H and O^6-H bonds are oriented in such a way that both should give rise to parallel dichroism in the O-H stretching region of the infrared spectrum.

Pervaiz & Abdul Haleem (1975) in a further examination of the β-chitin structure have found that the most favorable position of the O^6 atom should be in the range 120 - 140°, however it is possible that there are two positions. β-Chitin is easily swollen in water, and it is suggested that the two crystalline hydrate structures recognized contain one and two water molecules per residue. In these hydrated structures the hydrogen bonded sheets remain intact but are moved apart. The intrasheet hydrogen bond model is consistent with this behavior whereas an intersheet bond would tend to hold the sheet together. The intrasheet $O^{6'}_a-H\cdots O^7$ bond can be considered as an additional stabilizing factor for the sheets of chitin chains also linked by the amide $N-H\cdots O=C$ bonds.

There are numerous points of analogy between the structures of β-chitin and cellulose I. Both structures contain extended parallel chains and can be visualized as an array of hydrogen bonded sheets. According to the description given by Gardner & Blackwell (1974), in cellulose I the O^2-H group forms an intramolecular $O^{2'}-H\cdots O^6$ bond, and since there is also an $O^3-H\cdots O^{5'}$ intramolecular bond, the ribbonlike chain has approximately parallel hydrogen bonds on both sides of the glycosidic bridge. In β-chitin the same $O^3-H\cdots O^{5'}$ intramolecular bond is present, but the N-H is involved in bonding with the amide group in the next chain.

Upon conversion of cellulose I to cellulose II, the $O^{2'}-H\cdots O^6$ intramolecular bonds are broken and $O^2-H\cdots O^2_d$ bonds are formed between antiparallel chains; this is analogous to the conversion of β-chitin to α-chitin and maybe additional refinements of α-chitin might show additional intersheet hydrogen bonds.

CHITIN MICROFIBRILS.

In living systems chitin occurs in the form of microfibrils: the microfibrils have been clearly brought to evidence in recent years by electron microscopists who have called them crystallites as well.

Arthropod cuticle is a composite material like fiber-glass, with crystallites or microfibrils dispersed in a matrix. In the arthropod cuticle (Neville, 1975), microfibrils can be resolved with particular clarity because their edges are defined against the protein matrix in which they are embedded. Moreover, when they are arranged in an helicoidal superstructure the component planes of microfibrils can be seen in electron microscope sections to rotate around, appearing in transverse and longitudinal sections with intervening oblique patterns. The best technique according to Neville & Luke (1969) for revealing them is to stain the surrounding matrix with potassium permanganate and lead citrate: this makes the microfibrils evident as white points in transverse sections against a dark background.

Rudall (1967) described the faster degradation of the matrix by molting fluids than the chitin, leading him to the hypothesis of the occurrence of chitin microfibrils in a protein matrix. However, even in early reports on the preparation of samples for x-ray diffraction studies, it was remarked (Fraenkel & Rudall, 1947) that partly digested cuticles give a chitin x-ray diffraction spectrum sharper than the untreated cuticle spectrum, because of the increased orientation introduced by the treatment.

Further confirmation that chitin occurs as a microfibrillar component has been obtained by Weis-Fogh (1970) on locust prealar arm ligaments which had been partially digested with subtilisin at neutral pH, so it can be said that the rubber-like cuticle consists of chitin microfibrils embedded in a resilin matrix.

Chitin microfibrils in cuticle samples exhibit diameters around the value of 2.8 nm, as reported by Neville (1975 and 1976) for various insects, a crustacean and an arachnid. Rudall (1967) has also reported microfibril diameters of about 2.5 nm on the basis of electron micrographs and x-ray diffraction spectra.

As the dimensions of the unit cell of α-chitin are 1.885×0.476 nm and it contains two anti-parallel molecular chains, the cross-sectional area is 0.448 nm^2. Assuming that the microfibril (which is equivalent to the crystallite according to x-ray diffraction results) is square in cross-section, there could be up to three sheets ($3 \times 0.942 = 2.826$ nm) and six chains per sheet ($6 \times 0.476 = 2.856$ nm) (Neville, 1975). This is in agreement with the results of Rudall & Kenchington (1973) who estimated three sheets of seven chains.

Chapman (1966) advanced the hypothesis for the in vivo control of the fibril diameter, based upon the minimum energy of the constituent molecular chains: the cylindrohelical arrangement of molecules would include a tilt respective to the microfibril axis, increasing with the radius of the growing structure. Beyond a certain limit, the addition of more chains is energetically unfavorable and thus the diameter is limited.

Chitin microfibrils have been recently synthesized in vitro by Ruiz-Herrera & Bartnicki-Garcia (1974); an electron micrograph showing the chitin microfibrils is in Fig. 2.4. These authors incubated a membrane fraction from the yeast cell of *Mucor rouxii* which contains most of the chitin synthetase activity with the substrate uridine diphosphate-N-acetylglucosamine and the activator N-acetyl-D-glucosamine. By so doing, they were able to have about 20 % of the total chitinsynthetase activity buoyant after centrifugation at 81,000 *g* for 30 min. When this soluble supernatant was brought to room temperature, turbidity appeared in about 20 min, followed by a precipitation of a fibrous material. Using radioactive substrate in the incubation mixture, polymerization was detected at the extent of 30 % after 1 hr. Chitin was identified by partial acid hydrolysis, N-acetylation and paper chromatography. Electron microscopy revealed that the product thus synthesized consisted of networks of highly regular, long microfibrils about 20 nm in diameter. Some of the networks

Fig. 2.4. Networks of chitin microfibrils synthesized in vitro. Magnification × 56,800. Ruiz-Herrera, Sing, Van Der Woude & Bartnicki-Garcia, 1975.

showed a striking resemblance to pieces of incipient cell walls formed in vivo and light microscopy showed mostly pellicles of many different shapes. Also observed were thick fibers and numerous isolated particles (1 to 2 μm). X-ray diffraction analysis proved definitely that the biosynthesized polymer was α-chitin of a crystallinity comparable to that of highly purified chitin from *Cancer magister*, with reflections slightly sharper.

The results by Ruiz-Herrera & Bartnicki-Garcia (1974) are the first results concerning the formation of cell wall microfibrils in vitro. Further research by Ruiz-Herrera, Sing, Van Der Woude & Bartnicki-Garcia (1975) has pointed out that chitin microfibrils are synthesized in vitro by granules of chitinsynthetase. The association of single granules with the terminal part of microfibrils suggests that microfibrils originate from the granules. These fibrils were 12 - 18 nm wide and up to 2 μm long; their x-ray diffraction spectrum characterized them as α-chitin (Fig. 1.5).

The molecular chains of chitin are therefore synthesized and assembled collectively and simultaneously into a microfibril by a large multienzyme aggregate or a chitinsynthetase granule.

Muzzarelli (1976) has presented small-angle x-ray diffraction measurements on chitosan (deacetylated chitin) from crab shells. The chitosan powder, exposed in a mylar bag, gave a spectrum whose analysis revealed the presence of clusters of average diameter $\sim$ 25 nm in addition to larger clusters beyond the instrument resolution set at 60 nm. When the chitosan powder was exposed in a Lindeman glass capillary 0.5 mm in diameter, the spectrum revealed a sharp orientation of the clusters in a direction normal to the capillary axis. While confirming the values of 25 and > 60 nm, the spectrum indicated a regular packing of the particles in parallel bundles, analogous to that of stretched polyehtylene and polypropylene fibrils. These data indicate that the inter-actions of the chitosan macromolecules along an axis give rise to a fibrous structure.

Large microfibrils, 25 nm in diameter, have been shown in crustacean cuticles by Bouligand (1965). Neville (1970) has resolved the substructure of 50 nm components of decalcified cuticles from *Carcinus maenas*; these components consist of bundles of microfibrils approximately 2.5 nm in diameter which have been called macrofibrils. Their cross-section is variable; they resemble the bundles of 3.5 nm cellulose microfibrils from the alga *Valonia*, whose diameter is in excess of 20 nm according to Sarko & Marchessault (1969). Crustacean macrofibrils can in fact be seen under the phase contrast light microscope as reported by Dennel (1973). The helicoidal change of direction is clearly shown by crustacean microfibrils, so that in a section cut normal to the layers some macrofibrils are cut in LS and some in TS, while the intervening ones are sectioned obliquely.

Plant biophysicists and polymer morphologists do not easily distinguish between elementary protofibrils of 2.8 nm diameter and the larger bundles that they form. Nozawa, Kitajima & Ito (1973) isolated cell walls of *Epidermophyton floccosum* in high purity after mechanical breakage, sonication and sodium dodecylsulfate extraction. Glucosamine was present in the proportion of 30.9 %. These cell walls, from examination of shadowed preparations at the electron microscope showed a fibrous texture with random orientation. Short and thick fibrils in bundles were observed at the outer surface, whilst long and thin fibrils were visible on the inner surface. Similar types of chitin fibrils were reported by Manocha & Colvin (1967), Troy & Koffler (1969) and Wang & Bartnicki-Garcia (1970).

The extra-cellular chitin spines of the euryhaline diatom *Thalassiosira flu-viatilis* have been investigated by Dweltz, Colvin & McInnes (1968) and by Gardner & Blackwell (1971). When the diatom spines were suspended in water by hand shaking, long, rigid needles were seen with widths of approximately 100 nm and variable lengths up to $\sim$ 60 μm. Smaller fragments were present in these suspensions, consisting of ribbons with a thickness of 2.0 nm and widths of 20 - 30 nm, indicating that the spines can be broken into fragments in a variety of ways. These ribbons appeared to be constituted of a number of fibrils with cross-section 2.0 × 3.5 nm. Striated ribbons were visible only in sonicated specimens, indicating that harsh treatment is needed to disrupt the bonds between the 3.5 nm units. The 3.5 nm fibrils are analogous to the elementary fibrils occurring in *Valonia ventricosa* cellulose. Diatom chitin has been shown to have a β structure by Blackwell, Parker & Rudall (1967) since all the chains run in a parallel fashion.

The chitin chains are linked in sheets through hydrogen bonds and are part of a structure which is easily swollen in the direction perpendicular to the sheets. X-ray diffraction data obtained on mats of spines with uniplanar orientation by Falck, Smith, McLachlan & McInnes (1966) show that the plane of hydrogen bonding corresponds to the width of the spines, and hence to the width of the 2 × 20 ÷ 30 nm ribbons. The thickness of 2.0 nm corresponds to two chains arranged side by side. Unit cells containing two such chitin chains have been seen in the hydrate structures of pogonophore tube chitin, a system very similar to the diatom spines. Chitin layers from blow-fly larval cuticle have been reported by Rudall (1967) to contain chitin layers 2.5 nm thick.

Helicoidal architecture.

A characteristic feature of cuticles is the parabolic pattern which consists of a lamellar repetition of rows of parabolae. Bouligand (1965) derived a model to explain the parabolic patterning and described an helicoidal archi- tecture to represent the way in which the chitin microfibrils are organized in the protein matrix. According to Bouligand (1965) microfibrils are arranged in parallel in each plane and with the planes parallel to the cuticle surface. In successive planes the direction of orientation of the microfibrils rotates through a small angle always in one direction of rotation.

Parabolic patterning showing up in oblique sections of helicoidal cuticle, is of universal occurrence in arthropods (Neville, 1967). It is observed in egg, larva, pupa and adult and in all the main arthropod classes. It is found in all types of cuticle (solid, rubber-like and arthrodial membrane). These patterns are large enough to be seen under the light microscope and have been reported by many authors since long ago. They were not reported in insect cuticle until later (Locke, 1960) because the electron microscope is necessary to make them visible in this group.

Weis-Fogh (1970) has proposed a refinement of the helicoidal model; instead of having the planes of microfibrils separate from each other, she considers a continuous "screw-carpet" structure in which microfibrils are continuously secreted on to the ends of existing ones. In this way, the microfibril taking part in one region of the helicoid would gradually sweep around, changing its vertical position in the stack. In this way the single microfibril contributes to other parts of the helicoid in turn. This idea is supported by experimenta- tion on enzyme digested insect membranes.

Therefore, two systems have been described: the first system is made of layers in which the chitin microfibrils are helicoidally arranged in the protein matrix; the second is made of layers in which the microfibrils are all uni- directionally oriented to form a preferred layer. Either type may occur exclu- sively or both may be present in successive layers of the cuticle. The trans- ition between the two systems is smooth and progressive (Neville & Luke, 1969; Neville, 1975). In fact, it is known that the daily changes in microfibril orientation in locust are controlled by a circadian clock (Neville, 1965). By uncoupling this control, locusts can be induced to grow endocuticles which are either lamellate or non-lamellate throughout their thickness. Most exocuticles are helicoidal (lamellate throughout their thickness) with the exception of a few which have single unidirectional layers of microfibrils sandwiched between two helicoidal layers.

The two system model has important implications for x-ray diffraction studies of cuticles. The two components are analogous to the molecular architecture of cholesteric and nematic liquid crystals.

The above data concerning the stereochemical model of chitin and its architecture in some of the most important living organisms are part of the fruit
of the research activities of a large number of scientists. The picture that
we have of chitin at this point is necessary for the correct understanding of
its properties, and for the most appropriate treatment of chitin in its production, refinement and applications.

Liquid crystals.

A state of matter intermediate between that of a liquid and a solid is called
liquid crystal. Liquid crystals combine the ability to flow of a liquid with
the order of a solid and are formed by oriented long molecules. Liquid crystals
are anisotropic and exist over rather narrow ranges of temperature depending on
their nature.

Nematic liquid crystals are given by molecules arranged with an average unidirectional orientation. Cholesteric liquid crystals are constituted by molecules whose orientation is in parallel planes, with the orientation rotation
to form an helicoid. The two system model for cuticle is equivalent to nematic
and cholesteric liquid crystals.

The initial substances constituting secretions of numerous tissues, are
liquids or anisotropic gels able to undergo various chemical alterations without losing their initial geometrical properties. The order existing in the
liquid crystal is normally kept by the tissue originating from it. The meso-
morphic states show numerous faults whose distribution is not random, and
normally they form patterns. These insoluble and hydrated gels become extremely
hard fibrillar systems when some side chemical reactions make the fibrils more
coherent, for instance, when the insect cuticle is tanned by quinone derivatives after each molting. Such fibrillar systems are often slightly hydrated
solids, whose geometrical stereochemical and optical characteristics are those
of liquid crystals or their colloidal analogues.

Therefore, it is not surprising that the decalcified crab shells and insect
cuticles have properties similar to those of liquid crystals. The cholesteric
liquid crystals, normally reflect a given wavelength in the visible spectrum
with maximum intensity. Incident light is resolved into two polarized circular
and opposite vibrations, one of which is reflected and the other which pene-
trates into the cuticle or the liquid crystal. This explains the special color
effects of the cuticles of insects whose main color in certain regions is
shifted to a sharply different color, for instance green to red.

Liquid crystal systems including fibrillar cellulose and partially deacetylated
chitin have been reported by Marchessault, Morehead & Walter (1959) and by
Bouligand (1974). A suspension of crystallite particles of chitin was prepared
by treating 20 g of purified chitin from crab shells for one hr in 750 ml of
2.5 N hydrochloric acid under reflux. The excess acid was decanted and
distilled water was added. The chitin hydrolysate was still mostly a sediment,
and it was homogenized when still acid. From this treatment, a stable isotropic
suspension was obtained and the pH raised to 3.5 due to unacetylated amino
groups at the crystallite surface. This suspension and a similar one of cellulose were the starting materials for the preparation of the liquid crystals.
The concentration was always less than 1 %.

The formation of a permanent birefringent gel was first observed when a suspension of cellulose crystallites was heated on a steam bath. A soft, reddish-brown gel, formed on the surface of the heated suspension, was found to be birefringent but without extinction directions, like a powder of a birefringent crystal.

In the birefringent gel, low-angle x-ray measurements have shown that the interparticle distance varies as the square root of the concentration. For a 15 % gel it is about 40 nm. The properties of this system were found to be similar to the well-known behavior of tobacco mosaic virus particles.

Colloidal chitin.

Colloidal chitin has been prepared since a long time ago by many authors. Smirnoff (1975) describes its preparation as follows: 10 kg of crude chitin was soaked for 24 hr in 60 1 of diluted hydrochloric acid obtained by diluting 1:10 concentrated acid, and then for 24 hr in 60 1 of 10 % sodium hydroxide solution. This soaking was repeated twice more. After filtration and water removal (washing with acetone) the chitin was dissolved or dispersed in 60 1 of concentrated hydrochloric acid. This highly viscous suspension was filtered through a fritted glass filter into a container holding 300 1 of distilled water. On dilution, the chitin precipitates as a colloidal suspension which is again filtered and diluted to final pH 6.4. The colloidal chitin can then be recovered in solid form by vacuum drying.

Along with amylose, xylan, mannan and cellulose, chitin is mentioned in the text of a patent granted to Battista (1966) dealing with a method of preparing aqueous dispersion-forming aggregates. Cellulose crystallite aggregates have been found to be extremely useful as non-caloric food additives, as blend components for cosmetics and pharmaceutical preparations and for a variety of other uses. Insoluble polysaccharides, like chitin and cellulose, can form stable colloidal dispersions and thixotropic gels on attrition by means of a high speed shearing action in the presence of an aqueous medium. At consistencies above 25 % by weight of cellulosic material, attrition produces a material which resembles a shiny oleaginous paste and, as the concentration increases, the material acquires a progressively firmer and wax-like appearence and consistency. The patent does not include any example dealing directly with chitin but it is understood that chitin can form similar gels.

Chitin gels have been prepared by Hirano, Kondo & Ohe (1975) and by Hirano & Ohe (1975) by deacetylating chitin with 40 % sodium hydroxide in the presence of 1 % $NaBH_4$, to produce chitosan: 500 mg of the latter were dissolved in 25 ml of 10 % acetic acid. To the solution was added 12.5 ml of acetic anhydride. The excess acetic anhydride added is important to the gel formation and is considered to be used for the acetylation of both amino and hydroxyl groups present in chitosan as well as for the reaction with water. The mixture was kept at room temperature for about 30 min to afford a rigidly solidified gel to the whole solution in a flask. The solidified gel was kept at room temperature overnight and then suspended in a large volume (about 2 1) of distilled water, which was changed several times, at room temperature for 3 days or dialysed against running water. This treatment exchanged free acetic acid present in the gel for water and removed unreacted chitosan as its acetate salt which is soluble in dilute acid. Thus, the chitosan gel free of acetic acid was obtained by filtration as about 31.4 g yield. The gel was colorless, transparent and rigid, and gave no smell of acetic acid and almost no taste. The gel was soluble in formic acid but insoluble in water, 50 % formic acid, 10 % acetic acid, alcohols and acetone. The gel did not melt by heating over a flame in a test tube and could be sliced with a knife without distruction. The product had a degree of substitution of 2.36 per monosaccharide residue. The dried gel was soluble in formic acid and 50 % resorcinol but insoluble in other solvents examined. The solubility is similar to that of a peracetylated chitin prepared with acetic anhydride by passing dry hydrochloric acid.

SOLUBILITY AND VISCOSITY.

Chitin.

Hexafluoroisopropanol and hexafluoroacetone sesquihydrate are solvents for chitin, as reported by Capozza (1975), who prepared chitin films from 2.0 % and 1.4 % solutions, respectively. While these solvents were proposed a few years ago as solvents for polymers and biological materials, they have been used with chitin only recently: data on chitin solutions in these solvents are therefore scarce but it is foreseen that hexafluoroisopropanol and haxafluoro-acetone sesquihydrate will be of assistance for special preparations.

It has been found by Austin (1975) that chloroalcohols in conjunction with aqueous solutions of mineral acids or with certain organic acids are effective systems for dissolving chitin in any form, such as native, reprecipitated powders, or crystalline conformations. These solvent systems give relatively low viscosity solutions of chitin, they dissolve chitin rapidly at room or mildly elevated temperatures and hydrolytic degradation proceeds relatively slowly in them.

The chloroalcohols that may be employed include 2-chloroethanol, 1-chloro-2-propanol, 2-chloro-1-propanol and 3-chloro-1,2-propanediol. Certain of these materials are also called chlorhydrins and a commercial mixture of propylene chlorhydrins comprising 1-chloro-2-propanol and 2-chloro-1-propanol may be conveniently employed. However, the simple 2-chloroethanol is generally preferred.

The choice among these solvents is always a compromise among such aspects as rate of solution, rate of chitin degradation, viscosity of the acid and viscosity of the resulting chitin solution. Mixtures of 2-chloroethanol with hydrochloric, phosphoric , sulfuric or nitric acid have their compromise properties. However, all the acids have a reduced degree of ionization in the chloroalcohol which leads to greater stability of the chitin in such solutions. With formic acid solutions of chitin, a limited amount of chloroalcohol may be used. 2-Chloroethanol in conjunction with aqueous sulfuric acid appears to facilitate solution of the chitin, and gives high chitin concentrations at workable viscosities, a low rate of hydrolysis and the ability to precipitate chitin in crystalline fibrillar form. For example, twelve parts of 2-chloroethanol and sixteen parts of 73 % sulfuric acid mixed together and heated moderately below the boiling point of the 2-chloroethanol readily dissolve one part of chitin to give a relatively low viscosity solution. The chitin can be precipitated by addition of water, metahnol or aqueous ammonium hydroxide.

Table 2.1. Solubility of chitin from crab shells and *Loligo* pen and beak.

Chitin origin	Solvent	Effect
Crab shells	hexafluoroisopropanol	dissolves
id	hexafluoroacetone	dissolves
id	1,2-chloroalcohols	completely dissolves
Loligo pen β	88 % formic acid	swells and dissolves
id	98–100 % formic acid	completely dissolves, does not re-dissolve when precipitated by water
id	an. trifluoroacetic ac.	no
id	98 % formic + 2 % acetic anhydride	no
id	dimethylformamide	no
id	dimethylsulfoxide	very slight swelling
id	picric acid + dimethylformamide	very slight swelling
id	8 M urea	no
id	8 M guanidine HCl	no
id	0.1 M LiCNS, boiling	no
id	2-aminoethanol	no
id	Cadoxen	no
Loligo beak α	88 % formic acid	bleaches
crab shells α	88 % formic acid	bleaches, sample rubbery
id	98–100 % formic acid	bleaches, sample separates into layers
both	polyethyleneimine	colored "dispersions"
both	cyclohexanone oxime	colored "dispersions"

The solubilities of 1 g of polymer in 50 ml of dimethylformamide, with a dinitrogen tetroxide to polymer ratio of 3:1 are 5 % for chitin and 100 for chitosan (Allan, Johnson, Lai & Sarkanen, 1971). Infrared spectral analysis of the starting materials and products indicate the absence of chemical modification and that their solubilities in appropriate solvents were identical. The highly ionic polymers alginic acid, chitosan and the mucilage of *Ulva lactuca* gave non-viscous solutions presumably owing to tight chain coiling induced by common ion effect. On the other hand, the essentially non-ionic cellulose, agar and propyleneglycol alginate gave viscous solutions. For the mechanism by which solution occurs, it was suggested that hydrogen bonds are disrupted by NO^+ ions.

While chitin is known to dissolve in the concentrated mineral acids and in anhydrous formic acid, it is not clear whether all chitins are soluble in the latter (Lee, 1974). The observations on solubility are given in Tab. 2.1. The ease of dissolution evidently depends on the degree of crystallinity, as only β-chitin dissolves in anhydrous formic acid; γ-chitin was not examined. In 88 % formic acid, however, deproteinizing the *Loligo* pen rendered it less soluble than an untreated pen. This bears out the observation that large quantities of protein are intimately associated with chitin: chitin crystallites of insect cuticles, originally very small in lateral dimensions, showed large crystal-

lites on removing the protein and the disappearance of the 3.1 nm axial
periodicity on deproteinization in 5 % potassium hydroxide. In the case of
α-chitins, anhydrous formic acid was effective in bleaching and penetrating
into the cuticular interlayers to separate the integument into its component
parts. The polar aprotic solvents such as dimethylsulfoxide are only able to
swell β-chitin very slightly, and are totally ineffective on α-chitin. Protein
denaturants such as urea, guanidine and the chaotropic agent lithium thiocyanate are also ineffective. The cellulosic reagents do not dissolve the
polysaccharide, probably because the acetamido group at C^2 prevents formation
of the required complex. The polymer polyethyleneimine and the industrial deacetylating agent cyclohexanone oxime, were also tried as solvating agents.
Samples refluxed in these solvents gave a highly viscous, deeply colored
concoction. Whether they are true solutions is difficult to judge and were
described as dispersions by Lee (1974).

Chitin precipitates out as formic acid solutions of the material are slowly
diluted with water. However, its behavior and stability in anhydrous formic
acid are measured by viscosity. The polymer shows a non-linear dependence on
concentration: above certain concentrations, the polymer becomes more
entangled and there may be formation of internal mesh-work. By extrapolation
to infinite dilution, however, Lee (1974) was able to measure the intrinsic
viscosities. Fig. 2.5 shows some viscosity measurements of chitin dissolved in
anhydrous formic acid.

Generally speaking, the viscosity of a solution of a macromolecular compound

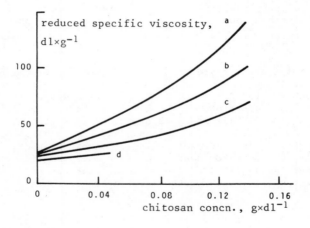

Fig. 2.5. Extrapolation to zero concentration of the reduced viscosity of chitin samples solubilized in anhydrous
formic acid at 2° under nitrogen. Measurements done at 25°.
The ordinate intercepts of the lines give the intrinsic
viscosities. a = 4th day; b = 5th day; c = 6th day; d =
14th day. Lee, 1974.

is related to the degree of polymerization through the constant K_m that depends on the nature of the solute and the solvent, the type of bond and the molecular shape:

$$\lim_{c \to 0} (\eta_{sp} / c) = K_m \times P^n$$

where n is between 0 and 2.

Currently, the intrinsic viscosity η is related to the average molecular weight M_w in the Staudinger equation:

$$\log|\eta| = \log K + a \times \log M_w$$

where a is unity for a long molecule kinked in a random fashion, and approaches zero for a chain coiled into a ball. Lee (1974) obtained the values $K = 8.93 \times 10^{-4}$ and $a = 0.71$ for chitin.

K values for cellulose derivatives ($11.0 \times 10^{-4} \div 11.8 \times 10^{-4}$) and linear, unbranched molecules having β-anhydroglucosidic bonds are about ten times bigger than for branched polysaccharides, like methylstarch (1.0×10^{-4}) and arabogalactan (0.3×10^{-4}) or with α-anhydroglycosidic bonds like acetylamylose (0.14×10^{-4}). The glycolchitin value was found to be 12.0×10^{-4} by Okimasu & Senju (1958).

Viscosity data relevant to chitin ethyl ether having molecular weight 66,000 daltons, obtained by Danilov & Plisko (1958) are in Fig. 2.6. The intrinsic

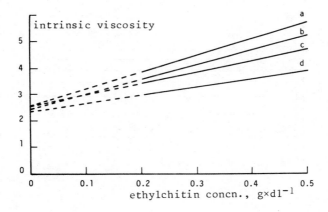

Fig. 2.6. Effect of solvent on the slope of the curve and on the intrinsic viscosity of chitin ethyl ether solutions in: a = _o_-xylene, benzene and toluene; b = methylethylketone; c = 1:4 mixture of ethanol and benzene; d = chloroform. Danilov & Plisko, 1958.

viscosity in different solvents varies between 2.3 and 2.6. The viscosity
curves have different slopes which are characteristics of the interactions
between solvent and solute: the best solvent among those studied is the 1:4
mixture of ethanol and benzene.

Chitosan.

Chitosan is currently prepared by deacetylation of chitin in concentrated
sodium hydroxide solution (see page 94). The viscosity variation as a function
of the deacetylation time under usual deacetylation conditions is shown in
Fig. 2.7. It is possible to remark that deacetylation does not need to be prolonged for hours or days, as reported by several authors in the past, as 30
min are enough to obtain a chitosan soluble in acetic acid solutions; however,
during this period, the viscosity shows a dramatic change and this enphasizes
the problems involved in the production of high molecular weight chitosan.
When the deacetylation extent is enough to permit dissolution of the product

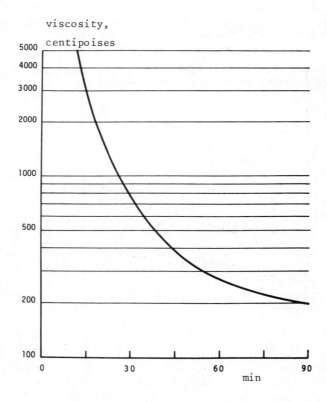

Fig. 2.7. Viscosity as a function of deacetylation time in
50 % NaOH solution at 118°. Muzzarelli, original results.

in dilute acetic acid, the name "chitosan" is preferred to "deacetylated chitin". Formic and acetic acid are widely used for research and applications of
chitosan.

Chitosan in solution exhibits the polyelectrolyte effect: in the absence of
salt, there is an abnormal increase in the viscosity of the more dilute
solutions because of an enlarged effective volume due to charge repulsion and
stretching out of the molecules. When sufficient salt is added to neutralize
this charge effect, the viscosity behavior is normal. Fig. 2.8. shows the
viscosity behavior of chitosan in 0.2 M acetic acid and various acetate buffers. In the presence of the acetate salt, the pH changes from pH 3.0 (at 0.1
M salt) to pH 5.5 (at 0.5 M salt). At 0.2 M salt and above, the solutions
become turbid when chilled, and the turbidity is non-reversible on warming.
For this reason, the solvent system of 0.2 M acetic acid + 0.1 M sodium
acetate should be preferred for viscosity measurements.

Viscosity in acetic acid tends to increase with increasing acid concentration
(decreasing pH), while the hydrochloric acid tends to decrease with increasing
acid concentration. Viscosity curves for three grades of chitosan obtained
after various deacetylation times, are in Fig. 2.9. They were obtained by dis-
persing chitosan in water and by adding the required amount of acetic acid.

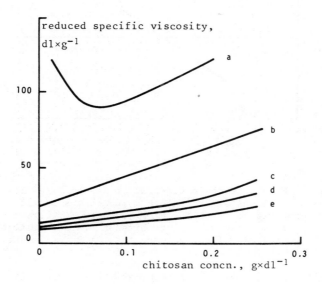

Fig. 2.8. Dependence of the reduced specific viscosity on
concentration of chitosan in the absence and presence of
salts. a = 0.2 M acetic acid; b = 0.2 M acetic acid + 0.01
M sodium acetate; c = 0.2 M acetic acid + 0.1 M sodium
acetate; d = 0.2 M acetic acid + 0.2 M sodium acetate;
e = 0.2 M acetic acid + 0.5 M sodium acetate. Lee, 1974.

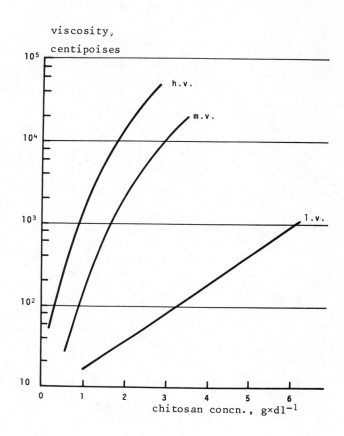

Fig. 2.9. Effect of the concentration on the viscosity of
chitosan solutions at 25° and pH 4 in aqueous acetic acid.
h.v. = high viscosity chitosan; m.v. = medium viscosity
chitosan; l.v. = low viscosity chitosan. This terminology
is related to molecular size. Courtesy of Hercules Inc.

Solutions of chitosan decrease in viscosity as their temperatures are raised.
The solutions regain their viscosities when cooled to initial temperatures.
This nearly linear negative dependence of intrinsic viscosity on temperature
has also been found for cellulosic materials and was attributed to the
enhanced chain flexibility and reduced root-mean-square unperturbed end-to-end
distance of the chains with temperature increase. Solutions continually
saturated with oxygen show a slightly greater rate of degradation.

SHEARING.

Shear degradation is an efficient and economical process for selectively degrading macromolecules and producing a relatively narrow molecular weight
distribution. Such an approach to obtaining selected monodispersed samples,
with sharpening of the molecular weight distribution in general is highly
desirable when working with polymers. Important physical properties of polymers depend to a significant degree on the molecular weight distribution and
superior properties can be obtained when the molecular weight distribution is
sharp.

In shearing of long molecules by hydrodynamic forces, the actual force is not
uniform over a given chain but is parabolic in form with the maximum in the
center and with the height of the maximum proportional to the average force.
A general view is that chain scission results from cavitation. When high-
frequency sound waves are irradiated into a solution, or when there are
swiftly moving bodies, cavities of partial vacuum or bubbles are formed, and
the energy of the field is carried locally. These cavities then collapse
violently and release their stored energy by radiating intense shock waves
through the surrounding medium. As the solvent rushes through a collapsing
cavity, a strong velocity gradient is imposed across a macromolecule located
near the cavity and, if forces exceed bond strength, breakage occurs. In such
model, the rate of degradation is a function of the chain length.

In fact, longer chains are more resistant to flow; these macromolecules are
less able to rearrange and relieve shear stresses, so that they break more
frequently, with the breaks located at or near the center. There is a minimum
chain length for which the shear forces can not exceed the bond strength and
the chain can not break. Thus, in qualitative terms, for an initial broad
distribution of chain sizes, shearing would be expected to repeatedly break
the largest molecules until resulting fragments are below the size at which
the stress is sufficient to break them. However, there are inherent difficulties in quatitative shear degradation and applying theory to this hydrodynamic phenomenon.

Experimentally, degradation is often difficult to control in that the
rheology of the many devices preclude obtainment of meaningful results.
Chitosan suspensions were placed in two thin-cylinder Couette devices which
were rotated in opposite directions. The experiments carried out by Lee (1974)
indicate that particles in high velocity gradients undergo translation and
rotation with a tensile stress developed at small angles with respect to
streamlines of flow, so that shear-induced deformation on the rotational
motion can occur.

In shearing chitosan, two features of the degradation process stand out: the
molecular weight fall-off is steepest with the initial passes, and the molecular weight rapidly approaches a limiting size with repeated passes (Fig. 2.
10). Ultracentrifugation reveals that the molecular weight distribution
narrows considerably. Repeated degradation, reduces the size of the molecules
to the limit produced by the critical forces applied per molecule, so that,
unless a greater force is applied, no further degradation occurs. When the
same chitosan was sheared in a solution containing 0.2 M acetic acid + 0.1 N

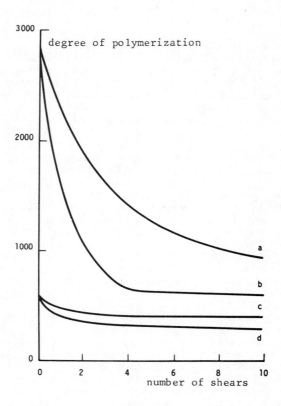

Fig. 2.10. Shear degradation of chitosan. a = 0.2 M acetic
acid; b = 0.2 M acetic acid + 0.1 N sulfuric acid; c =
commercial chitosan in 0.2 M acetic acid; d = commercial
chitosan in 0.2 M acetic acid + 0.1 N sulfuric acid. Lee,
1974.

sulfuric acid, initial molecular weight fall-off was more rapid, and the
limiting degree of polymerization of 660 was substantially lower, suggesting
that solvolysis may contribute to lowering of the activation energy of bond
scission.

A commercial chitosan preparation (initial degree of polymerization 581) could
be sheared to final degree of polymerization 420 in 0.2 N acetic acid and to
350 in 0.2 acetic acid + 0.1 N sulfuric acid (Fig. 2.10).

The low sheared value of the commercial chitosan preparation which had an
initial degree of polymerization lower than the leveling degree of polymerization found for sheared chitosan can be explained in that apparent degree of

polymerization values can mask the actual molecular weight distribution: while
the bulk of the commercial preparation contained short molecules well below
the critical length, shearing reduced the sizes of those longer molecules to
below the critical length, until further shearing did not affect the molecular
weight distribution. The shortening of this high molecular weight "tail" would
then result in a slightly decreased degree of polymerization. For a chitosan
sample, while the bulk of the chains were longer than those in the commercial
preparation, there was also a much larger proportion of longer chains well
above the critical length. Degradation of these to or below the critical
length had a larger effect on the molecular weight redistribution. Hence, for
the same force applied, the apparent limiting degree of polymerization would
be different for the different preparations.

HYDROPHILICITY AND SWELLING.

It was shown by Danilov & Plisko (1954) that by repeatedly freezing and de-
freezing chitin in alkali solution is not only swelled but could also be
wholly dissolved in the alkali. The authors' explanation of this process was
that the structure of chitin became friable, analogous to the phenomenon observed in cellulose fibers.

The study of cellulose shows that their hydrophilic properties could give an
idea of the state of the fiber structure, density of packing of the chain-like
cellulose molecules and characterize the possible reactivity in a number of
cases. Determinations of the hygroscopicity at relative humidities of 65 and
100 %, of the amount of unfrozen water in chitin after its saturation with
water vapor, of the heat of wetting with water and the heat of swelling in
sodium hydroxide solution were carried out. As the value of the specific heat
capacity of the material investigated needed to be known for calculating the
amounts of unfrozen water in it, it was determined calorimetrically for chitin
as well. For this, a sample of chitin was dried in vacuum over phosphorus
pentoxide, cooled to a definite temperature and then placed in the calorimeter
in a sealed tube. The specific heat capacity of chitin was derived from the
heat effect of the calorimeter by the appropriate calculations. The measurements carried out showed that the specific heat capacity of chitin is somewhat
higher than that of cellulose and is 0.373 ± 0.03 cal$\times$g$^{-1}\times$degree.

Chitin has a hygroscopicity close to that of mercerized cellulose fibers and
considerably greater than that of ramie fibers. However, the moisture absorbed
by the chitin is almost wholly frozen at $-6°$, while ramie fibers still retain
12,5 % of the water and the mercerized wood cellulose fibers 26,2 %. As water
molecules are most securely retained on an active inner surface of a hygroscopic material, the results given by Danilov & Plisko (1954) indicate that
the chitin surface is less active and permeable to water in comparison with
cellulose fibers. The moisture absorbed by chitin is probably in coarser
capillaries, from which is readily frozen when the temperature is lowered to
$-6°$. The values of the heat of swelling of chitin and cellulose fibers in
water and in alkali solutions are reported in Tab. 2.2.

Air-dried chitin at $105°$, very readily becomes wet with water and reacts with
solutions of sodium hydroxide in concentrations of up to 23 %. Cellulose
fibers behave similarly; then readily become wet and swell. The alkali reaction

Table 2.2. Heats of swelling of chitin and cellulose fibers in water and in solutions of sodium hydroxide at 20°. Klenkova & Plisko, 1957.

Material investigated	Moisture %	Heat of swelling, $cal \times g^{-1}$						
		in water		in NaOH soln. at NaOH % concn.				
		sample dried at 105°	sample dried in air	8.0	12.0	17.5	23.0	34.5
Chitin	8.90	12.4	3.3	–	3.8	6.7	8.6	79.9
Ramie fibers	5.50	11.9	3.7	8.8	21.5	28.1	–	63.3
Cellulose viscose fibers	8.96	17.3	4.5	13.9	22.5	26.8	–	54.7

for both chitin and cellulose proceeds very slowly in higher concentrations of sodium hydroxide and the equilibrium in the calorimeter is established 30 min after immersion in the alkali solution for cellulose fibers, while it takes 70 min for chitin. Dried chitin does not become wet readily even in the lower concentrations of sodium hydroxide and due to this, the calorimetric data cannot be reliable. Taking the above into consideration, the heats of swelling were measured without drying the chitin, i.e., with some moisture absorbed.

The data in Tab. 2.2. show that the heat of wetting of chitin by water is almost the same as the value for ramie fibers and is less that that for mercerized cellulose. However, the air-dried cellulose fibers swell considerably in the reaction with sodium hydroxide and the heat effect of swelling in alkalis, even in low concentrations, is much greater than the value for the heat of wetting in water. Chitin, however, reacts very little with solutions of sodium hydroxide in concentrations of up to 17.5 %; the heat effect simply corresponds to the values of its wetting by water in the air-dried state. This effect increases slightly in 17.5 and 23 % solutions of sodium hydroxide and increases sharply in a 34.5 % solution, even exceeding the values for cellulose fibers. It is possible that this is due to splitting out of acetyl groups, which occurs in concentrated solutions of sodium hydroxide. Independent experiments were carried out with a chitin sample from which some of the acetyl groups had been removed with 30 % sodium hydroxide solution: this chitin reacted somewhat more readily. Thus, for example, its heat of reaction with 17.5 % sodium hydroxide solution was 11.0 $cal \times g^{-1}$, while with a 34.5 % solution it was 101.3 $cal \times g^{-1}$. The freezing of the chitin in alkali in preparing this sample probably made its structure more friable and more permeable to sodium hydroxide.

The hygroscopicity of ethylchitin after 24 hr at room temperature at relative humidity of 42, 62 and 96.5 % is 0.51, 2.05 and 5.55 respectively. Ethylated chitin is highly hygroscopic, evidently because of the partial removal of acetyl groups. Ethylated chitin, containing 33.83 % of ethoxyl, dissolved in various organic solvents and gave clear 5 % solutions in alcohols (methanol, ethanol, butanol, isobutanol, hexanol, cyclohexanol and others), ketones (acetone, cyclohexanone, methylethylketone), halogen derivatives (chloroform), esters (diethyl and dibuthyl phosphates, butylacetate, isoamylacetate) and in benzene, toluene, xylene and their 20 % mixtures with ethanol.

MEMBRANE FORMATION.

Several attempts have been made in the past toward the production of chitin membranes. Films or membranes can be more easily obtained when chitosan is used as starting material; chitin membranes can then be obtained by acetylating chitosan membranes.

Capozza (1975) has found that 2 % solutions of chitin in hexafluroisopropanol or 1.4 % solutions in hexafluoroacetone sesquihydrate are suitable for the preparation of chitin membranes; they are transparent, flexible and do not show any symptom of hydrolysis when kept under water for several days. The chitin membranes thus obtained are degraded by lysozyme and have been used for medical applications.

Brine & Austin (1975) developed solvent systems for chitin and produced some data on chitin membranes obtained directly from chitin. To obtain chitin films of sufficient high quality to be cast, coagulated and renatured in crystalline form, and sufficiently ordered to be cold-drawn to high tenacity products, several conditions should be met simultaneously. The solvent should be anhydrous: trichloroacetic acid + methylene chloride was tried and was supplemented with chloral hydrate as a critical component. High molecular weight chitin should be made available by short solution times. An anhydrous non--solvent, usually acetone should be used for washing. Neutralization should be done with anhydrous alkali such as potassium hydroxide in 2-propanol. Renatured chitin should be oriented by cold drawing to enhance its properties and to approximate the structure of natural fibrous chitin. Extraction and/or decomposition of residual solvents should be done to reduce them to a low level. X-ray diffraction was used by Brine & Austin (1975) to demonstrate the fiber-type orientation realized.

Chloral hydrate in the trichloroacetic acid + methylene chloride system speeded solution and dissolved more chitin, up to $2 \div 3$ %. It also aided the cold drawing operation, but contributed markedly to the chlorine content of the film (see Tab. 2.3 and Tab. 2.4). Chloral may react with acetamido groups or free amino groups to form an aldehyde ammonia or Schiff's base type of structure, which would account for the difficulty of removing halogenated impurities.

More extended research work has so far been done on chitosan and chitosan derivative membranes; Izard (1941) and Mima, Yoshikawa & Miya (1975) produced membranes after addition of polyvinyl alcohol solutions to chitosan solutions. Nud'ga, Plisko & Danilov (1974) produced membranes of sulfoethylchitosan for medical applications (see Chapter 7), and also Kochnev, Moldkin, McHedlishvili & Plisko (1973) experimented with sulfonated chitosan membranes in medicine. If chitosan and cellulose acetate are mixed in the presence of solvents, like acetone and ethanol, transparent articles can be molded from the material after removal of the solvents. According to Du Pont de Nemours & Co. (1936), ten parts of cellulose acetate and one part of chitosan acetate are mechanically mixed and molded at at 150° under pressure: the product is hard, translucent and iridescent.

Table 2.3. Description of renatured chitin films. Brine & Austin, 1975.

Film preparation	Analysis N %	Cl %	Spher- ulites	Cold draw	Path diff.	X-ray
40 % trichloro acetic in $CH_2\bar{C}l_2$ 4 hr	6.52	0.58	no	no	–	–
idem, 0.5 hr	6.52	0.58	++	15 %	25 125	sharp rings
chloral	5.71	10.06	+++	100 %	100 125	sharp rings
chloral, ext. 12 hr with acetone+CH_2Cl_2	5.03	9.45	+++	85 %	125	rings

Table 2.4. Mechanical properties of some chitin films.

Film	Tensile strength $kg \times mm^{-2}$	Elongation, %	Remarks and references
Natural	35	–	Kunike (1926)
Natural (oriented fiber)	58	–	Clark & Smith (1936)
Regenerated (xanthate)	9.49	–	Thor & Henderson (1940)
Renatured	32.5	11	Non crystalline; refers to first line of Table 2.3.
Renatured	52 – 58	125	Crystalline; extracted; refers to fourth line of Table 2.3
Renatured	75 – 95	4	Pre-oriented by partial cold drawing. Brine & Austin (1975)

In addition to a simple technique based on the evaporation of formic acid from a chitosan solution spread on a glass plate, and neutralization of the dry membrane in ammonia, there is another approach to the preparation of a chitosan membrane. A chitinous membrane was reported to occur in the shell of the cuttlefish *Sepia officinalis*; to isolate the chitin membrane, the shell of the cuttlefish was first decalcified and the resulting chitin-protein complex was deproteinized. As soon as a chitin membrane is available, it can be deacetylated to yield a chitosan membrane. The form and area of the membrane thus obtained are those of the shell.

To produce a chitosan membrane from the *Sepia officinalis* shell, the isolation procedure included a 10 % hydrochloric acid treatment for three or four days, an ether extraction and a deacetylation with potassium hydroxide in ethanol + monoethyleneglycol mixture. The chitosan identity was checked by infrared spectrophotometry. Both procedures, involving chitosan powder from crab shells or the *Sepia officinalis* membrane, yielded membranes perfectly transparent and of high mechanical strength.

In fact, the tensile strength was found to be 7 $kg \times mm^{-2}$ on both directions. No appreciable elongation was observed. These data are in agreement with data incidentally reported by Nud'ga, Plisko & Danilov (1971) in a paper on the preparation of chitosan. The burst strength was found to be 1.8 $kg \times cm^{-2}$.

The permeabilities to gases and vapor are reported in Tab. 2.5 and compared to data for a cellophane membrane. The outstanding ability of the chitosan membrane to prevent the passage of oxygen, nitrogen and carbon dioxide is evident and its permeability to water moisture is comparable to that for cellophane (Muzzarelli, Isolati & Ferrero, 1974).

Table 2.5. Permeabilities of chitosan membrane and cellophane. Muzzarelli, Isolati & Ferrero, 1974.

Membrane	Thickness, μm	Permeabilities			
		$g \times m^{-2} \times 24\ hr^{-1}$ water D = 90% at 100°F	$ml \times m^{-2} \times 24\ hr^{-1}$, 21°, 760 mm		
			O_2	N_2	CO_2
Chitosan	20	1200	3.6	0.7	2.7
Cellophane	22	1200	11.3	3.7	264.0

Man-made chitosan membranes as well as chitosan membranes obtained from *Sepia officinalis* show x-ray diffraction lines. This finding indicates a certain ordered structure derived from that in chitosan powder: it is interesting to note that this order survives after dissolution and casting treatments. It was also observed at the microscope that alcohols make the membrane crystallize upon wetting.

The viscosities of chitosan solutions obtained from chitin deacetylated at 100° for various periods of time with 50.9 % sodium hydroxide, and the strength of films obtained from these solutions are given in Tab. 2.6. The viscosity of a chitosan solution may be decreased by the addition of oxidizing agents such as hydrogen peroxide, chlorine, bromine, hypochlorous acid, perborates, permanganate, dichromates, oxygen, air etcetera. The addition of 20 parts per million of 30 % hydrogen peroxide reduced the viscosity from 568 to 213 poises in one hr, or from 510 to 80 poises in 3 hr, but no further drop in viscosity is observed after three weeks. The viscosity is also altered by allowing one of the said agents to act upon the solid chitosan before dis-

Table 2.6. Strength of membranes obtained from chitin deacetylated for
different lengths of time. Du Pont de Nemours, 1936.

Deacetylation, hr	Viscosity, poises	Tensile strength, $kg \times mm^{-2}$
1	630	6.767
2.5	63	5.768
6	46	5.722
16	36	5.357

solution in acetic acid or other solvents.

A film was cast by Danilov & Plisko (1958) from a concentrated solution (5 ÷ 8
%) of ethylchitin in alcohol + benzene mixture, whose viscosity at 20° was 484
centipoises. The film was cast on a carefully cleaned surface. After the solvent had evaporated, the film was allowed to dry in air before drying out at
60 ÷ 80° in a vacuum desiccator for 5 ÷ 6 days. The films were tested for
yield strength and extension (Tab. 2.7); samples did not soften on heating.

It appears from Tab. 2.7 that the tensile strength varied between 5.5 and 7.3
$kg \times mm^{-2}$ and the extension between 5 and 23 %. Films, made from ethylated chitin
prepared in ampoules, showed a lower tensile strength and a lower extension,
since under these conditions there was a greater degradation of the alkali
derivative of chitin.

Table 2.7. Properties of ethylchitin films. Danilov & Plisko, 1958.

Sample no	Tensile strength, $kg \times mm^{-2}$	Extension at rupture, %	$|n|$
1	6.8 - 7.3	15.1 - 22.8	2.30
2	6.4 - 7.1	23.0 - 23.9	1.40
3	5.5 - 5.9	5.1 - 8.1	1.73

General characteristics.

Chitosan membranes have been prepared by De Chirico & Gallo (1958) by mixing
chitosan and polyhexamethyleneadipamide both in 99 % formic acid solution. The
resulting homogeneous solution was poured on crystal plates and evaporated at
40 ÷ 50°. The membrane was then washed in 1 N sodium hydroxide solution and
rinsed with water until neutral. When the polymers are mixed in the ratio 3
parts of chitosan to 7 parts of polyhexamethyleneadipamide, quite water imper-

meable membranes are obtained with excellent mechanical properties. The membrane potentials were measured on $10 \div 20$ μm thick membranes by using a potentiometer connected to a membrane through saturated calomel electrodes, saline bridges and sodium chloride solutions of variable concentrations c_1 and c_2:

SCE | sat. KCl | c_1 | membrane | c_2 | sat. KCl | SCE

The ionic potentials in cells:

0.1 N NaCl | membrane | 0.1 N NaX

where X was thiocyanate, iodide, nitrate, bromide, chloride, formate, iodate and acetate were also measured. Fig. 2.11 shows how the membrane potentials vary with the logarithm of the ratio a_1/a_2.

When c_1/c_2 was varied up to 8, the variation of the potential with the activity ratio was close to the theoretical, according to the equation:

$$E = (1 - 2\tau^-) \frac{RT}{F} \ln \frac{a_1}{a_2}$$

where τ^- is the anion transport number. Negative membrane potentials were recorded for uni-univalent salts, whilst positive potentials were recorded for sulfate, chromate and hydrogen phosphate. Even though these relations are linear, it is doubtful that the Nernst equation applied to these salts. De Chirico & Gallo (1958) advanced the hypothesis that this was due to the fact that the membrane behaves like a base. Further research in this direction was carried out by Costantino & Vitagliano (1964) with chromate, sulfate and chloride.

The osmotic flow through chitosan + polyhexamethyleneadipamide membranes $12 \div 18$ μm thick was found to be between 100 and 250 $mm^3 \times 100$ $cm^{-2} \times hr^{-1}$.

The divalent anions taken under examination by De Chirico & Gallo (1958) were all capable of interacting with chitosan to form insoluble chitosan salts; this information was also available in the previous literature, but apparently was not noticed by those authors. In any case the data thus made available confirm the formation of insoluble salts of chitosan.

The chitosan membrane is the first chelating membrane so far obtained with a natural polymer. There is interest in making chelating membranes available for preconcentration of trace transition elements, for selective isolation of certain elements from saline solutions or for special sample preparations (Muzzarelli, Isolati & Ferrero, 1974). Of course, a chelating polymer membrane is expected to exhibit superior characteristics as it would be quite selective, would not release any ion to solution and would bind the metal of interest by dative bonds instead of ionic bonds (see also pages 139 and 193).

Crystallization occurs upon collection of transition metal ions, which is accompanied by color appearance, for example yellow with hexavalent chromium, green with trivalent chromium and blue with copper. In each case particular crystallization patterns can be observed. Chitosan membranes generally show lower capacity than chitosan powder, due to the reduced contact surface; however, collection is good, especially for molybdenum, chromate and mercury.

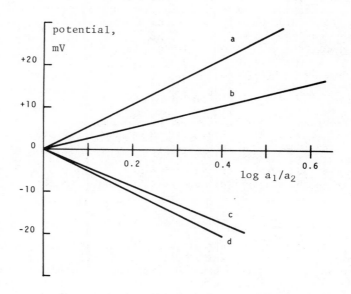

Fig. 2.11. Chitosan + polyhexamethyleneadipamide membrane
potentials as functions of $\log(a_1/a_2)$. a = chromate; b =
sulfate; c = nitrate; d = chloride, all as sodium salts.
De Chirico e Gallo, 1958.

The chitosan powder surface area is 1.64 $m^2 \times g^{-1}$ while for the reprecipitated
chitosan is 76.6 $m^2 \times g^{-1}$ as reported by Nud'ga, Plisko & Danilov (1974).

The metal ions, once collected on the membrane, do not undergo the usual
reactions currently used for their identification: for instance chromate does
not react with diphenylcarbazide, and the membrane keeps its yellow color
while the solution remains colorless.

Semipermeable membranes were prepared by Matsuda & Shinoda (1975) from glycol
chitosan and/or heparin and/or chondroitin sulfate. Unlike Loeb membranes,
these preparations are stable at high pH values, and do not compress under
high pressure. Thus, 0.2 % chondroitin sulfate solution and 0.2 % glycolchitosan were mixed and adjusted to pH 7.5. The mixture was spread on glass plates
to produce 20 μm membranes after evaporation under reduced pressure.

Reverse osmosis membranes.

The traditional reverse osmosis membrane is a film of cellulose acetate.
However, the film is made extremely thin so that a high water permeability is
achieved. Moreover, another porous membrane is necessary to support the film.
Since a cellulose acetate film is not alkali resistant, it cannot be used in

the high pH range: these are limitations of the traditional reverse osmosis
membranes.

To produce a reverse osmosis membrane, chitosan is dissolved in dilute solution of acetic, formic or propionic acids. The solvent is water or a mixture
of water and methanol, ethanol or acetone. The ratio of water to organic
solvent is 6:4 and excess water is preferably used. The resulting $0.1 \div 2$ %
solution is coated on a suitable surface, such as a glass plate and is subsequently dried to produce a film $0.5 \div 10$ µm thick. The film produced is
neutralized with diluted sodium hydroxide or pyridine so that a reverse
osmosis membrane of chitosan possessing free amino groups is obtained.

Permeability testing of a 6 µm chitosan membrane was carried out by Yaku & Yamashita (1973) using 0.2 % sodium chloride under a pressure of 40 kg×cm^{-2}: the
water permeability was 0.55 tons×m^{-2}×day^{-1} and the rejection of sodium chloride was 72 %. Under the same pressure, with a 3 µm membrane, the water permeability for an aqueous solution of 0.2 % calcium chloride, was 1.01 tons×m^{-2}
×day^{-1} and the rejection of calcium was 81 %.

For a 10 µm thick chitosan membrane, an acetylating bath containing 100 parts
metahnol, 3 parts acetic acid, and 3 parts dicyclohexylcarbodiimide at room
temperature for 24 hr, was used to prepare an acetylated chitosan membrane
(chitin membrane); the dicyclohexylcarbodiimide acts as a dehydration condensing agent. The water permeability was 0.33 tons×m^{-2}×day^{-1} and the rejection
of calcium chloride was 93 %. This reverse osmosis membrane is insoluble in
0.25 N acetic acid and 0.2 N hydrochloric acid. The acetylation degree is
72 %.

According to Yaku & Yamashita (1973) the solid phase acetylation of a water-
swelled chitosan film is carried out in a solvent such as dimethylformamide,
dimethylsulfoxide, pyridine, alcohol or methylpyrrolidone. All of these
solvents can be combined with water and acetic acid. A solvent must be used to
provide for the penetration of the acetylation agent into the interior of the
film so that interior amino groups may be sufficiently acetylated. Many experimental results have demonstrated that the swelling of film by water is
indispensable to this reaction. When an unswelled film is acetylated, acetylation occurs only on the surface of the film (acetyl content $3 \div 4$ %) and the
film produced is soluble in dilute acid and exhibits scarce rigidity. The
acetylation of the free amino groups in chitosan produces the aminoacetyl
structure found in chitin which increases insolubility in dilute acid. Moreover, intermolecular hydrogen bonding may occur which increases the rigidity
and a tight film is thus produced.

The acetylating agent is acetic acid and dicyclohexylcarbodiimide dehydration
condensing agent. When dicyclohexylcarbodiimide is used, the solid raw
material can be reacted at room temperature in a water containing solvent. The
solvent may be an anhydrous organic solvent or an organic solvent containing
less than 40 % water (preferably $0 \div 20$ % water). Since an O-acetyl group is
unstable under these conditions, it will be decomposed and N-acetylation is
selectively carried out to produce the original chitin structure. A water
content greater than 40 % would prevent acetylation. A chitosan hydrochloride
film was soaked in 80 % pyridine to free the amino groups. The thermal decomposition of this film was 230°. The film was rinsed and swelled with water and
then soaked in 95 % pyridine. This solution was then combined with 5 equi-
velents of dicyclohexylcarbodiimide and 10 equivalents of acetic acid relative

to the chitosan film. The solution was allowed to stand for 48 hr in order to carry out the acetylation. The film was then washed with methanol and alkali and subsequently rinsed with water and dried. The produced film was colorless, transparent and insoluble in dilute acid. The acetyl content of the film was 17.8 %; the tensile strength 412 kg×cm^{-2}; the percentage elongation 10.7 %; the Young's modulus 10,313 kg×cm^{-2} and the thermal decomposition point was 256°.

The dicyclohexylcabodiimide reaction has been widely applied in the synthesis of peptides as a carboxyl activating agent: anhydride formation or acyl iso-urea derivative formation is expected at low or high concentrations of di-cyclohexylcarbodiimide, respectively. The stoichiometry of this reaction has been discussed by Tometsko & Comstock (1975).

The chitosan and the acetylated chitosan membranes do not possess any small pores which can be observed by a light microscope or an electron microscope; they are homogeneous and have high mechanical strength, especially tensile strength. Moreover, they have high permeability toward water. Due to this fact, a membrane does not need to be processed in order to form a porous structure. A membrane whose thickness makes it self-supporting can be used as a good reverse osmosis membrane. Because of this, it can be used as a cover layer on top of a porous material or it can be applied as a coating on a porous material or it can be impregnated into a porous material.

The chitosan membrane possesses a high permeability toward water even when the thickness (several μm) of the membrane makes it self-supporting. The Loeb membrane which is a cellulose acetate membrane and whose effective layer thickness is about 0.2 μm exhibits a water permeability of 0.8 tons×m^{-2}×day^{-1} under a pressure of 40 kg×cm^{-2}. On the other hand, the chitosan membrane possesses a water permeability of 2 tons×m^{-2}×day^{-1} under conditions identical to those mentioned above. The high water permeability of this membrane may be a result of the chitosan molecules being swollen with water and possessing a stereospecific structure. Moreover, since the rejection of bivalent metal salts is higher than the rejection of monovalent metal salts, the softening of hard water can be accomplished effectively by a reverse osmosis membrane made with chitosan.

The chitosan membrane allows permeation of urea and low molecular weight organic compounds and rejects high molecular weight compounds. That is, it is permeable to urea, aminoacids and creatine and is impermeable to serum proteins in the blood. Since it also has a high mechanical strength, it can be used as an artificial kidney membrane. The chitosan can be combined with carboxymethylcellulose, gelatin, alginic acid, polyvinyl alcohol and high polymers (Mima, Yoshikawa & Miya, 1975).

A chitosan membrane does not need to be stored in a wet state, although cellulose acetate membranes require such storage.

Chitin xanthate membranes.

Thor (1936) has found that by converting chitin into a suitable compound such as chitin xanthate, it can be regenerated. Thus, by extruding the chitin compositions through nozzles or orifices of the desired shapes and sizes into a suitable coagulating bath, regenerated chitin articles, such as filaments, threads, films, ribbons, tubes, straws etcetera can be produced.

Zimmermann (1952) prepared chitin from fungi and obtained the xanthate derivative that he used to prepare chitin xanthate membranes. If membranes are to be
prepared by spreading, these steps should be followed, according to Thor &
Henderson (1940):

1) The filtered and deaerated xanthate dispersion is spread smoothly on a
glass plate.
2) The plate is immersed in a coagulating bath consisting of 40 % ammonium
sulfate and 5 % sulfuric acid in water. It is left in the bath for several min
until regeneration is complete.
3) The film is washed in running water until free of acid and until the initial
cloudiness has disappeared. This may be facilitated by a final wash in 1 % ammonium hydroxide.
4) If the film is to be dried, it is desirable to treat it with a softening
agent such as glycerine. To do this, the film is immersed for about 30 min in
15 % aqueous glycerine, is blotted free of excess solution and then allowed to
dry, preferably after being stretched on a frame to prevent shrinkage.

The coagulating and regenerating bath, preferably contains sulfuric acid and
ammonium sulfate at various concentrations. Best results have been secured
when the aqueous coagulating bath consists of 35 % ammonium sulfate and 7.5 %
sulfuric acid. Various compounds and substances may also be incorporated in
the bath to impart certain characteristics to the product.

After the chitin has been regenerated, the articles, preferably while in the
gel state, are subjected to treatments to remove undesirable substances. For
example, the membranes are given several water washes including a very dilute
solution of an alkali, such as 0.5 ÷ 1 % ammonia, to expedite and facilitate
the removal of the sulfuric acid. The articles, while still in the gel state
and after washing, may be subjected to desulfuring and bleaching operations.
Softening or conditioning agents may be incorporated into the membranes in an
operation continuous with the process; for instance, the regenerated chitin
membrane is passed through an aqueous bath containing 10 ÷ 20 % glycerin prior
to drying.

To produce an opaque or low luster product, titanium oxide, barium sulfate,
zinc oxide, antimony oxide and other compounds may be incorporated. Likewise,
if a colored product is desired, a suitable dyestuff or pigment may be incorporated in the spinning or casting composition.

Chitin xanthate dispersions are miscible in all proportions with regular
viscose and such mixtures can be used to cast formed articles consisting of an
intimate mixture of regenerated cellulose and regenerated chitin in any
desired proportion.

Regenerated chitin films, prepared in the manner described above, behave very
much in the same way as the original chitin toward swelling media and solvents;
they are insoluble in water, dilute acids, dilute alkalies and all neutral
organic solvents, and do not gelatinize in any of these media. Determinations
of the nitrogen content of several regenerated chitin films and of the original chitin flake indicated that no significant deacetylation occurs through
the whole process of dispersion, xanthation and regeneration.

When treated with iodine + potassium iodide in dilute sulfuric acid, films of regenerated chitin become brown in color as does the original chitin flake. Deacetylated chitin, on the other hand, gives a deep violet color with this reagent. Films prepared from a chitin xanthate dispersion which has been stored at room temperature for several days, instead of at 0°, do give a violet color with the reagent. They, also, show other properties of deacetylated chitin, such as greatly decreased wet strength and the property of swelling in dilute acetic acid.

The tensile strength of regenerated chitin films in the dry condition is comparable with that of regenerated cellulose films prepared in the same manner from viscose. The wet strength, however, is considerably lower.

The elastic modulus for regenerated chitin films, as determined by Joffe & Hepburn (1973) is $1.97 \pm 0.07 \times 10^9$ N×m^{-2}; the tensile strength is $5.10 \pm 0.19 \times 10^7$ N×m^{-2}. Rupture occurred for elongations of about 7 % and 14 % for stress applied parallel or perpendicular to the preferred axis of the specimen, respectively.

Regenerated chitin membranes retain both acidic and basic dyestuffs. In view of the fact that regenerated chitin has been used in the manufacture of washable wall-papers and colored fabrics resistant to laundering (Richards, 1941) dye retention is an important quality of this material. The acidic dyes were fast green, orange G, Biebrich scarlet, Heidenhain's hematoxylin and Evans blue; the basic dyes were acridine orange, methylene blue, basic fuchsin and toluidine blue (Joffe & Hepburn, 1973).

MOLECULAR WEIGHT.

Hackman & Goldberg (1974) reported on molecular weight determinations of chitin by light scattering. They used chitin from clean shells of fully grown, freshly caught crabs *Scylla serrata*. The shells were air-dried and powdered in a mill. The 150 mesh fraction was decalcified with cold 1 M hydrochloric acid and the residue was extracted with 1 M sodium hydroxide for 24 hr at 100°. The alkali treatment was repeated twice. Chitin was collected, acidified with HCl and dialyzed against water until free of chloride. Their derivative O-carboxymethylchitin and |O-(2-hydroxyethyl)chitin| were prepared according to methods described in Chapter 3.

Carboxymethylchitin sodium salt and |O-(2-hydroxyethyl)chitin| were dissolved in aqueous sodium chloride at approximately 1 mg×ml^{-1} and the concentration range of sodium chloride investigated was 0.01 ÷ 2.5 M. Chitin was dissolved in aqueous lithium thiocyanate saturated at room temperature and brought to 95°, the cold solution was centrifuged at 26,000 g for one hr and the supernatant was diluted with an equal volume of water. The solutions were equilibrated by dialysis, each with its respective solvent, and measurements were taken on the dialyzed solutions, the dialyzate being taken as reference solvent. Solvents and solutions were clarified by filtration through 0.2 and 3 µm filters, their concentrations were determined by refractometry. The specific refractive index increment for chitin was assumed to be 0.164, the same as that of a lyophilized sample of hepta-N-acetylchitoheptaose. The average molecular weights, the Z-average radii of gyration and the second virial coefficients were calculated from the Zimm plots.

The results given in Tab. 2.8. show that chitin in 5.55 M aqueous lithium thiocyanate and carboxymethylchitin in 2.5 aqueous sodium chloride both have a weight average degree of polymerization of about 5200, i.e. each contains the same number of sugar units in a chain (there would be no association of molecular chains in these solvents): thus chitin has an average molecular weight of 1.036×10^6 daltons and carboxymethylchitin 1.33×10^6 daltons. Since the repeating unit of the chitin derivative has a higher molecular weight than has that of chitin, carboxymethylchitin must have the higher molecular weight unless degradation of the chain occurred during its preparation. In sodium chloride solutions of lower concentrations, e.g. 0.5 M, association of the molecular chains of carboxymethylchitin occurred so that, in these solutions, its measured average molecular weight is higher. $|O-(2-hydroxyethyl)chitin|$ gave similar results.

The results presented in Tab. 2.8 show that the degree of polymerization of carboxymethylchitin and $|O-(2-hydroxyethyl)chitin|$ in 0.5 M sodium chloride are similar, which is further evidence that these two derivatives did not degrade the chitin chain. For carboxymethylchitin, as the concentration of the sodium chloride in the solution was decreased, so the intrinsic dissymmetry increased, pointing out that the size and shape of the scattering particle increased. On the other hand, as the concentration of the sodium chloride was increased from 0.5 M, there was little change in the intrinsic dissymmetry but in 2.5 M sodium chloride the average molecular weight dropped from 1.896 to 1.338 Mdalton. The change in the radius of gyration was also small, so it is concluded that the dissociation occurred, a chain (or chains) of molecular weight 0.558 Mdalton was removed and the packing was essentially side-by-side.

Therefore, the average molecular weight of α-chitin from the *Scylla serrata* is 1.036 Mdaltons: this value approaches the value given for cellulose. It is lower than the value for β-chitin calculated by Lee (1974). In fact, assuming that the constants for the Staudinger equation obtained for chitosan in 0.2 M acetic acid + 0.1 M sodium chloride + 4 M urea are not significantly different for chitin in anhydrous formic acid, the size of native chitin was calculated by Lee (1974) using the value at 2° (see page 61):

$$|n| = K \times M_w^a$$

$$31.7 = 8.93 \times 10^{-4} \times M_w^{0.71}$$

Therefore, the molecular weight of chitin is 2.5 Mdaltons and the degree of polymerization is 1.38×10^4, according to this author.

The second virial coefficient for the three polymers in the solvent used were low: this indicates little interaction between these polymers and the solvent. For carboxymethylchitin, $|O-(2-hydroxyethyl)chitin|$ and chitin the values were 6.64×10^{-5}, -4.36×10^{-6} and -2.65×10^{-4} cm$^3 \times$mol$\times$g^{-2}, respectively.

Hackman & Goldberg (1974) concluded that chitin and its two derivatives are random coil structures, polydisperse in molecular weight with a polydispersity parameter of one. The molecular weight distribution for each compound corresponds to one characterized by $M_z:M_w:M_n = 3:2:1$. Thus, the effective bond length b can be calculated by using the formula:

$$b^2 = \overline{r}^2 \times N^{-1}$$

Table 2.8. Molecular parameters of carboxymethylchitin, glycolchitin and chitin. Hackman & Goldberg, 1974.

Compound	Molecular weight rep. unit, daltons	Molecular weight; Av. M_w, Mdaltons	Average degree polymer.	Intrinsic dissymmetry	Z-average radius of giration, nm	Effective bond length, nm
Carboxymethylchitin sodium salt in 0.5 M NaCl	252.7	1.896	7503	3.35	143.49	3.31
Carboxymethylchitin sodium salt in 2.5 M NaCl	252.7	1.338	5295	3.18	131.69	3.62
Glycolchitin in 0.5 M NaCl	221.6	1.819	8207	4.31	251.57	5.57
Chitin in 5.55 M LiSCN	199.0	1.036	5206	1.93	64.14	1.80

and substituting the appropriately averaged-degree of polymerization N and root-mean-square end-to-end dimension $\bar{r}^2$. The latter value is calculated from the radius of gyration. The bond length given in Tab. 2.8 confirms that rotation about the C-O bond is restricted. In sodium chloride solutions, $|$O-(2-hydroxyethyl)chitin$|$ has a more extended coil configuration than has carboxymethylchitin; chitin in lithium thiocyanate has a comparatively compact coil configuration. The effective bond length of carboxymethylchitin in sodium chloride solution is of similar magnitude to that of carboxymethylcellulose in sodium chloride solution.

In an early work, Van Duin & Hermans (1957) studied a couple of samples of chitosan obtained from *Loligo* by a method not specified: they reported that the radius of gyration was between 90 and 145 nm while the average molecular weight was between 145,000 and 185,000. In that work chitosan was brought into solution with 0.1 M hydrochloric acid and the pH adjusted to 3 with sodium hydroxide. Those chitosan samples showed a steadily increasing value of molecular weight with decreasing ionic strength. This phenomenon was attributed to association of the polymer, due to the presence of carboxyl groups or to the presence of metal ions.

Muzzarelli, Ferrero & Pizzoli (1972) determined the molecular weight of chitosan by light-scattering in 8.5 % formic acid + 0.5 M sodium formate. The refractive index increments were measured at four distinct concentrations of chitosan at the two wave-lengths 436 and 546 nm and from the results the value dn/dc = 0.174 ml×g^{-1} was obtained. The protonated amino groups of the polymer are responsible for the polyeletrolyte effect, the formate counter-ion being free to move.

At high concentrations, the chitosan molecules are close together and the counter-ions do not leave the molecular domain; at lower concentrations the counter-ions diffuse away from the polymer molecules. The effective charge increases and the polymer structure becomes more extended: sodium formate was added to the solvent to reduce the polyelectrolyte effect.

For each chitosan concentration, the light-scattering was measured and the Zimm plots were obtained. The extrapolated average molecular weight was 1.2×10^5. This chitosan sample of a commercially available chitosan from crab shells showed an average molecular weight whose value is very close to the other value reported by Lee (1974) for commercial chitosan. Nagasawa, Tohira & Inoue (1971) reported exactly the same value 1.2×10^5.

These results confirm that the commercially produced chitosans have similar average molecular weights. The fact that many results were obtained for chitosan or chitosan derivatives, whilst for many years since the publication by Meyer & Wehrli(1937) no determination of the average molecular weight of chitin was published, induced many authors to believe that the average molecular weight of chitosan and chitin were of the same order of magnitude, until the publication of the above cited data by Hackman & Goldberg (1974) and by Lee (1974).

It is now understood that the molecular weights of chitin and chitosan are very different, and that a severe degradation of the chain takes place during the production process. This happens during the decalcification step, when the crab or shrimp shells are submitted to the action of acid solutions, at high concentrations and at room temperature, or during the deacetylation step which involves harsh treatment with alkali at high concentration and at high temperature. In fact, the samples of α-chitin on which the highest values of average molecular weight (1.036×10^6) have been obtained, were decalcified in the cold with dilute hydrochloric acid, and were extracted with 1 M NaOH at 100° for 24 hr, to avoid degradation. The samples of β-chitin studied by Lee (1974) were from *Loligo* pens and were deproteinized with pronase and three washes with 5 N NaOH, and the chitosan samples obtained from part of this chitin were deacetylated in 45 % NaOH under nitrogen for 40, 60 and 80 min at 140°. For these samples, Lee (1974) obtained average molecular weights as follows: chitin, 2.5×10^6, chitosan A, 7.25×10^5; chitosan B, 4.92×10^5 and chitosan C, 2.35×10^5.

It is therefore clear that the chitin average molecular weight is much higher than was supposed until 1974 and that extended degradation occurs during the preparation of the samples and during the preparation of the derivative unless precautions are taken.

The acid treatment surely shortens the chain length: the β-chitin samples reportedly had an average molecular weight more than double that of the α-chitin obtained after decalcification; no information is available on the differences between the α- and β-chitin average molecular weights other than this.

Also the alkali treatment used to perform deacetylation degrades the chains because the value for chitosan decreases with increasing deacetylation time, even under nitrogen.

THERMAL DEGRADATION.

Chitin and chitosan at high temperature in air undergo degradation. Thermal analysis performed with a derivatograph showed that these polymers cannot withstand temperatures higher than 100 ÷ 120°. The data recorded by Bihari-Varga, Sepulchre & Moczar (1975) refer to N-acetylglucosamine, poly-N-acetyl-glucosamine (chitin obtained from chitosan by acetylation) and chitosan, and are presented in Fig. 2.12.

The thermal decomposition of N-acetyl-D-glucosamine takes place with maximum rate at about 200° followed by a second process at 400 ÷ 450°. Polymers show increased thermostability: water loss takes place at 60° and the main thermal process takes place at 275° and 280° for chitin and chitosan, respectively. The peak at 200° is due to the presence of components having a low degree of polymerization.

Chitosan has been utilized as an adhesive for tobacco particles in reconstitu- ted tobacco sheet formulation (Moshy & Germino, 1966). A low tar yield in smoke of chitin led to its trial as a tobacco extender (Austin, 1975) and prompted a study of chitin pyrolysis. It was determined that chitin and chito- san could be added to tobacco blends without altering significantly such phy- sical properties as packing ability, burning rate and ash retention, or grossly affecting aroma, taste or mildness. Since identification of the major pyrolytic products of chitin would provide additional evaluation of this ma- terial as a tobacco extender, chitin and tobacco were pyrolyzed separately and in admixture under identical conditions; the pyrolyzates were analyzed by gas

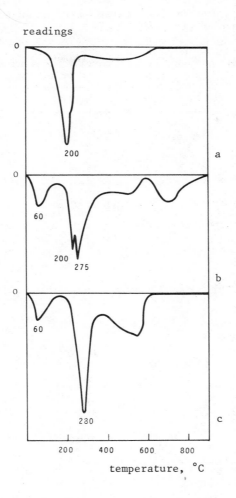

Fig. 2.12. Differential thermal degradation curves of:
a = N-acetyl-D-glucosamine; b = chitin obtained after
acetylation of chitosan; c = chitosan. Bihari-Varga, Sepulchre & Moczar, 1975.

chromatography for volatile, ether-soluble neutrals, bases, phenols and
carboxylic acids. Chitosan was also pyrolyzed to determine the effect of deacetylation on the carboxylic acid profile of the chitin pyrolyzate (Tab. 2.9).

The major basic components of tobacco pyrolysis were pyridine and nicotine,
while those from chitin were picolines and pyrazine. The presence of pyrazine
is of special interest, since pyrolysis of glucosamine yields pyrazines, which
are reportedly responsible for the "roasted" aroma of certain cooked foods.

Table 2.9. Components identified in chitin and chitosan pyrolyzates.
Schlotzhauer, Chortyk & Austin, 1976.

Components identified	Product distribution, % of GC volatiles from	
	chitin	chitosan
in basic pyrolyzate fractions:		
pyridine	0.91	
picolines	16.87	
pyrazine	15.49	
quinoline	8.53	
nicotine	–	
unidentified	49.20	
in phenolic pyrolyzate fractions:		
phenol	59.62	
o-cresol	10.45	
m,p-cresols	13.60	
2,6-xylenol	4.12	
p-ethylphenol	3.06	
3,5-xylenol	–	
unidentified	7.13	
in acidic pyrolyzate fractions:		
formic acid + acetic acid	92.10	41.34
propionic acid	3.84	4.25
isobutyric acid	0.33	<1
butyric acid	0.28	19.88
isovaleric acid	0.16	2.99
valeric acid	<1	<1
isocaproic acid	1.03	<1
caproic acid	0.52	<1
unidentified	1.74	31.53

On pyrolysis, tobacco yielded a series of lower fatty acids, where the C_2, C_3 and C_6 predominate. Chitin produced mostly acetic acid (92 %). In addition to C_2, C_3 and C_5 acids, chitosan yielded relatively large amounts of butyric acid and less than half of the acetic acid obtained from chitin, but more than 3 times that from tobacco. Chitosan also produced about 70 times as much butyric acid as chitin. Since tobacco and chitin yielded similar quantities of acids (0.5 %) on pyrolysis, production of acetic acid by chitin was more than eight-fold that of tobacco. Compared to chitin, chitosan yielded half as much acetic acid. The data reported by Schlotzhauer, Chortyk & Austin (1976) are in Tab. 2.9.

ELECTRICAL PROPERTIES.

Fukada & Sasaki (1975) reported the piezoelectricity of α-chitin; piezoelectricity depends markedly on the mechanical and dielectric properties of chitin. The small values of the dielectric constant are probably due to the structure of the polymer which contains many microvoids. The adsorbed water should exist in the crystalline surfaces, voids and/or amorphous regions: in such phases around -95° the dielectric constant increases. Ionic conduction increases above 60°. Muzzarelli, Isolati & Ferrero (1974) reported the electrical conductivity of chitosan membranes.

OPTICAL PROPERTIES.

The infrared spectra of chitin and chitosan have been published by many authors on both naturally occurring chitins and purified or modified chitins and chitosan. The most significant parts of the spectra are those showing the amide bands at 1665, 1555 and 1313 cm^{-1}, all of which show perpendicular dichroism and which are, respectively, assigned to the C=O stretching, to the N-H deformation in the CONH plane and to CN bond and CH_2 wagging.

As shown in Fig. 2.13, the chitosan spectrum differs from that of chitin in that the new band at 1590 predominates over the one at 1665 cm^{-1} and the band at 1555 is absent. This is the most important feature for practical purposes.

When chitosan has chelated copper, the following shifts were observed in the infrared absorption bands of O=C-NHR: 1650 → 1620 cm^{-1}; $-NH_2$: 1590 → 1575 cm^{-1}; $-NH_3^+$: 1560 → 1510 cm^{-1}. Further, 1650 → 1625 cm^{-1} in each of mercuric and nickel complexes, and 1650 → 1640 and 1560 → 1530 cm^{-1} in ferrous ion complex of chitosan were also observed shifts.

These results imply that not only $-NH_2$ or $-NH_3^+$ groups but also the N-acetyl-amino groups remaining in the chitosan macomolecule have considerable interaction with the metal ions or metal salts themselves.

The metal ions introduce modifications which depend on the nature of the metal ion, on its concentration and on its counter-ion. A quite new spectrum is recorded for chitosan polymolybdate (see curve d in Fig. 2.13 and page 140).

The reaction of chitosan with aldehydes is also revealed by infrared spectroscopy. Formaldehyde and amidodialdehyde makes the 1590 cm^{-1} band disappear while a broad band at 1650 cm^{-1} persists; glutaraldehyde, 2,5-esandione and glyoxale practically cancel the bands in the region 1500 ÷ 2800. N-methyl-chitosan shows bands at 1460, 1410, 1380, 1180 and 800 cm^{-1} only, in addition to the usual broad band at 2900 cm^{-1}.

Chitin

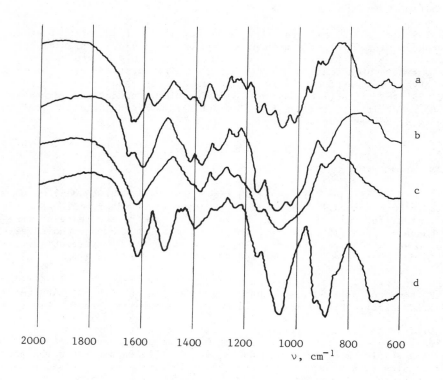

Fig. 2.13. Infrared spectra of: a = chitin; b = chitosan;
c = chitosan with chelated copper or cadmium; d = chitosan
polymolybdate. Muzzarelli, original results.

CHAPTER 3
CHITIN CHEMISTRY

Most of the research work carried out on chitin concerns the amino group, which is of course the most important function of the macromolecule. In chitin the amino group is acetylated, thus chitin is an amide of acetic acid; in chitosan the amino group is free and therefore chitosan is a primary amine. However, as it appears from the previous Chapters, chitin can not be sharply distinguished from chitosan, because fully acetylated and fully deacetylated chitins do not normally occur in nature and are difficult to prepare.

The term chitin currently refers to a polymer of N-acetylglucosamine where a minority of the acetyl groups has been lost, while the term chitosan currently refers to a deacetylation product obtained from chitin, where most of the acetyl groups have been removed. Experimentally, chitosan can be distinguished from chitin because of its solubility in dilute acetic or formic acid; chitin is also a product that contain less than 7 % nitrogen, while chitosan contains 7 % or more nitrogen.

The amino groups of chitin and chitosan are exceptionally stable in 50 % sodium hydroxide, even at 160°, at which most amines liberate ammonia or yield degradation products.

Ideally, the monomer of chitin is 2-acetamido-2-deoxy-β-D-glucose, while the monomer of chitosan is 2-amino-2-deoxy-β-D-glucose; for practical purposes, however, the extent of deacetylation in both the natural and the modified polymers has to be measured.

The official name of chitin (a fully acetylated product) is:
| (1→4)-2-acetamido-2-deoxy-β-D-glucan|
and the official name of chitosan (a completely deacetylated product) is:
| (1→4)-2-amino-2-deoxy-β-D-glucan|.

Giles, Hassan, Laidlaw & Subramanian (1958) made an untraditional approach to the preparation of chitin for chemical analysis. One of the main difficulties experienced in the characterization of the components of chitin has always been the degradation and the chemical alterations induced by harsh preparation treatments, so they removed protein by enzymic hydrolysis rather than with sodium hydroxide and obtained chitin in as unmodified a condition as possible. A comparison of the estimated values of the elemental analysis in purified chitin specimens with the theoretical values for acetylglucosamine suggested that chitin is assembled with acetylglucosamine, glucosamine and water molecules. Several authors have estimated that chitin contains 82.5 % of acetylglucosamine, 12.5 % of glucosamine and 5 % water.

Glucosamine occurs as an essential part of the polymer structure. Runham (1961) reported that newly formed chitin in *Patella* reacts positively to the PAS test, but on tanned chitin the test was negative. He suggested that newly formed chitin is less aggregated, so that the amino groups are accessible to chemical reagents, whilst quinone tanning and association with proteins prevent the older and more crystalline parts from reacting. Similar reports came from Wigglesworth (1956) concerning the reaction of the *Rhodnius* chitin from the conjunctiva of the limbs and the PAS negative test on the endocuticle chitin. Sannasi (1969) remarked that the termite *Odontotermis obesus*, when infected with fungus *Aspergillus flavus* shows changes in the staining and the histochemical reactions of the cuticle, indicative of changes in its chemical composition. One of the deviations from the normal histochemical reactions is the positivity of the endocuticle to PAS reagent; a disruption of chitin following fungal infection may result in the breakup of the bonds with the protein, thereby rendering it reactive to the PAS reagent. Wigglesworth (1957) pointed out that the cuticle of *Rhodnius*, when subjected to the action of molting fluid is rendered positive to the PAS test. Therefore, while it is agreed that some polymer units are deacetylated, the proportion of acetylglucosamine and glucosamine is not known in general.

The presence of bound water is another point open to debate. On exposure of chitin to deuterium oxide vapor, the infrared absorption band at 1656 cm^{-1} disappears, thus suggesting that this band is due to bound water that is susceptible to deuteration when the chitin is less aggregated. The density value for chitin reported on the basis of the assumption that it consists of units of acetylglucosamine is in agreement with the theoretical values calculated in the light of the newer concept of chitin as partially deacetylated poly-N-acetyglucosamine, if it is assumed that for every missing acetyl group there are two water molecules. The assumption of bound water reconciles the data obtained by different approaches.

The stoichiometry of chitin is $(C_8H_{13}NO_5)_n$. Elemental analysis on chitin samples reveal them to bind water tenaciously. This is quite in line with the general picture of chitin structure as a chain of N-acetylglucosamine punctuated by free glucosamine units with considerable amount of trapped water as part of the molecule. For the β-chitin in *Loligo* pens, it would appear that the frequency of free glucosamines is about one every five or six residues, a structure slightly less acetylated than the one Giles, Hassan, Laidlaw & Subramanian (1958) found for the α-chitin of lobster shells. The theoretical and observed values agree rather well if one includes bound water as part of the structure, as indicated in Tab. 3.1. One can see that, with increasing cooking time in strong alkali, the content of the acetyl group decreases as the nitrogen content increases.

In preparing chitosan samples, Lee (1974) observed that whole pens cooked for less than 30 min swelled in acid but did not dissolve as a result of chromogenic decomposition.

In addition, for these β-chitins, the short cooks are in sharp contrast to the rigorous deacetylation times required for α-chitins. This lends credence to the findings that β-chitins are more accessible.

The composition for chitin and chitosan was ascertained by chromatographic separation of the acid hydrolysates. When microquantities of the sugars are hydrolyzed and separated, there is only the glucosamine band for both chitin and chitosan samples. However, when large quantities of the polymers were hydrolyzed and chromatographed, chitin showed trace amounts of aminoacid con-

Table 3.1. Composition of *Loligo* chitin and chitosan samples. Experimental data are in brackets. Lee, 1974.

Sample	C	H	N	CH_3CO	N/C	Monomer
Poly-N-acetylglucosamine $C_8H_{13}O_5N$ = I	47.3	6.4	6.9	21.2	0.146	203
Polyglucosamine $C_6H_{11}O_4N$ = II	44.7	6.8	8.7	0.0	0.195	161
82.5 % I + 12.5 % II + 5 % H_2O	44.6	6.5	6.8	17.5	0.153	198
Chitin 77.0 % I + 13.7 % II + 9.3 % H_2O	42.5 (42.66)	6.88 (6.5)	6.50 (6.25)	16.3	0.153	181
Chitosan A 58.5 % I + 35.7 % II + 5.8 % H_2O	43.7 (41.51)	6.80 (6.54)	7.15 (7.51)	12.4	0.164	177
Chitosan B 43.0 % I + 50.9 % II + 6.1 % H_2O	43.3 (42.31)	6.88 (6.55)	7.40 (7.87)	9.1	0.171	170
Chitosan C 29.7 % I + 59.7 % II + 10.6 % H_2O	41.3 (40.45)	7.12 (5.88)	7.25 (7.41)	6.3	0.178	158

tamination. The nature of the contaminating protein should not be of the collagen type, as no hydroxyproline was detected.

Another aspect of the chitin composition problem is in regard to the sugar components of chitin. The chitin of the cuticle of insects yields on hydrolysis glucosamine and glucose, but the chitin present in the outermost layer of the cuticle of *Palamneus swammerdami* was different in that the end product of hydrolysis was galactose (Krishnan, Ramachandran & Santanam, 1955). Stegemann (1963) also reported that galactose and mannose and not glucose are the constituent sugars of the chitin from the pen of *Sepia*. In the chitin of the polyzoan *Scrupocellaria berthelotti*, the end products of hydrolysis are rhamnose and galactose. Therefore, in special cases, chitin comprises an array of macromolecules; they are polyaminosaccharides, but differ in the respect of the nature of the monosaccharides. A similar situation is verified for cellulose: in certain seaweeds of the Caulerpales group the cell walls yield xylene and in the Codiaks group, mannose and fructose upon hydrolysis. Sundara Rajulu & Gowri (1967) have pointed out that the chitin of the coenosteum of Millipora contains galactose and mannose in addition to glucosamine.

PREPARATION OF CHITIN.

Harsh chemical treatments are usually required to remove calcium carbonate and protein from raw chitinous material. The raw material most abundantly available includes crab shells, shrimp shells and prawn wastes (see Chapter 6).

Several procedures have been developed through the years to prepare chitin and chitosan; they are at the basis of the chemical processes for industrial production of chitin and chitosan: various methods are reported below. In addition it should be recalled that pigments, melanins and carothenoids can be eliminated with 0.02 % potassium permanganate at 60°, or sodium metabisulfite, hydrogen peroxide or sulfuric acid. Purification is normally achieved by doing extractions with boiling water, ethanol and ether.

Method of Hackman (1954).

Shells are cleaned by washing and scraping under running water, and dried in an oven at 100°. 220 g of the shells are digested for 5 hr with 2 l of 2 N hydrochloric acid at room temperature, washed, dried and ground to a fine powder. The finely ground material (91 g) is extracted for 2 days with 500 ml of 2 N hydrochloric acid at 0°; the content of the flask being vigorously agitated from time to time. The collected material is washed and extracted for 12 hr with 500 ml of 1 N sodium hydroxide at 100° under occasional stirring. The alkali treatment is repeated four more times. The yield is 37.4 g corresponding to 17 %; there are no ashes and the nitrogen is 6.8 %.

Method of Whistler & BeMiller (1962)

Shells are cleaned by washing and dried in a vacuum oven at 50°. 500 g of the ground shells are soaked for three days in 10 % sodium hydroxide solution previously deaerated, at room temperature. Fresh hydroxide solution is used each day. The deproteinized chitin is then washed until free of alkali, washed with 95 % ethanol, 6 l being necessary to clean the product from pigments. Further, the chitin is washed with one l acetone, 2.5 l absolute ethanol, 500 ml ether. The white product is then dried under reduced pressure and introduced into a 37 % hydrochloric acid solution at -20° for 4 hr. The swelling in cold hydrochloric acid and following washings may be repeated. The yield is 100 g (20 %) and the nitrogen 7.1 %.

Method of Horowitz, Roseman & Blumenthal (1957).

Ten grams of decalcified lobster shells are shaken for 18 hr with 100 ml of concentrated formic acid (90 %) at room temperature. After filtration the residue is washed with water and treated for 2.5 hr with 500 ml of 10 % sodium hydroxide solution on a steam bath. The yield is 60 ÷ 70 % on the basis of decalcified shells, nitrogen 6.95 %.

Method of Foster & Hackman (1957).

The product obtained followed the methods described above is termed chitin by the above mentioned authors, but the drastic treatments, especially the prolonged extraction in hydroxide solutions often at high temperature, surely affect the chitin structure. It removes protein and peptides as desired, but it also removes acetyl groups and may well lead to fragmentation. The method by Foster & Hackman is a milder method for the isolation of chitin, based on the use of ethylenediaminetetraacetic acid on the cuticle of the crab *Cancer pagurus*. Large cuticle fragments were attacked slowly (2 or 3 weeks) by EDTA at pH 9.0. Powdered shells having particle size 1 ÷ 10 μm were decalcified more rapidly, in 15 min under the same conditions. No carbohydrate material was extracted in this process. After a further treatment with EDTA at pH 3.0, the residue was extracted with ethanol for pigment removal and with ether for lipoid removal. Spectrographic results are as follows (the values before treatment in parenthesis): silica, 1.07 (5.6); magnesium oxide 0.08 (4.5); calcium oxide (77.0); phosphoric anhydride (8.20) % with an error of 10 ÷ 20 %. Elemental analysis gave the following data, corrected for 2.6 % ash as oxide: carbon 43.5; hydrogen 6.6 and nitrogen 7.6. As indicated by the high nitrogen percent, this chitin contained protein and this was demonstrated by hydrolysis in 6 N hydrochloric acid for 24 hr, followed by paper electrophoresis. Protein was roughly estimated to be 5 % of the product. This protein could not be removed at room temperature by extraction with dimethylformamide or phenol + water mixture. Extraction with 98 ÷ 100 % formic acid gave a soluble and an insoluble fraction, both containing the same proportion of protein. The soluble fraction could be reprecipitated by dilution with water, and looks like chitosan. The insoluble chitin could be dispersed in aqueous lithium thiocyanate saturated at room temperature and then brought to 100°.

The treatments presented in the three methods above include prolonged digestion in hydroxide solutions, which are known to deacetylate chitin and not only to eliminate protein. Therefore, in all cases, a partially deacetylated product is obtained. These data are also of interest in connection with the production of chitosan: the ability of hydroxide solutions in eliminating proteins should allow one to care less for a preliminary protein elimination step because the deacetylation process is based on hydroxide solutions at high temperature, and therefore remaining aliquots of proteins are taken away at that stage.

Method of Takeda & Abe (1962) and Takeda & Katsuura (1964).

King crab shells are decalcified with EDTA at pH 10.0 at room temperature and then digested with a proteolytic enzyme such as tuna proteinase at pH 8.6 and 37.5°, papain at pH 5.5 ÷ 6.0 at 37.5° or a bacterial proteinase at pH 7.0 and 60° for over 60 hr. The protein still present in chitin was about 5 %. To remove the remaining protein, sodium dodecylbenzensulfonate or dimethylformamide were particularly effective.

Method of Broussignac (1968).

Decalcification is carried out by a simple treatment with hydrochloric acid containing about 50 g hydrochloric acid per 1 at room temperature. This can be done in any container of plastic or wood; metals should be carefully avoided. The acid solution is recycled from various containers at various decalcification stages. This operation takes about 24 hr: one should pay attention to the important gas evolution at the beginning, which stops after one day. Before ending it is suggested to check the ash content which should be in the range 0.4 ÷ 0.5 %. After completion of the decalcification treatment, proteins are to be eliminated and for production of chitin deacetylated as little as possible, papain, pepsin or trypsin should be used.

When chitosan is the final product, the enzymatic treatment can be conveniently replaced by an alkali treatment at high temperature. The alkali treatment can be performed in a steam heated steel container.

Other methods.

Rigby (1936 and 1937) treated crustacean wastes with hot 1 % sodium carbonate solution followed by dilute hydrochloric acid (1 ÷ 5 %) at room temperature and then 0.4 % sodium carbonate solution. Blumberg, Southall, Van Rensburg & Volckman (1951) prepared chitin from shells with hot 5 % sodium hydroxide solution, cold sodium hypochlorite solution and warm 5 % hydrochloric acid. Many other authors have reported on the preparation of chitin, most of them have followed the scheme of the acidic decalcification after removal of proteins and removal of pigments if desired.

Lovell, Lafleur & Hoskins (1968) reported on the preparation of chitin from crayfish waste; Kishi (1943) obtained chitin from silkworm pupae: dried silkworm pupae were powdered and moistened, hydrogen peroxide was added under stirring until the chitin and the oily substances floated on the surface to be skimmed off and the protein was recovered by filtration and drying. Lagunov, Kryuchova & Nikolaeva (1970) obtained chitin from krills *Euphausia superba*, simply by pressing; the solid part remaining was mostly chitin, while the liquid fraction contained proteins: in fact, Yanase (1975) reports that the content of chitin in the exoskeletons of the Antarctic krills is 38.7 %. Peniston, Johnson, Turril & Hayes (1969) described a process for the production of chitin from crustacea after removal of protein with 1 ÷ 2 % sodium hydroxide at 60°. This process led to the installation of a pilot plant for industrial production. The by-products of the South Africa fishing industry were considered by Molteno (1951) for the production of various substances including chitin.

A few authors remarked that viscosity determinations on chitosan produced from chitin obtained in various ways depend on the demineralization scheme followed and Lusena & Rose (1953) called attention to this fact and also reduced the deacetylation time.

Madhavan & Ramachandran (1974) obtained chitin from prawn *Matapenaeus dobsoni* in a processing factory. The fresh waste is washed in water and heated to boiling with 0.5 % aqueous sodium hydroxide in the ratio 2:3 for 30 min. The alkali is drained off and kept separately for the recovery of protein. The residual protein is removed by heating the residue to boiling with an equal weight of 3 % sodium hydroxide solution, draining off the alkali and repeating the process once again. The residue is immersed in cold hypochlorite solution containing 0.3 ÷ 0.5 % available chlorine for about 30 min, when most of the pigments contained in the prawn waste are bleached. The liquor is drained off, the residue is washed and demineralized by immersing in 1.25 N hydrochloric acid at room temperature for 1 hr. The residue after draining off the acid and washing with water is subjected to the final process of deacetylation by dipping in 1:1 sodium hydroxide solution for 2 hr at 100°. The alkali can be used for deacetylation of subsequent batches. The deacetylated mass is washed several times until free of alkali, dried, pulverized to the required size and stored. Shorter deacetylation periods and bleaching with hydrogen peroxide are preferred by Moorjani , Achutha & Khasim (1975).

Recovery of protein from the prawn waste using a lower concentration of the
alkali necessitates only a proportionately low amount of acid for neutralization which further helps to bring down the final concentration of salt in the
recovered protein. Use of bleach liquor for bleaching of pigments dispenses
with the use of costly hydrogen peroxide.

Demineralization of the protein-free mass is an important step in the preparation of chitosan from prawn waste as the degree of demineralization deter-
mines to a great extent the characteristics of the product, particularly the
viscosity of its solutions. With progressive increase in the concentration of
hydrochloric acid, the degree of demineralization is increased, however, use
of acid of concentration above 1.25 N adversely affects the viscosity of the
final product. A comparative account of the concentration of acid used, time
of treatment, content of acid soluble ash in chitin and viscosity of the final
product are given in Tab. 3.2. Demineralization using hydrochloric acid at
higher concentration and temperature is not a good practice, as the viscosity
of chitosan solutions would be reduced.

Depolymerization and deacetylation can take place when conditions are not
properly controlled and lead to low quality chitin and chitosan production.

Table 3.2. Effect of demineralization conditions on the viscosity of chitosan
solutions. Madhavan & Ramachandran, 1974.

Concn. HCl, N	Treatment length, min	Ash in chitin, %	Viscosity of 1 % chitosan in 1 % CH₃COOH, centipoises
1.25	30	24.34	106.85
	60	18.82	97.07
	120	6.33	58.05
	180	2.97	46.44
	∞	1.31	40.89
1.50	30	15.34	49.28
	60	7.90	43.95
	120	3.14	40.06
	180	1.46	38.84
	∞	1.31	34.58
2.00	30	2.71	37.66
	60	1.76	31.52
	120	1.03	26.94
	180	0.65	17.79
	∞	0.54	17.20

Radioactive chitin.

Radioactive chitin was prepared by Bade (1974) who injected ^{14}C-glucosamine dissolved in insect Ringer's solution through the muscular footpad of *Manduca sexta* larvae 10 hr prior to ecdysis from 4-th to 5-th instar. The radioactive cuticles were washed, milled and treated with 1 M sodium hydroxide to solubilize protein at 80°. Radioactive chitin may also be prepared by acetylating chitosan with labeled acetic acid in the presence of dicyclohexylcarbodiimide.

DEACETYLATION OF CHITIN.

Amide linkages are more difficult to cleave under basic conditions than ester groups. Under vigorous basic conditions, acetamido groups adjacent to *cis* related hydroxyl groups may undergo N-deacetylation, but *trans* related analogs are much more resistant.

The amino sugar moiety in amino sugar-containing polysaccharides is generally N-acetylated; heparin is exceptional in being N-sulfated, and it may be readily N-desulfated by treatment with dilute hydrochloric acid in methanol. N-Acetyl groups can not be removed by acidic reagents without hydrolysis of the polysaccharide, and alkaline methods must be employed for N-deacetylation.

Anhydrous hydrazine at 100° can be used to cleave amide linkages in proteins, and this reagent has been suggested for N-deacetylation of chondroitin sulfate preparations. Other workers have used this procedure for the N-deacetylation of blood group A substance from hog gastric mucus. This reagent gives extensively degraded products and N-deacetylation is usual incomplete (Matsushima & Fujii, 1957; Wolfrom & Juliano, 1960).

Chitin possesses the 2,3-*trans* arrangements of substituents in it monosaccharide unit, and it is remarkably stable to most reagents, including aqueous alkali. Chitosan was also studied by Von Furth & Russo (1906): they concluded that three out of four acetyl groups can be removed from chitin. Lowy (1909) considered that chitosan similarly prepared contained one acetyl group per disaccharide unit. Extended treatment of chitin with hot concentrated sodium hydroxide solution gives an almost completely N-deacetylated but degraded product. Rigby (1936) presented a treatment of chitin with 40 % aqueous solution of sodium hydroxide, for 4 hr at 110°. Evidence of the chitosan identity was put forward by Clark & Smith (1936) who registered x-ray diffraction data. However, this treatment should not be prolonged. Infrared spectrophotometry can be used for checking the removal of acetyl groups (Darmon & Rudall, 1950) because of the disappearance of the amide carbonyl band.

Wester (1909) and Meyer & Wehrli (1937) considered chitin a stable polymer in acidic solutions because it could be recovered from solutions by using concentrated strong acids. However, viscosity measurements of chitin in 45 % nitric acid and 30 % hydrochloric acid showed that extensive degradation occurs even at 0°.

It should be kept in mind, therefore, that any acidic treatment carried out on chitin or on chitinous raw materials, leads to a partial or extended depolymerization. When acidic demineralization of crab shells or other chitinous material is unavoidable, Lusena & Rose (1953) suggest the use of hydrochloric acid at pH not lower than 3 and of ethanol for extraction of lipids. Thus, satisfactory chitosan production depends largely on the acidic treatment to which the raw material has been submitted prior to the deacetylation step.

For instance, Holan (original results) determined that in 2 % hydrochloric acid at 98° during 14.5 hr, the deacetylation is 29.5 % and the hydrolysis to glucosamine is 4.4. %. After 3 hr in 3 % sodium hydroxide, the deacetylation is 3.4 % at 60° and 9.8 % after 2 hr at 100°.

The effect of the alkali treatment on the macromolecule length, during the production of chitosan from chitin, is less pronounced than that of the acid treatment. The size of chitin particles within the range 20 ÷ 80 mesh has no effect on the extent of deacetylation and viscosity of the chitosan solutions. Deacetylation in an inert atmosphere yields chitosan of higher viscosity than deacetylation in air. Prolonged time or higher temperature increase the percentage of deacetylation and reduce molecular size. Lusena & Rose (1953) reported that one hour's deacetylation in two half-hour stages separated by washing and drying, is as effective in terms of the degree of deacetylation, as one continuous 15-hr deacetylation and the product more viscous. Deacetylation can rarely extend beyond 80 % removal of acetyl groups, unless the alkali fusion procedure is applied (Horton & Lineback, 1965), along with fractionation.

More recent data by Nud'ga, Plisko & Danilov (1970) and by Numazaki & Kito (1975) are in agreement with those reported above. These authors point out that, according to the results in Table 3.3, prolonged deacetylation does not affect the degree of deacetylation but lowers the viscosity (both in air and nitrogen) down to one tenth of the maximum observable value. According to these authors, the alkali treatment combines the deacetylation and the protein removal in one step. Deacetylation at 50° for 15 min in 30 % sodium hydroxide gives chitosan with 1650 centipoise viscosity (as a 0.5 % solution in 0.1 % hydrochloric acid at 20°) and M_W 365,000.

The combined degradation effects produced by the acid and alkali treatments

Table 3.3 . Chitin deacetylation with aqueous 49 % sodium hydroxide solution. Nud'ga, Plisko & Danilov, 1970.

Sample number	Warming time to reach 140° hr		Time 140° hr	Nitrogen content, %		Deacetylation degree	$|\eta|$ in 2 % acetic acid
				total	amine		
1	air	1.0	1.0	6.85	6.83	0.79	8.8
2	air	1.0	2.0	–	6.79	0.78	6.8
3	air	1.0	6.0	7.66	7.35	0.84	6.85
4	N_2	0.0	1.0	6.55	6.38	0.73	31.5
5	N_2	0.5	2.0	7.34	7.13	0.82	13.8
6	N_2	1.0	2.0	7.55	7.23	0.83	13.4
7	N_2	1.5	2.0	6.86	6.82	0.78	7.4
8	N_2	2.0	2.0	7.64	7.58	0.87	6.8
9	N_2	1.0	6.0	7.88	6.40	0.74	3.65
10	N_2	0.0	1.0	8.21	6.37	0.73	12.8
11	Ar	0.0	1.0	8.14	7.61	0.87	18.9

lead to a polydisperse product that can be fractionated by dissolving the chitosan in 2 % acetic acid, rising the pH value to about 6.5 with sodium hydroxide, and precipitating with acetone and washing free of sodium acetate. Fractionation concerns the molecular weight only and not the percentage of deacetylation. The quality of the fractionation can be verified according to:

$$K = \frac{\sum W_i |\eta|_i}{|\eta|}$$

where W_i is the weight of the fraction i, $|\eta|_i$ is the intrinsic viscosity of the fraction i and $|\eta|$ the intrinsic viscosity of chitosan before fractionation. When K is close to unity, the fractionation is satisfactorily performed. Integral and differential fractionation curves are reported in Fig. 3.1.

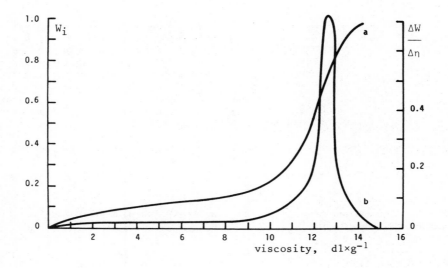

Fig. 3.1. Integral (a) and differential (b) curves for the chitosan fractionation. Nud'ga, Plisko & Danilov, 1970.

Method of Horowitz, Roseman & Blumenthal (1957) and Horton & Lineback (1965).

Thirty g of chitin can be converted to chitosan by fusion with 150 g solid potassium hydroxide in a nickel crucible while stirring in a nitrogen atmosphere. After 30 min at 180° the melt is poured carefully into ethanol and the precipitate is washed with water to neutrality. Purification can be done by dissolution in 5 % formic acid and sodium hydroxide precipitation. A 95 % removal of acetyl groups was claimed; however, the chain length after dialysis was found to have about 20 units only, which is not very much for a polymer to be used in chromatography, flocculation etcetera.

Method of Rigby (1936); Wolfrom, Maher & Chaney (1958) and Wolfrom & Shen-Han.

Fifty g of chitin are treated with 2.4 l of a 40 % aqueous solution of sodium hydroxide at 115° for 6 hr under nitrogen. After cooling, the mixture is filtered and washed with water to neutral reaction. An 82 % removal of acetyl groups can be obtained.

Method of Broussignac (1968).

A systematic survey by Broussignac led to the following deacetylation mixture that is nearly anhydrous: potassium hydroxide 50 % + 96° ethanol 25 % + mono-ethyleneglycol 25 % by weight. To prepare this mixture, the two solvents are first mixed and then potassium hydroxide is added in small portions under stirring. The dissolution is exothermic and the temperature can reach 90° during this step. This mixture can be used in both glass or stainless steel reactors. A stainless steel reactor consisting of a steam heating system, a stirrer and reflux has been in operation for a certain time. It contained 360 kg of the deacetylation mixture and 27 kg of chitin. The temperature was 120° and corresponded to the boiling of the mixture. Chitin should be dried in advance, of course. The treatment is continued for the desired length of time, and after filtration chitosan is washed with water until neutral reaction. Chitosan can be dried at moderated temperature. In Fig. 3.2 viscosity and percent of free amino groups are given vs deacetylation treatment time. One can observe that the viscosity range of the obtained chitosan is very broad, and depends on the length of treatment. Amino groups are deacetylated to the extent of about 83 % in 16 hr, which is very satisfactory. The Broussignac method yields 7 % chitosan of the crab shells weight. Muzzarelli (1973) has remarked that, when doing filtration, the alkaline mixture contains certain quantities of chitosan in dissolved form. Upon dilution with water, a white flucculent precipitate can be obtained from the spent deacetylation mixture. This amount of chitosan should be added to the yield mentioned. In order to reduce the amount of ethanol necessary and to lower the amount of chitosan dissolved in the deacetylation mixture, the latter has been modified: 500 g of potassium hydroxide dissolved in 450 ml of diethyleneglycol and 76 g of chitin added, the mixture kept at 170° for 6 hr or less.

Method of Fujita.

Ten parts of chitin are mixed with ten parts of 50 % sodium hydroxide, kneaded, mixed with 100 parts of liquid paraffin, and stirred for 2 hr at 120°; then the mixture is poured into 80 parts of cold water, filtered and thoroughly washed with water; the yield is 8 parts of chitosan. The free amino group content is 0.92 per glucose residue. This method is simple and requires much less hydroxide than the other methods reported. This method has been seldom applied.

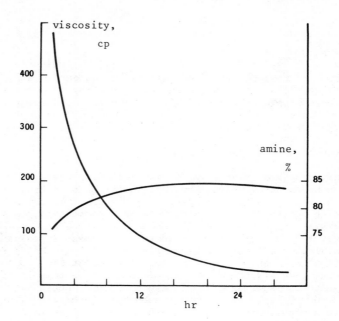

Fig. 3.2. Deacetylation of chitin and formation of free
amino groups of chitosan *vs* time. The viscosity of the
chitosan obtained after deacetylation is also shown.
Broussignac, 1968.

Method of Peniston & Johnson (1975).

Chitosan is produced directly from the shellfish wastes and permits recovery
of proteins, sodium acetate and lime or calcium carbonate as by-products,
providing nearly complete conversion of shellfish wastes into marketable
commodities. According to this method, a 30 ÷ 50 % alkali solution is prepared
at 120 ÷ 150°, the chitin being simultaneously deacetylated and the calcium
carbonate being converted to calcium hydroxide in situ according to the
equilibrium equation:

$$Na_2CO_3 + Ca(OH)_2 \rightleftarrows CaCO_3 + 2\ NaOH$$

While this equilibrium is well known in the technology of caustic soda manufacture, it has been operated substantially in the forward direction only, to
obtain increased yield of caustic soda.

This equilibrium was studied at 95° for concentrations of total equivalent
sodium hydroxide up to 240 $g \times l^{-1}$: this is approximately 20 % sodium hydroxide
by weight. The concentration of 240 $g \times l^{-1}$ was limiting due to precipitation of

the double carbonate $CaCO_3.Na_2CO_3.2 H_2O$. With excess calcium hydroxide, about 4 hr were necessary to attain equilibrium. For the reverse reaction, it is found that the percent conversion of sodium hydroxide to sodium carbonate is approximately linear for the range studied and can be expressed by the equation:

$$percent\ conversion = 0.1475\ x - 12.9$$

where x represents the sodium hydroxide concentration in $g \times l^{-1}$. At x = 240, the conversion is 22.6 %. The equilibrium for conversion of calcium carbonate to calcium hydroxide is favored by higher sodium hydroxide concentration. In a countercurrent process any reasonable degree of conversion would permit carrying the reaction to completion ; i.e., complete conversion of calcium carbonate to calcium hydroxide, since carbonate is removed from the reaction zone as sodium carbonate in the extraction.

At 20° a 39.8 % solution of sodium hydroxide will dissolve 0.3 % of sodium carbonate and the solid phase at equilibrium is sodium carbonate monohydrate. At higher temperatures, solubilities are considerably higher. It is thus possible in the process to extract carbonate as sodium carbonate from the shell and to remove this as a solid phase on cooling the extract liquor. This avoids the necessity for dilution of the sodium hydroxide and treatment with lime to recausticize the liquor before reuse in the process.

Sodium acetate shows solubility characteristics similar to those of sodium carbonate in strong hydroxide solutions. At 20° a 40 % solution of sodium hydroxide will dissolve 3.5 % sodium acetate. This can be removed from the sodium carbonate by fractional crystallization or alternatively the mixture can be dissolved and treated with lime to precipitate calcium carbonate. The sodium acetate can then recrystallized from the concentrated sodium hydroxide solution.

Referring to Fig. 3.3, shellfish waste coarsely ground to the particulate size of 3 ÷ 6 mm, is shown applied at 1 to a protein extraction apparatus (2) where the shell is moved countercurrently to the flow of dilute sodium hydroxide, applied at 3, and obtained from a later step of the process. The dilute hydroxide solution may be 0.5 ÷ 2 %. The amount of extraction by alkali solution applied at 3 is controlled to maintain a residual of alkalinity needed to form proteinate, shown extracted at 4. The time of the extraction step can extend from 1 to 4 hr depending upon the porosity of the shell and the temperature which is best kept in the range 50 ÷ 70°.

The sodium proteinate solution at 4 is then clarified by centrifugation or filtration at 5, and may also be treated with refining agents to remove lipids or pigments shown ejected at 6. The clarified product is then neutralized with the help of added hydrochloric acid to the pH of minimum solubility, which will be in the range 4.5 ÷ 3.4, depending upon the shellfish species and extraction conditions. The resulting precipitated protein at 7 is collected at 8 and it is washed and dried at 9 by reslurrying and spray-drying to provide the protein by-product at 10.

Following protein removal, the shell is again extracted countercurrently in a further series of extraction cells 11. Strong sodium hydroxide is shown applied at 12. The effluent solution from this operation (13) contains excess sodium hydroxide, sodium acetate and sodium carbonate, and is passed to a crystallizer (14) to precipitate sodium acetate as a useful by-product at 15,

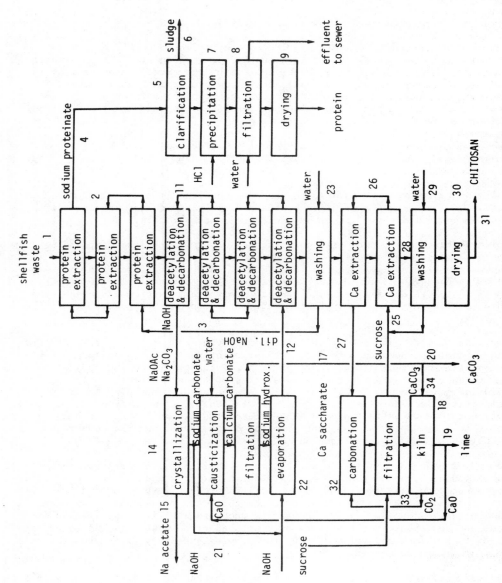

Fig. 3.3. Chitosan production process scheme. Peniston & Johnson, 1975.

the same being removed by filtration or centrifugation, washed and purified by
conventional means. Most of the sodium carbonate is also precipitated at this
stage.

The mother liquor is diluted with water and treated with lime (16) in order to
convert the remaining sodium carbonate back to sodium hydroxide. Sodium carbonate recovered from the crystalline sodium acetate + sodium carbonate mixture
is also lime treated for sodium hydroxide recovery. The precipitated calcium
carbonate is collected as at 17. The regenerated sodium hydroxide solution
resulting at 21 is combined with added strong alkali to replace that removed
at 15 as sodium acetate, and is evaporated at 22 to the desired strength for
application at 12 to the extracting process.

Following deacetylation and decarbonation, the residual shell, consisting of
chitosan and calcium hydroxide, is washed at 23 with carbonate-free water to
remove residual sodium hydroxide. This washing is the source of the dilute
sodium hydroxide initially used at 3 for the protein extraction.

The chitosan + calcium hydroxide mixture is then extracted with an aqueous
solution of sucrose applied at 25. The calcium hydroxide becomes dissolved as
calcium saccharate, withdrawn at 27, leaving at 28 pure chitosan that may be
washed to neutrality at 29, dried at 30 and made available at 31. The
saccharate is then decomposed by carbonation and precipitates calcium carbonate at 34 which is washed and passed to the lime kiln 18. The sucrose solution is evaporated to the desired concentration and reused at 25 in the
process. Other substances capable of chelating calcium, such as glycols, EDTA,
sorbital or gluconates may also be used.

ALKALI CHITIN.

Alkali chitin was described by Thor & Henderson (1940) who prepared it by
steeping chitin flakes (150 g) in 43 % sodium hydroxide (3 1) for 2 hr at 25°.
Penetration of the caustic into the chitin may be facilitated by conducting
the steep under vacuum. The resulting alkali chitin weights approximately 3
times as much as the original chitin. In fact, more than 0.75 equivalents of
sodium can combine with each N-acetylglucosamine unit as shown in Fig. 3.4. It
is however recommended to operate at low temperature (0 ÷ 15°) to avoid deacetylation and degradation (Noguchi, Arato & Komai, 1969; Haga, 1971). The
excess steep liquor is separated by means of a basket type centrifuge or by
filtration on a suitable screen followed by adequate pressing. The resulting
cake should then be broken down or shredded to a state of subdivision at least
equal to that of the original chitin.

When working with sodium chitin, reductions in temperature in so far as
practical, would always be an advantage from the chemical standpoint. If, for
example, a sodium chitin containing 0.75 or more equivalents of combined
sodium per acetylglucosamine unit is desired, the maximum temperature is 25°
when the caustic concentration range of steep liquor is 40 ÷ 50 %, and progressively below 25° as the caustic concentration drops to 30 % (Thor, 1939).

When alkali chitin is extracted with water, it regains the appearance of the
original chitin. That the chitin is not deacetylated to any appreciable degree
by the treatment with the caustic, is indicated by the fact it remains in-

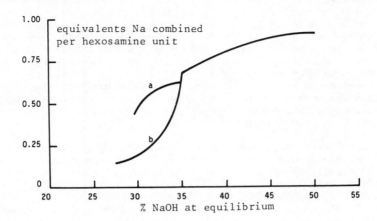

Fig. 3.4. Effect of steep caustic concentration on the
combination of sodium hydroxide with chitin, a = at 5°;
b = at 25°. Thor & Henderson, 1940.

soluble in dilute acetic acid, the nitrogen content remains unchanged and it
does not give a violet color when treated with iodine + potassium iodide in
dilute sulfuric acid.

If the alkali chitin is kept at room temperature for more than a few hr after
it is prepared, then deacetylation becomes detectable. This can be minimized
by storing at 0° or lower. As far as the time of steeping is concerned, it was
found that chitin steeped in 38 ÷ 48 % sodium hydroxide for 18 hr at 25° was
not sufficiently deacetylated to become soluble in dilute acetic acid.

Shredded alkali chitin, prepared as above, was found to react with materials
such as carbon disulfide, ethylene oxide, monochloroacetic acid, and dimethylsulfate to give products with different properties. These reactions were
neither smooth nor complete due to the unfavorable physical state of the
chitin used as large flakes.

Sodium chitin can be dispersed in water by mixing it with the desired quantity
of crushed ice, which, in contact with sodium hydroxide in sodium chitin provides a freezing mixture (Thor, 1939). During the process, the temperature of
the mixture drops to several degrees below 0°, which is undoubtely an important factor in bringing about the dispersion. However, when unsteeped chitin
is added to a similar mixture of sodium hydroxide and ice, it is not attacked.
A dispersion of alkali chitin prepared in this manner is stable for several
days at 0°; at room temperature, chitin begins to precipitate after several
hours and, if allowed to stand, the chitin obtained will be deacetylated to a
considerable degree. Noguchi, Arato & Komai (1969) demonstrated that no deacetylation occurs below 15°.

Sannan, Kurita & Iwakura (1975) noticed that alkali chitin films allowed to stand at room temperature for two days became soluble in water due to the chain degradation and destruction of the secondary structure. They confirmed that prolonged treatment with alkali leads to viscosity decrease; chitin films were produced by casting alkali chitin on glass plates and by regenerating chitin by washing them free of alkali with methanol in a Soxhlet extractor.

Chitin can be recovered from the dispersion of alkali chitin by diluting it with a large volume of water, by salting out with sodium sulfate or by neutralizing the sodium hydroxide with dilute acid. The dispersion of alkali chitin is a good reaction medium for the preparation of a number of derivatives.

CHITOSAN SALT FORMATION.

The pK_a value for chitosan is 6.3 according to Nud'ga, Plisko & Danilov (1973), to Noguchi, Arato & Komai (1969) and to Muzzarelli (1971), while a higher value was reported by Doczi (1957). The pK_a value for 2-amino-2-deoxy-β-D-glucose, as determined by Tamura, Miyazaki & Suzuchi (1965) is 7.47; lower values, 7.15, 7.10 and 6.92 were found for its methylglucopyranoside, the 3-methylether and the 3,4,6-trimethylether, respectively. The determinations illustrated in Figs. 3.5 and 3.6 include the calculation of free amino groups. Chitosan forms numerous salts, most of which are listed in Table 3.4.

Anions

From chitosan salt solutions, chitosan can be reprecipitated by several precipitating agents including several strong acids, especially sulfuric acid, and also ortophosphoric acid, chromic acid, nitric acid, salicylic acid, hydrochloric and hydrobromic acids.

Other precipitating agents are phosphotungstate, iodomercurate, iodobismuthate, molybdate, tungstate, picrate, tannin, iron ferrocyanide (Brunswik, 1921). Tab. 3.4 lists many salts of chitosan; a number of them are insoluble and the corresponding acids are therefore suitable for precipitation. Some of them deserve particular mention.

To prepare chitosan perchlorate, Wolfrom, Maher & Chaney (1958) suspended chitosan in a mixture of 132 ml of glacial acetic acid and 33 ml of 60 % perchloric acid. The suspension was stirred at 8° for 16 hr prior to filtration and drying with ether; the yield of chitosan perchlorate was 3.2 g.

From formic acid solutions, chitosan can be precipitated by 0.1 N sulfuric acid, by copper sulfate and ammonium sulfate, as white crystalline chitosan sulfate. Concentrated hydrochloric acid also precipitates chitosan. Sodium sulfite, sodium thiosulfate and triethanolamine also precipitate chitosan from formic or acetic acid solutions.

It is worth while to note that the precipitate is white when the precipitation is carried out with copper sulfate, i.e. copper ions are left in solution by chitosan sulfate, while the controlled precipitation of the free base chitosan leads to the collection of copper ions, as described in further sections.

Table 3.4. Chitosan salts. s = soluble; ss = slightly soluble; ins = insoluble.

Acetate	s	Lactate	s
monochloro-	s	laurate	ss
dichloro-	s	levulinate	s
monoiodo-	s	linoleate	ss
phenyl-	s	Malate	s
trimethyl-	s	maleate	s
acetacetate	s	malonate	s
acrylate	s	diethyl-	s
ethacrylate	s	mandelate	s
furacrylate	s	molybdate	ins
hydracrylate	s	phospho-	ins
adipate	s	Naphtenate, M_w 186	s
anthranilate	s	M_w 450	s
azelaate	s	Nitrate	ss
Benzensulfonate	s	Oxalate	ss
benzoate	s	Palmitate	ss
o-benzoyl-	ss	phosphate	ss
borate	s	phtalate	ss
iso-butyrate	s	picrate	ins
α-chloro-	s	propionate	s
hydroxy-	s	α-bromo-	s
n-butyrate	s	α-chloro-	s
α-bromo-	s	α-iodo-	s
Caproate	s	pyruvate	s
chromate	ins	Salycilate	ss
cinnamate	ss	sebacate	s
citrate	s	stearate	s
crotonate	s	succinate	ss
cyanate	ss	sulfanilate	s
Dilactate	s	sulfate	ins
dithiocarbonate	ss	sulfite	s
Formate	s	sodium bisulfite	ss
fumarate	ss	sulfosalycilate	s
Glycinate	ss	Tartrate	s
phenyl-	ss	terephtalate	ss
glycolate	s	tetrachloroaurate	s
di-	s	tetraiodomercurate	
thio-	s	thiocyanate	ss
Hydrobromide	s	tungstate	ins
hydrochloride	s	phospho-	ins
hydroiodide	s	meta-Vanadate	ss
hypoc..lorite	ss		

A process for the purification of chitosan from extraneous material such as inorganic salts, proteinaceous material and gums, has been published by Doczi (1957). It consists of adding an excess of a soluble salicylate, preferably sodium salicylate, to an aqueous solution of chitosan salt, followed by chilling the resulting mixture in an ice-water bath whereupon the chitosan

salicylate precipitates. The high solubility of sodium salicylate makes it
particularly useful in this application. The precipitate is separated by
centrifugation, redissolved in water and the resulting solution filtered and
adjusted to a pH of about 9, by the addition of a water-soluble base. The
resulting precipitate is collected, washed by a water-miscible organic solvent
and dried, yielding the chitosan in purified form. The base used in regenerating the chitosan may be any organic or inorganic base, provided that its
strength is greater than that of chitosan. Doczi (1957) has preferred diethyl-
amine as a convenient reagent of adequate basic strength. Among the process
variables the ratio of equivalents of salicylate to equivalents of chitosan
may vary widely, and appear to control the product yield in the range 2 to 3,
as the yield for 2.28 is 32.3 and for 3 is 89.6.

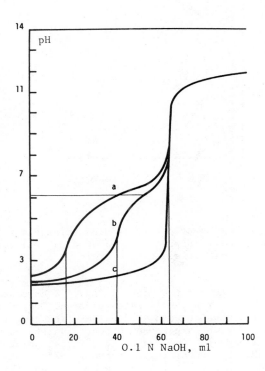

Fig. 3.5. Potentiometric titrations of chitosan dissolved
in 20 ml of 0.3 N HCl, a = 1.0 g chitosan; b = 0.5 g
chitosan; c = reference curve for 0.3 N HCl. The inflection
points are respectively at 16.2, 39.0 and 64.0. The formula
of Broussignac gives: 16.2 × (64 − 39) : 0.5 = 79 % or
16.2 × (64 − 16.2) : 1.0 = 79 % amino groups. The pK_a value
is 6.3 in both cases. Muzzarelli, original results.

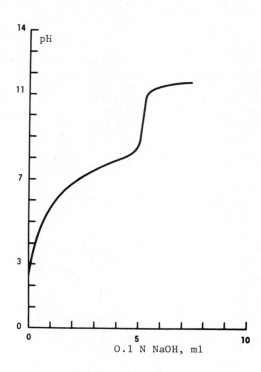

Fig. 3.6. Potentiometric titration of triethylchitosan
with a substitution degree of O.55. Nud'ga, Plisko &
Danilov, 1973, (modified).

Chitosan polymolybdate precipitates upon mixing an acetic acid solution of
chitosan with a molybdate or paramolybdate solution. Due to the tendency of
molybdate to form polyanions at acidic pH values, the amount of molybdate that
reacts with chitosan is much larger than expected on the basis of calculations
like those above. Chitosan polymolybdate and chitosan polytungstate have
fibrous aspect and show a reversible photochromic effect when exposed to sun
light: the metavanadate, which is yellow, becomes green, the molybdate and
tungstate which are white, become light blue, whilst dichromate, which is
orange, undergoes oxidative degradation and becomes brown. A red derivative of
chitosan obtained by reaction with hexachloroplatinic acid (but not with potassium hexachloroplatinate) has also been observed. These derivatives can be
decomposed using concentrated carbonate solutions (Muzzarelli, 1973).

The isolation of the molybdate and tungstate derivatives is not only important
from the standpoint of a deeper insight in the retention mechanisms of metal
ions on natural polymers, but also for further developments in the application
of these derivatives as chromatographic supports. In fact, derivatives of

chitosan with metavanadate, molybdate or paramolybdate and tungstate may contain more metal by weight than other derivatives, to the point that the inorganic portion exceeds the organic one, as can be seen in Fig. 3.7.

Chitosan polymolybdate can be expected to be a quite stable chromatographic support in salt solutions and in acid solutions, and to be a selective agent for the collection of phosphate because of the well-known reaction of phosphate with molybdate.

Chitosan polymolybdate can be prepared as follows: 6.5 g of chitosan powder are suspended in 500 ml of water to which 2.5 ml of glacial acetic acid are added; the resulting pH is 3.5. In another flask, 20 g of ammonium para-molybdate tetrahydrate are dissolved in one 1 of distilled water, pH 6.5. The second solution is slowly introduced into the chitosan solution through a separating funnel and a large magnetic rod is then used to provide very strong stirring. The white precipitate is then filtered and washed with distilled water until the stannous chloride + thiocyanate test for molybdenum is negative. It is best to lyophilize the product, or to keep it under water to avoid the formation of a crust.

Chitin adsorption of acids was tested by sealing 100 mg of finely powdered chitin with 20 ml of acid solution. The titration curves obtained on the solutions after shaking for 12 hours showed a shoulder corresponding to an

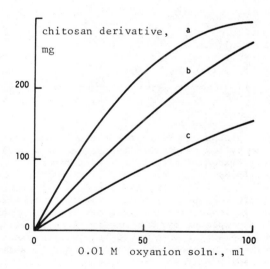

Fig. 3.7. Amounts of chitosan derivatives obtained when adding measured amounts of oxyanion to a 100 mg chitosan solution in acetic acid. The precipitates were lyophilized before weighing: a = molybdate; b = tungstate; c = vana-date. Muzzarelli, 1972.

acid binding capacity of 4.8 eq×kg^{-1} at about pH 2; at lower pH values a
sudden further rise in absorption occurs and proceeds without apparent limit.
Once the acetamido and amino groups are saturated, no other acid combining
group would be available unless the ether groups form oxonium salts. Even if
this were so, the total quantity of acid could not be accounted for; the
general reaction would be:

$$R-NHCOCH_3 + H^+X^- \rightleftharpoons R-\overset{+}{N}H_2COCH_3X^-$$

The adsorption of acids by unground chitin at 50° is given in Table 3.5, where
the values for hydrochloric acid and sulfuric acid are the highest.

Table 3.5. Adsorption of acids by chitin. Giles, Hassan, Laidlaw & Subramanian
1958.

Acid	mmol×kg^{-1}, at 50°	
	pH 3.0	pH 4.5
Hydrochloric	3800	450
formic	385	–
acetic	255	–
chloroacetic	320	–
sulfuric	3870	150
nitric	235	85
benzenesulfonic	200	100
aniline-2,5-disulfonic	100	55
naphtalene-2-sulfonic	195	–
2-naphtol-6,8-disulfonic	105	50

The acetamido and amino groups of chitin are the only operative groups toward
hydroxylic solutes; the hydroxy and ether groups on the glucosidic rings are
probably protected by solvated water and thus rendered inactive (Giles, Hassan,
Laidlaw & Subramanian, 1958).

Chitin can absorb anionic compounds by salt formation with its substituted
amino groups, but this action can be suppressed without preventing the ability
to form hydrogen-bonded complexes by working in alkaline buffer solutions.
The following compounds were tested: phenol, resorcinol, methanol, benzene-
sulfonic acid, naphthalene-2-sulfonic acid and its sodium salt, antraquinone,
sulfonic acid and its sodium salt and sulfonated alizarin.

Simple hydroxy compounds like phenol and methanol in iso-octane, benzene and
water are adsorbed by chitin by hydrogen bonding; simple sulfonic acids are
not absorbed by chitin when salt formation with the acetamido groups is prevented by an alkaline buffer, while free sulfonic acids of different molecular
sizes and without hydrogen bonding substituent groups are adsorbed on chitin
to approximately equal extents from unbuffered solutions. The effect of

hydogen-bonding groups on different solutes of similar molecular areas can be
studied in the case of applications at pH 9 of 2-hydroxyanthraquinone, anthraquinone sulfonic acids, sulfonated alizarin and azo dyes: the hydroxy or amino
groups of chitin promote strong adsorption (Giles, Jain & Hassan, 1955; Giles,
Hassan & Subramanian, 1958).

Dyes.

The adsorption of aniline-naphthionic acid was compared to that of the
corresponding "double" molecule Congo Red. The two dyes have identical hydrogen bonding groups but the second one would have much enhanced van der Waals
affinity by virtue of its larger molecular area, linearity and higher conju-
gation; at pH 9 ion-exchange by chitin being suppressed, the dye with the
larger molecule is adsorbed a little more that the simpler dye, but in the pH
range of 7 to 9 it is adsorbed to a considerably greater extent by cellulose.
This is another indication that cellulose can exert stronger van der Waals
attractions than chitin can, by virtue of the more regular outline of its
molecular chain. If hydrogen bonding alone were the adsorptive forces, both
dyes should be adsorbed to a similar extent on cellulose. In neutral solutions
both dyes can be adsorbed by an ion-exchange mechanism.

Ciba Ltd. (1962) has produced data on the interactions of proteins and polyaminosaccharides with compounds of the general formula shown in Fig. 3.8.

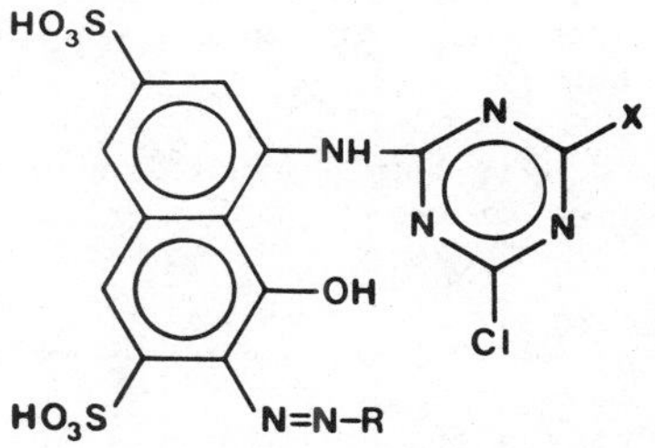

Fig. 3.8. General formula of dyes with which colloidal
chitosan gives red products. R = $2.NaO_3SC_6H_4^-$ and X =
$3.NaO_3SC_6H_4NH^-$; R = $2.HO_3SC_6H_4^-$ and X = $C_6H_5NH^-$. Ciba
Ltd., 1962.

Methylene blue, tartrazine, Victoria blue and phloxine have also been studied
on chitin and cellulose thin layers; some results are in Tab. 3.6.

The determining factor in adsorption on chitin is the ability of the acetamido
group to act as an ion-exchanger. Cellulose can adsorb solutes by hydrogen
bonding only in absence of water, whereas chitin can do so whether water is
present or not.

Table 3.6. Comparison of adsorption activity of chitin and cellulose for coloring matters at 30°. Takeda & Tomida, 1969.

Coloring matter	Content of nonionic N, %	Adsorption activity, $mmol \times g^{-1}$	
		chitin	cellulose
methylene blue (basic)	11.2	0.2	(0.01)
tartrazine (acidic)	10.5	1.6	0.0
victoria blue (basic)	2.4	2.5	2.3
phloxine (acidic)	0.0	2.1	1.0

In water, chitin can form hydrogen bonds with some solutes and can also take part in ion-exchange reactions, in both cases through its acetamido groups. Its hydroxyl groups appear to be protected by water in the same way as those of cellulose; its molecular chain exerts van der Waals attractions in a weaker way compared to cellulose.

A qualitative comparison of the affinities for chitin of a certain number of dyes was made by Giles, Hassan & Subramanian (1958). The affinity rises with increase in the number of benzene nuclei, i.e. with increasing planar area of the anion. The affinity falls with the increase of the degree of sulfonation of the dyes. Each sulfonate group after the first has a negative affinity for chitin. This would mean that owing to steric effects, the second and subsequent sulfonate groups in an anion are unable to interact with cationic sites and remain in the water.

The adsorption of sulfonated dyes can be regarded as an ion-exchange process at the cationic sites; the reaction with a dye in sulfuric acid medium can be written:

$$R-\overset{+}{N}H_2COCH_3 \tfrac{1}{2}SO_4^{2-} + NaDye \rightleftarrows R-\overset{+}{N}H_2COCH_3Dye^- + \tfrac{1}{2}Na_2SO_4$$

The disulfonated dyes with large anions have relatively high apparent heats of adsorption but in general the heat values are as low as those of alumina, silica, graphite and some resins.

Hackman & Goldberg (1964) found that the apparent heat of adsorption of water soluble protein on chitin is very low: the temperature changes have little effect on the adsorption which is nil on the alkaline side of pH 9 and rises to a maximum at about pH 5. Powdered chitin from cuttlefish was heated 5 min with a 2 % solution of Procion Red or Remazol Brilliant Violet in 1 N sodium hydroxide. After enzymatic hydrolysis the total amount of chitin which had gone into solution could be determined on the basis of absorbance of the digest. This method allows the use of a fine chitin powder instead of colloidal suspensions to be determined by nephelometry. In fact, colloidal chitin is known to be considerably degraded during preparation to the extent that chitinase capable of hydrolyzing this substrate may eventually fail to attack chitin in its naturally occurring formations. Chitins precipitated with concentrated mineral acids are converted to α-chitin from the naturally

occurring form; degradation, orientation and chemical modifications may alter
the results of the enzymatic tests; therefore the dyed chitins can be useful
for enzymatic degration studies.

Krichevskii & Sadov (1961) studied the kinetics of the reaction of Procion
Yellow RS with the amino groups of chitosan. Dichlorotriazine dyes react with
the amino groups of chitin and chitosan in both alkaline and acidic media, the
hydroxyl groups taking part in the reaction in alkaline media only.

The relative intensity of fluorescence and the change in adsorption spectra of
chitosan and glucosamine hydrochlorides, when added to uranin, eosin and
erythrosin, were examined by Mataga & Koizumi (1955). There occurs no change
either in absorption or fluorescence until the concentration of chitosan
hydrochloride reaches about 10^{-5} M. Beyond this concentration, the fluorescence
is quenched gradually and the absorption spectrum shifts toward red and somewhat broadens, accompanied with the decrease of extinction (see Fig. 3.9).

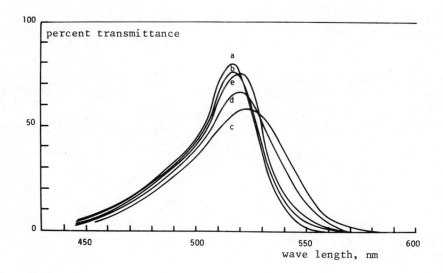

Fig. 3.9. Change of absorption spectrum of aqueous eosin
solution by chitosan hydrochloride. Dye concentration
10^{-5} M; temperature 27°. Quantity of added chitosan
hydrochloride, a = 0 ÷ 10^{-5} M; b = 6×10^{-5} M; c = 10^{-4} M;
d = 4×10^{-4} M; e = 10^{-3} M. Mataga & Kozumi, 1955.

Such a change becomes most prominent near 10^{-4} M, but over this point a marked
phenomenon of recovery comes to be observed both in absorption and fluorescence and when the concentration reaches 10^{-3} M, the shape of the absorption
curve becomes quite similar to that in the pure aqueous solution.

OXIDATION.

Oxidative deamination of chitosan.

Deamination of chitosan can be performed with barium hypobromite. D-Arabinose
was produced from glucosamine by the action of silver nitrate (Shoruigin &
Makarova-Zelyanskaya, 1935). Matsushima & Fujii (1957) deaminated chitosan in
hydrochloric acid solution, by using silver nitrate. The evolution of nitrogen
and the decrease of viscosity were immediately observable; silver chloride was
removed by filtration and residual nitrous acid was removed by a small amount
of urea. Chitosan may be also deaminated in water suspension with gaseous dinitrogen trioxide.

Oxidation of chitosan by periodate at 25° and at a pH above 5 and below 3 does
not show a definite end point, whereas at pH 4.1 a constant end-point is
reached. The amount of ammonia liberated is 0.6 ÷ 0.7 mol per glucosamine unit
according to Jeanloz & Forchielli (1950). Periodic acid was also used by
Neuberger (1941) on glucosamine derivatives. Hypochlorite, chorine and lead
dioxide have been proposed by Matsushima (1951) as oxidation agents for
glucosamine. All these procedures bring about extended degradation of the
polymer chain.

The half-time of the deamination of chitosan by the action of nitrous acid is
one minute, as measured by changes in the optical rotation of the solutions.
The reaction of chitosan with nitrite is a cleavage reaction as demonstrated
by Peniston & Johnson (1975) and entails a deamination at the point of
cleavage, the products being chitosan fragments consisting of 2-deoxy-2-amino-
glucose for all units in the chain, except for the unit at one end of the
chain, which is an amine-free reducing sugar. Such products are not formed by

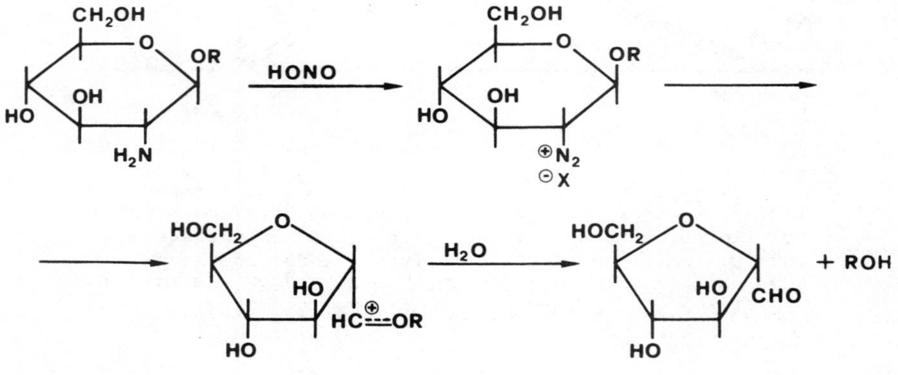

Fig. 3.10. Scheme of reaction of the glucosamine unit at
the end of the chitosan chain, with nitrous acid.

current hydrolysis reactions which cleave glycosidic linkages without deamination. The extent of deamination is greater in small fragments, from 50 %
in the dimer, to 3 % in a 33-unit fragment.

The deaminative cleavage of chitosan can be carried out in dilute aqueous
solutions under mild conditions, the nitrite being converted to nitrogen gas.
For instance, chitosan (2 g) in 6 % acetic acid (100 ml) was treated at 20°
with a solution containing 85.5 mg of sodium nitrite, i.e. 10 % of the theoretical amount for complete deamination. The sodium nitrite was added in small
increments over a 30 min period with stirring and the mixture was then neutralized. While chitosan precipitates at ∿ pH 6.5, the nitrous acid reaction
products precipitate partially at pH 7.4. The reducing value of this precipitate was 188 mg cuprous oxide per gram, equivalent to 84.3 mg glucose,
demonstrating an average chain length of about 13 units; the intrinsic viscosity was greatly reduced.

Consumption of nitrite, deamination and depolymerization of chitosan proceed
simultaneously and rapidly, and the reaction goes to completion when equimolar
quantities of chitosan and sodium nitrite are used.

The deamination of chitin is mentioned briefly by Katz, Fishman & Levy (1975)
who used deaminated chitin as a chromatographic support to isolate lysozyme
(see page 196). However, the deaminated chitin was not characterized from the
physico-chemical point of view: presumably, deamination and depolymerization
occurred where deacetylated amino groups were present. A review article has
been published by Williams (1975) on the subject of deamination of carbohydrate amines and related compounds.

Carboxyl group formation.

Today, oxidation of chitin has importance as a means of introduction of new
functional groups into the polymer chain. An example is the oxidation of the
C^6 as carried out by Horton & Just (1973). They achieved a simulation of the
gross structure of heparin (whose key functional groups are the 2-sulfoamino
and the 6-carboxyl moieties) by converting chitosan into its 6-carboxyl analog
and then N-sulfating this polymer to afford a product having moderate anticoagulant activity. Although partial C^6 carboxylation of chitosan has been
achieved by Whistler & Kosick (1971) by use of dinitrogen tetraoxide or
oxygen-platinum, only low degrees of substitution by the carboxyl group were
obtained, and there was some concomitant degradation of amino functionality
and partial oxidative depolymerization.

In order to oxidize the chitosan to the 6-carboxyl analog, it was necessary
to devise conditions that would obviate the need for temporary substitution at
the functional positions 2 and 3; give specific reaction at C^6 without oxidation at C^2 or C^3, and allow the polymeric structure to be maintained with
the minimum of degradation. Chromium trioxide in acetic acid was selected as
the oxidant, and the amino group was protected from oxidation by protonation
to the ammonium ion by use of a strong acid (perchloric acid). This method of
protection follows the same rationale used in a successful synthesis of
α-amino acids through oxidizing β-amino alcohol precursors as their sulfate
salts by the action of permanganate.

As the subsequent oxidation of aldehydes to carboxylic acids appears to

require water, water was introduced at the terminal phase of the oxidation
step.

A solution of chitosan acetate in aqueous acetic acid was freed from most of
the water by evaporation and, by addition of perchloric acid, chitosan perchlorate was precipitated as a fine dispersion suitably surface-activated for
the oxidative step. The oxidation of the latter was performed under relatively
anhydrous conditions in acetic acid with an excess of chromium trioxide, and
water was introduced after 1 hr of the reaction at about 25° (see Fig. 3.11).

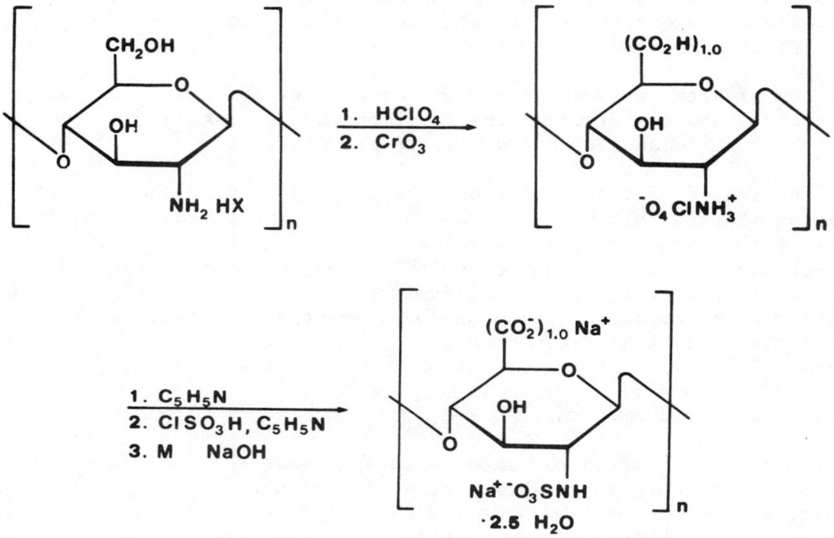

Fig. 3.11. Carboxylation and sulfation of chitosan.
Horton & Just, 1973.

The perchlorate salt of the 6-carboxylated polymer was obtained in good yield
as a water-soluble, methanol-insoluble powder. Microanalysis indicated that no
nitrogen had been lost from the product, and free amino groups were indicated
by a positive ninhydrin reaction. As chitosan has been reported to be an
inhibitor of tumor growth, this C^6 carboxylated analog was evaluated biologically and tested in vitro against leukemia L-1210 cells: it was observed to
inhibit cell growth by 50 % at a concentration of 0.6 μM (see Chapter 7).

Titration with base to the point of neutralization of perchloric acid and the
carboxylate groups indicated that the product had been fully converted into
the carboxyl derivative, with a substitution degree of 1.0. Titration to the
point of half-neutralization followed by dialysis yielded the free aminouronic
acid polymer, whose solubility in water was much lower than either the perchlorate salt or the sodium carboxylate salt of the free base.

By suspending it in pyridine, the perchlorate salt of carboxylated chitosan was obtained in a form suitable for heterogeneous sulfation. Sulfation with chlorosulfonic acid + pyridine, by the procedure of Wolfrom & Shen-Han (1959) for sulfation of chitosan gave a water-soluble polymer formulated as disodium (1→4)-2-deoxy-2-sulfoamino-β-D-glucopyranuronan of degree of substitution 1.0 by the sulfoamino group and the carboxylate group. This formulation was supported by elemental analysis, which indicated that equimolar amounts of sulfur and nitrogen were present, by a positive Toluidine Blue reaction for the sulfoamino group, and a negative ninhydrin reaction indicating the absence of free amino groups. The molecular weight (4.3×10^5) and the degree of polymerization (1,440) indicate that little depolymerization accompanies the sulfation. The infrared spectrum shows similarity with the spectrum of commercial sodium heparinate.

CHITIN AND CHITOSAN DERIVATIVES.

Sulfation.

A review of the esterification of chitin and cellulose under comparable conditions was published by Danilov, Plisko & Pyaivinen (1961). Nitration of chitin was attempted by Shoruigin & Hait (1934) who could introduce 1.5 nitrate groups per unit, by using concentrated nitric acid. Denitration could be performed with sodium hydrogen sulfide. To obtain chitosan nitrate, Wolfrom, Maher & Chaney (1958) stirred 200 mg chitosan into a mixture of 10 ml acetic anhydride, 10 ml of acetic acid and 13 ml of absolute nitric acid at 0 ÷ 5° for 5 hr. After centrifugation the residue was washed with acetic acid and ether; the yield was 330 mg of nitrate ester of chitosan nitrate. The product corresponded to the formula $C_6H_7O_2(ONO_2)_{1.65}(OH)_{0.35}(NHCOCH_3)_{0.15}(NH_3^+NO_3^-)$. Chitosan, with 85 % free amine) was found to dissolve in absolute nitric acid but the resulting chitosan nitrate was nearly identical to that obtained employing absolute nitric acid admixed with acetic acid and acetic anhydride which react heterogeneously with chitosan. Both reaction media afforded the nitric acid salt of chitosan nitrate ester in which approximately 85 % of the two available hydroxyl functions were esterified, corresponding to a degree of substitution of 1.7.

Several authors modifed chitin in view of obtaining heparin-like blood anti-coagulants: Cushing, Davies, Kratovil & MacCorquodale (1954) and Cushing & Kratovil (1956) reacted chitin in dichloroethane with chlorosulfonic acid. Complexes of sulfuric anhydride with pyridine, dioxane and N,N-dimethylaniline were used by Jones (1954) to direct the sulfonation to the hydroxyl groups of chitin:

$$R\text{-}OH \ + \ SO_3.NC_5H_5 \ \xrightarrow{\ C_5H_5N\ } \ R\text{-}O\text{-}SO_3NC_5H_6 \ \xrightarrow{\ NaCl\ } \ R\text{-}O\text{-}SO_3Na$$

The by-product , pyridine hydrochloride, can be easily recovered and reused, so sulfuric anhydride is quite a good sulfating agent. Preferred salts of chitosan are the formate, acetate, perchlorate, nitrate hydrochloride and hydrobromide; the chitosan salt is sulfated by dissolving it in formamide and the solution is treated with chlorosulfonic acid while cooling and stirring. The chitosan polysulfuric acid esters produced in this manner, show a relatively high content of free amino groups, especially in the case of perchlorate

and chloride salts of chitosan. The chitosan salt solution in formamide may be
introduced into a hot solution of pyridine and chlorosulfonic acid and the
sulfation can be carried out at 80°; the products obtained accordingly show
high nitrogen content but low amount of free amino groups.

In the previous section the oxidative carboxylation of chitosan according to
Horton & Just (1973) has been described (see pages 113 and 114).

The Upjohn Co. (1956) reported a process whereby chitosan is sulfated with a
mixture of sulfurous and sulfuric anhydrides in anhydrous conditions. The
sulfated chitosan is precipitated with barium ions and converted to a sodium
salt with sodium carbonate or other anions which form insoluble barium salts.

If chitosan formate in pyridine is used, a 9 % formyl content is obtained;
therefore, the product is reported as a polysulfuric acid ester of N-formyl
chitosan (Vogler, 1958). The formylation, which may be carried out either
before or after the sulfation step, is preferably effected by dissolving the
material to be formylated in 100 % formic acid on a steam bath, adding pyridine
dropwise while stirring and then warming the mixture; the proportions of
formic acid and pyridine are variable, preferably 1:1.

When the formylated chitosan has been sulfated, the product may be purified by
converting the washed and dried material into an alkaloid salt, usually the
narcotine salt by treatment with narcotine hydrochloride. The isolated
alkaloid salt may then be converted to an alkali-metal salt by treatment
with sodium hydroxide, for instance (Hoffmann La Roche & Co., 1957). If it is
desired to obtain end products of certain physical characteristics, the chitosan material can be submitted to a degradation step prior or subsequent to the
sulfation. This may be accomplished by oxidation, by means of aqueous hydrogen
peroxide or by hydrolysis, by means of acids or alkalis; N-formylated chitosan
sulfuric acid ester can be degraded to various extents by varying the peroxide
concentration. Viscosity measurements in an Ostwald viscosimeter in sodium
chloride solution, and nitrogen determination according to Van Slyke are
normally employed to characterize the products.

A fractionation is usually demanded, because the products of these reactions
are polydisperse: the polydisperse material is dissolved in water and it is
treated with increasing quantities of methanol or acetone, whereby portions of
the polymers precipitate according to their molecular weights and are separated. To precipitate the high molecular weight portion, a small amount of
precipitating solvent is required. The N-formylated chitosan sulfuric acid
polyesters are blood anticoagulants which are preferably administered in the
form of their alkali-metal salts (see Chapter 7).

Chitosan sulfate derivatives have been studied in connection with the anticoagulant activity of sulfated compounds, like heparin. The presence of acid
sensitive N-sulfate groups in the sulfated chitosan was demonstrated by
observing the rate of evolution of nitrogen in the Van Slyke deamination
apparatus: the sulfated chitosan reacts more rapidly under these conditions
than does heparin, in accordance with general results (Wolfrom & Shen-Han,
1959; Wolfrom, 1958).

A homogeneous sulfation of chitosan using sulfur trioxide-N,N-dimethylformamide
complex in excess of the latter is also feasible. The sulfur trioxide complex
can be formulated as a dipolar ion, whose reaction with a nucleophilic group

such as a primary amine should involve attack to the sulfur atom to effect
sulfation on nitrogen or oxygen with the concomitant release of N,N-dimethyl-
formamide.

The above procedure applied to chitosan yielded an amorphous product
containing one sulfoamino group and one sulfate acid ester group per monomer
unit. Thirty ml of the sulfur trioxide-N,N-dimethylformamide were for instance
necessary for 2 g of chitosan. Chitosan dissolves in this mixture and after 12
hr the crude product is isolated as the sodium salt by addition of sodium bicarbonate.

Transformation of chitosan in concentrated sulfuric acid was investigated by
Nagasawa, Tohira, Inoue & Tanoura (1971) and by Nagasawa & Tanoura (1972).
Finely powdered chitosan (average molecular weight 120,000; 92 % free amine)
was dried in vacuo over phosphorus pentoxide for 3 hr at 75°. Two g of the
dried chitosan were added in small portions during 15 min to 40 ml of concentrated sulfuric acid with vigorous stirring at the temperature indicated,
untill a gelatinous homogeneous state was reached. The reaction mixture was
then poured into 400 ml of ether at constant temperature and the precipitate
was collected on a glass filter, and after washing it was dissolved in water
and neutralized with cold 30 % sodium hydroxide. The neutralized solution was
dialyzed against tap water for 40 hr. The average molecular weight of the
sulfated chitosan was 31,000. Neither reaction temperature (0 ÷ 30°) nor time
(1 ÷ 10 hr) affected the degree of sulfation of chitosan, but it was markedly
influenced by the temperature of the ether used to separate sulfated products.

Chitosan is more stable than chitin to depolimerization due to the stabilizing
effect of free amino groups. Nagasawa, Tohira, Inoue & Tanoura (1971) studied
the relationships between depolymerization and sulfation of chitin and chitosan. They used sulfuric acid at 0°. On the basis of the molecular weights and
the sulfur content of the reaction products, they concluded that no structural
change in the monosaccharides occurred during reaction with sulfuric acid, but
drastic depolymerization and sulfation occurred immediately after dissolution
in sulfuric acid; the temperature and the length of the treatment influenced
the extent of depolymerization.

Nud'ga, Plisko & Danilov (1973) have produced further contributions to this
subject: they used a Schiff base derivative in order to protect the amino
group. Salicylidenchitosan was suspended in a mixture of propanol, isopropanol
and xylol. The suspension was poured into an organic alkali hydroxide solution
and a calculated amount of sodium β-chloroethansulfonate was added dropwise at
80°; two hr later the product was filtered and dialyzed against water. Chitosan was then regenerated from the Schiff base by the action of acids (see page
134). A maximum degree of substitution of 0.31 can be attained, while free
amino nitrogen is between 7.1 and 7.9.

O-Sulfoethylation is accompanied by a certain polymer chain degradation as
shown by the decreased viscosity of O-sulfoethylchitosan solutions in 2 %
acetic acid.

The conductometric titration curve includes three segments corresponding to
the titration of excess free hydrochloric acid, of sulfoethyl groups and amino
groups; the amount of sodium hydroxide in the reverse titration of the amino
and sulfoethyl groups corresponds to the expected on the basis of the above
data.

Table 3.7. Synthesis conditions for sulfoethylchitosan with sodium β-chloro-ethansulfonate; ¶ = dropwise addition. Nud'ga, Plisko & Danilov, 1974.

	Solvents	Mole ratio chitosan: NaOH: reagent: water	t°	time, hr	sulfur %	subst. degree	Amine N content found	calculated by substituting $-NH_2$	$-OH$
1	propanol	1:3:1:20	80	1	1.93	0.11	7.57	7.10	7.92
2	isopropanol + xylol 1:1	1:3:1:20	80	1	2.30	0.13	7.26	6.85	7.88
3	propanol + xylol 1:1	1:3:1:20	80	1	4.02	0.24	6.26	5.53	7.25
4¶	id	1:3:1:20	80	1½	4.79	0.30	6.69	4.80	7.00
5	id	1:3:1:20	50	1½	2.81	0.15	5.44	6.59	7.76
6¶	id	1:3:3:20	80	1½	5.32	0.35	5.76	4.42	6.78
7	id	1:3:3:20	130	1½	2.28	0.13	6.73	6.85	7.88

Table 3.7 reports data obtained by Nud'ga, Plisko & Danilov (1974) who prepared sulfoethylchitosan by reacting sodium β-chloroethansulfonate with chitosan in a suspension of alkali in alcohols under reflux. It has been noted that it is important to add the reagent in small portions at 30 min intervals, in order to obtain as high as possible degree of substitution. As shown in Tab. 3.7 the most satisfactory conditions are those corresponding to sample n° 6.

For the production of sulfoethylchitosan, reprecipitated chitosan is required because its surface is much larger than in chitosan powder. Sulfoethyl groups are attached to oxygen and nitrogen atoms in chitosan; this conclusion was reached on the basis of conductometric and volumetric analysis; data are shown in Fig. 3.11. No preferential substitution can be observed under the described experimental conditions. The sulfoethylchitosan samples thus produced are partially soluble in water and are soluble in 2 % acetic acid. From these acetate solutions, films have been cast for use as antitrombogenic surfaces.

Freshly prepared Schiff base (see page 131) is introduced into 40 % sodium hydroxide with a mixture of water, xylenol and isopropanol, and the reaction is carried out as indicated in the experiment n° 6 of Tab. 3.7. At the end, the Schiff base is decomposed in 50 % acetic acid and the O-sulfoethylchitosan is purified by dialysis, evaporated under vacuum and reprecipitated with acetone. O-Sulfoethylchitosan can be used as a polyampholyte.

The amino group protection through the Schiff base formation with salicyl-aldehyde has been used by Plisko, Nud'ga & Danilov (1972) for the preparation of O-carboxymethylated chitosan. In fact, the amino group protection allows selective carboxymethylation at the hydroxyl groups, by using sodium mono-chloracetate. In general, halo-substituted alkanethioacids or halo-substituted aliphatic carboxylic acids are suitable for this class of reactions. It is known that carboxymethylation of chitin leads to partial deacetylation and mixed carboxymethylation at both the hydroxyl and amino groups occurs.

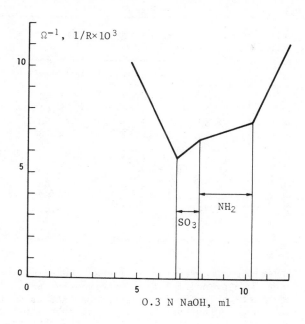

Fig. 3.11. Conductometric titration of sulphoethylchitosan
having a substitution degree of 0.35. Nud'ga, Plisko &
Danilov, 1974.

Cyanoethylation.

In the class of O-alkyl derivatives of chitosan, cyanoethylchitosan has been
investigated by Nud'ga, Plisko & Danilov (1975) who reacted chitosan in
various physical forms with acrilonitrile under alkaline, neutral and acidic
conditions.

Chitosan powder was added to a 10 % sodium hydroxide solution and after a few
minutes acrilonitrile in the ratio of up to 30 to 1 moles was added. Cyanoethylchitosan was washed with water and with methanol for the removal of
sodium hydroxide and of acrilonitrile, respectively. The acidic medium allowed
the reaction to be performed in homogeneous conditions. From data in Tab. 3.8,
it appears that the reaction at 20° is complete in about 6 hr and increased
mole ratio increases the substitution degree. At 40° monosubstituted cyanoethylchitosan is obtained in one hr, but after 6 hr the substitution degree is
lower, due to the side reactions including polymerization of the reagent. The
reaction is not feasible in neutral conditions. Temperatures higher than 20°
bring about carboxyl formation. The authors have demonstrated that at 20°
carboxyethyl groups are absent, by potentiometry and infrared spectroscopy,
the latter on thin films of cyanoethylchitosan.

Table 3.8. Cyanoethylchitosan formation at 20°. Nud'ga, Plisko & Danilov, 1975

	time, hr	mole ratio chitosan to acrylonitrile	total N, %	amine N, %		Substitution degree on monomer unit
				experimental	calculated	
1	6	1:30	15.08	4.47	4.48	1.87
2	24	1:30	15.20	4.48	4.41	1.91
3	48	1:30	14.67	–	4.90	1.70
4	24	1:15	14.33	4.18	4.78	1.54
5	6	1:30	14.54	4.03	4.70	1.63
6	24	1:30	14.15	4.20	4.83	1.47
7	6	1:15	14.43	–	4.70	1.60
8	24	1:15	14.29	4.90	4.87	1.54
9	48	1:15	14.19	–	4.82	1.50
10	3	1:6	9.71	5.89	6.60	0.25
11	6	1:6	11.45	5.49	5.90	0.64

Previous work on cellulose has demonstrated that the primary alcoholic groups can more readily undergo a cyanoethylation reaction. As shown in Tab. 3.8, the amine nitrogen content of the cyanoethylchitosan coincides with the calculated amino nitrogen; thus the conclusion concerning the substitution of the alcoholic groups is further supported by the fact that NMR spectra taken on cyanoethylchitosan obtained at 20° exhibits signals at τ 6.15 and 7.15, corresponding to protons in the $O-CH_2$ and $CH_2-C\equiv N$, whilst the signal at τ 7.5, corresponding to the $N-CH_2$ protons is absent. When performed at 70°, cyano-ethylation involves 30 % of the amino groups.

Carboxymethylation.

Carboxymethylchitin and carboxyethylchitin can be obtained by treatment of alkali chitin with the corresponding α-chloroacid. These chitin derivatives have been studied by Okimasu (1958), by Danilov & Plisko (1961), by Plisko & Danilov (1963), by Miyazaki & Matsushima (1968), by Hayashi, Imoto & Funatsu (1968) and by Trujillo (1968) among others.

All these authors take adavantage of the reaction mentioned above: typical reaction conditions are those listed in Tab. 3.9. Preliminary experiments established that the water solubility of carboxymethylchitin depends on the concentration of the sodium hydroxide bath used to pretreat chitin. For instance, with a 20 % sodium hydroxide bath the carboxymethylchitin obtained upon monochloroacetic acid treatment, is insoluble in water or dilute alkali solutions.

According to Okimasu (1958) the activation energy of carboxymethylation is 22.4 kcal, being the same as that of decomposition of sodium monochloroacetate with sodium hydroxide; therefore, it is impossible to inhibit the side reaction by controlling only the temperature of reaction. The concentration of sodium hydroxide in the reaction medium has greater effect on the velocity of

Table 3.9. Conditions for synthesis of carboxymethylchitin and its properties. Danilov & Plisko, 1961.

NaOH conc.	Time, hr	t°	Chitin taken, g	Reacting chitin, %	Sodium content, %		Substitution degree of soluble part	Viscosity of 1 % solution, cpoises
					soluble	insoluble		
50.15	14.0	40	5.2	85.5	6.88	4.10	0.84	40.43
49.76	18.5	60	3.0	87.3	7.32	4.13	0.89	37.19
49.76	18.5	40	3.4	88.2	7.25	3.91	0.88	22.84
49.76	12.0	60	2.7	88.8	6.13	2.30	0.75	30.20

carboxymethylation than on that of the side reaction. It is thus better to save reaction time by use of higher concentration of sodium hydroxide, which should be below 20 % to keep the medium homogeneous. The reaction of carboxymethylation is fast, being about 20 times as great as the velocity of the side reaction. When the initial concentration of sodium hydroxide is about 15 %, carboxymethylation is completed in about 12 hr at 30°.

Carboxymethylchitin is best prepared according to Trujillo (1968) when chitin (4 g) is mixed with dimethylsufoxide (20 ml) for 24 hr. The mixture is crushed to a pulpy consistency, and the excess solvent is removed. The chitin filter cake is then washed with ethyl alcohol, dried and ground to a powder before addition to 65 % sodium hydroxide (50 ml); the slurry is stirred for one hr. The mixuture is filtered and the cake is pressed to a wet weight of approximately 14 g.

This material is added to 2-propanol (50 ml) containing monochloroacetic acid (6 g) and the mixture is stirred for one hr. The material isolated on filtration is added to water (200 ml) and the pH adjusted to neutrality with concentrated hydrochloric acid. The resulting viscous solution is passed through glass wool, and then added slowly to one l of acetone. The sodium salt of carboxymethylchitin thus obtained is washed with absolute ethanol and dried (3.5 g).

The carboxymethylchitin may be dissolved again in water (120 ml) and after centrifugation at 4000 g slowly added to one l of acetone. The fibrous precipitate is washed with ethanol. The carboxymethylchitin is crushed to a fine powder and extracted with 95 % ethanol for 2 hr. The product is dried and stored over phosphorus pentoxide; the degree of substitution is one.

Carboxymethylchitin is a hygroscopic substance which normally contains 16 % moisture. The moisture uptake at 100 % relative humidity can reach 107 % after a few days.

Chitin glycolic acid is obtained by acidification of the carboxymethylchitin sodium salt with dilute hydrochloric acid at room temperature, followed by washing with methanol until negative reaction for chloride. Analysis of chitin glycolic acid for primary amino groups gave negative results which indicate that the acetamido groups are retained in the molecule (Danilov & Plisko, 1961).

Carboxymethylchitin containing 6.13 % sodium or more dissolves well in water. Heating the water solution does not cause coagulation, just as in the case of carboxymethylcellulose. In the study of dilute water solutions of carboxymethylchitin, it was shown that their properties are analogous to those of dilute aqueous solutions of carboxymethylcellulose, that is, with dilution of the solution there is an increase in specific viscosity.

Nud'ga, Plisko & Danilov (1973) have pointed out that hydroxyl groups only are involved in the alkylation process because the amino groups are mainly present in the acetylated form. According to previous works, the carboxymethylation process would bring about some deacetylation, and the carboxymethylchitosan could not be identified, strictly speaking, as O-carboxymethylchitosan. A substitution degree of one is in practice obtained with a ratio of sodium monochloroacetate to salicylidenchitosan of 1:9. After removal of the protection of the amino group, it has been verified with the Van Slyke method that the experimental amount of free amino groups coincided with the expected amount, and therefore hydroxyl groups only were interested in the carboxymethylation. Carboxymethylchitosan is soluble in 2 % acetic acid; if the substitution degree is close to one it is soluble in water.

Okimasu (1958) obtained carboxymethylchitosan as chloride or sodium salt, through carboxymethylation and deacetylation of chitin. The reduced turbidity *vs* concentration curves show the same upward tendency with dilution as that of ordinary polyelectrolytes in water, but some aggregation of polymer ions is observed in the presence of 0.1 M sodium chloride. Viscosity of carboxymethyl chitosan solutions falls gradually with time at room temperature and reaches an equilibrium after about 30 hr. The salts are coagulated from aqueous solutions by large amounts of ethanol and in the presence of neutral salts, such as sodium chloride or ammonium sulfate. Effects of neutral salts are more marked on the carboxymethylchitosan of higher carboxymethylation degree. Chloride salt is coagulated more difficultly than the sodium salt, whereas there is no difference in coagulability between them in neutral salt solutions. It is concluded from these results that the stability of carboxymethylchitosan is supported mainly by effective charges.

Viscosity of carboxymethylchitosan solutions varies with conditions in the same manner as that of ordinary polyelectrolytes, except that an isoelectric point is observed at pH 6.3. One of the remarkable characteristics of the solutions, is a very high viscosity due to a network structure among polymer ions as well as resistance to deformation. It seems that the network structure is unstable and changes gradually to the ion distribution of ordinary polyelectrolytes in a dilute solution. In a relatively concentrated solution, however, the structure is retained and affects the viscosity.

Effects of the concentrations of colloids, neutral salts and pH on the equivalent conductance of carboxymethylchitosan solutions are similar to those of ordinary polyelectrolytes. Results also coincide with those of coagulation measurements and viscosity data. In a relatively concentrated solution, however, the interference of conductance with a special network structure was observed by Okimasu (1958).

So far no information is available on the interactions of these interesting derivatives with metal ions and with other substances which are expected to easily react with them.

Hydroxyethylation (Glycolation).

Formyl-, propionyl-, butyryl-, benzoyl-, naphthalenesulfonyl-, benzenesulfonyl- and monomethylchitosan have been reported by Karrer & White (1930); however, at that time, those authors had no clear information about the type of combination of nitrogen in chitin and they even excluded that nitrogen was in the primary amino form. They also mentioned alkyl ethers of chitin.

Allylchitin has been used industrially to provide chitinous coatings; it was obtained by alkylation of the primary amino groups and etherification of hydroxyl groups with allyl bromide. The groups formed are $R-NH-CH_2-CH=CH_2$ and $R-O-CH_2-CH=CH_2$. Numerous other chitin derivatives of this particular type have been prepared by reacting unsaturated aliphatic halides containing more than two carbon atoms, like crotyl and methallyl bromides; since vinyl halides are not very satisfactory (Rust, 1949).

Senju & Okimasu (1950) prepared hydroxyethylchitin (glycolchitin) by reacting alkali chitin with ethylene oxide. Viscosity measurements on partially hydrolyzed mono(hydroxyethyl)chitin showed that the Staudinger equation holds for this derivative; however, the values of degree of polymerization and molecular weight, in the light of the current knowledge, were too low (Okimasu & Senju, 1950). Takeda & Tagawa (1960) solubilized glycolchitin into benzene at the maximum concentration of 0.013 $g{\times}l^{-1}$.

6-O-(Hydroxypropyl)chitin has been reported by Shimahara, Nagahata & Takiguchi (1974). Analysis in an autoanalyzer after hydrolysis in 4 N hydrochloric acid at 100° for 4 to 24 hr indicated that the glycolchitin preparation contained 3,6-di-O-hydroxyethylglucosamine, 3-O-hydroxyethylglucosamine, 6-O-hydroxyethylglucosamine and O-unsubstituted glucosamine residues in a molar ratio of 0.22:0.35:0.79:1.00.

Partially O-hydroxyethylated chitin labeled in N-acetyl groups was prepared as follows: partially O-hydroxyethylated chitosan (glycolchitosan, 40 mg) was treated at 4° with a mixture containing 400 mg of sodium bicarbonate and $|^3H|$- -acetic anhydride. After 24 hr, 200 µl of unlabeled acetic anhydride was added and the mixture was allowed to stand for another 24 hr at 4° (Araki & Ito, 1975).

Danilov & Plisko (1968) established the best conditions for synthesizing water soluble hydroxyethylchitin ethers, as follows: mercerization with 43 % caustic soda, cooling one to three times to low temperature, etherification at 70° for 30 hr with 16 mol of ethylene oxide to one of chitin and a chitin particle size of 1 mm. Other conditions gave ethers which only swelled in water or in dilute alkali, or did not even swell. The slightly broken down and etherified products dissolved like chitin in 60 % sulfuric, 78 % phosphoric and 50 % nitric acids. The more highly etherified chitin ethers dissolved in 4 ÷ 8 % caustic soda, in water and in dilute acetic acid. Analyses of samples of hydroxyethylethers of chitin are reported in Tab. 3.10.

Sample 1 was nitrated, with a nitration mixture containing 25 % nitric acid, 72 % sulfuric acid and 3.6 % water for 2 hr at 24°. The product contained 5.71 % nitrogen which corresponded to 1.34 hydroxyl groups. The intrinsic viscosity of chitin hydroxyehtyl ether (sample 3) in water and in 2.26 %

Table 3.10. Analyses of hydroxyethylethers of chitin. Danilov & Plisko, 1968.

Content				Composition	Number of CH_2CH_2OH groups
C	H	N	CH_2CO		
48.72	7.19	5.20	12.65	$C_{10.9}H_{19.4}O_{6.5}$	1.44
49.21	7.49	5.00	–	$C_{11.7}H_{21.4}O_{6.8}$	1.74
50.12	8.14	3.92	5.50	$C_{14.8}H_{20}O_{8.4}$	2.93

acetic acid is nearly five times as great in acetic acid as in water, while the slopes of the curves, characteristic of the interaction of solvent with solute are almost the same.

Hydroxyethylchitin is important because it has been used in a lot of research involving chitin: Araki & Ito (1975) have used it as a substrate to study the enzyme that catalyzes hydrolysis of acetamido groups in chitin derivatives. Yamasaki, Tsujita & Takakuwa (1973) developed a colorimetric method to determine lysozyme, based on the use of hydroxyethylchitin labeled with Remazol Brilliant Blue R. The latter was prepared as follows: glycolchitin (3 g) in water (150 ml) was labeled with Remazol Brilliant Blue R (300 mg) at 50°. Sodium sulfate (6 g) and sodium triphosphate (300 mg) were added to the mixture and after 2 hr methanol (600 ml) was used to precipitate the product. The colorimetric method is based on the release of the label by RBBR-glycolchitin in amounts proportional to the amount of lysozyme at 40° and pH 4.5. Acidified ethanol is used to precipitate the unreacted RBBR-glycolchitin before the readings are taken. Other applications are mentioned in further sections.

The degree of dissociation, the dissociation constant and the electrophoretic mobility of glycolchitosan have been studied by Kokufuta, Hirata & Iwai (1975) in water and in sodium chloride solutions by potentiometric titrations.

Noguchi, Tokura, Inomata & Asano (1965), Noguchi, Arato & Komai (1969) and Noguchi (1965) obtained anion-exchange polymers by epichlorohydrin treatment of chitosan and derivatives. The formation of epichlorohydrin derivatives is well-known in the preparation of modified celluloses. The epichlorohydrin treatment of chitosan brings about reactions with hydroxyl groups and some with amino groups of chitosan; the resulting epichlorohydrin chitosan is insoluble in both acidic and basic media. The pK_a for epichlorohydrin chitosan is 6.3.

Epichlorohydrin chitosan was prepared by Noguchi, Arato & Komai (1969) and by Haga (1971) by treating alkali chitin with epichlorohydrin first, and by carrying out the deacetylation in a further step, in order to protect the amino group. This approach to the preparation of epichlorohydrin chitosan is necessary to avoid secondary amine formation. Squeezed alkali chitin is added to epichlorohydrin in a proportion between 0.5 and 1.5 by weight and it is left 12 to 48 hr at 0° or, in any case, at a temperature lower than 15°. Deacetylation is carried out directly or after washing with water, according to one of the previously described methods. The resulting epichlorohydrin chitosan contains 6.93 % nitrogen, 93 % of which is present as primary amino group and 7 % is present as acetylamino groups.

A simpler procedure, leading however to secondary amine formation is as follows: epichlorohydrin (5.8) is added to a suspension of chitosan (10 g) in water (100 ml) and, after warming for 1 hr, 4 % sodium hydroxide (70 ml) is added and kept boiling for 2 hr. Washing is done with water, dilute hydro- chloric acid, dilute sodium hydroxide and water. The product contains 6.37 % nitrogen, 18.1 % of which is present as primary amino groups.

For the preparation of epichlorohydrin benzylchitosan, chitosan (20 g) is suspended in water (140 ml) and reacted with benzyl chloride (35 g) at 50° for 1 hr. 40 % Sodium hydroxide is added (28 ml) and after 3 hr the mixture is kept at the boiling for 1 hr. Washing is done with water and with alcohol. Benzylchitosan is insoluble in alkalis and soluble in dilute acids. If benzyl- chitosan is further reacted with epichlorohydrin, an insoluble product is obtained in both alkaline and acidic media.

Epichlorohydrin oxystyrene chitosan can be prepared by reacting chitosan (10 g) suspended in water (500 ml) with oxystyrene (10 ml). After warming, the temperature is raised to 115° for 3 hr. Oxystyrene chitosan is soluble in acids while epichlorohydrin oxystyrene chitosan is insoluble in acidic and basic media.

The description of the chromatographic properties of these polymers is to be found in Chapter 5.

Alkyl ethers.

Danilov & Plisko (1958) obtained the previously unknown chitin ethyl ethers by the action of ethyl chloride on the alkali derivative of chitin. The products obtained were almost completely soluble in organic solvents. Ethylchitin does not melt, but decomposes at 180 ÷ 190° in the presence of air, and at 200 ÷ 210° in vacuo.

A full analysis was carried out on a selected sample for functional groups and elements. Free hydroxyl groups were also determined by the Van Slyke method, since, under the conditions of the reaction, acetyl groups could split off from nitrogen. The results were: C, 54.98; H, 9.22; N, 6.19; C_2H_5O, 33.83; CH_3CO, 4.53; OH, 2.75 and NH_2 2.04. Ethylation was carried out in an autoclave in the presence of 16 mol of caustic soda and 15 mol of ethyl chloride for each mol of chitin over a period of 10 hr and in a temperature range of 60 ÷ 130°. The sample had the composition $C_{9.6}H_{17.53}N_{0.91}O_{4.14}$.

Shoruigin & Makarova-Semlyanskaya (1935) reported the production of monomethyl chitin, after an uneasy methylation process with dimethylsulfate and sodium hydroxide. Wolfrom, Vercellotti & Horton (1964) performed the methylation of chitosan with dimethylsulfate in alkaline medium. They obtained a product containing 29 % of the theoretical OCH_3 for fully methylated chitosan. A suspension of this material (10 g) in formamide (400 ml) was acetylated by shaking for one day with acetic anhydride (50 ml) and pyridine (25 ml) with addition of further acetic anhydride after 12 hr. The product was ninhydrin negative. The partially acetylated, partially methylated chitin (3 g) was further methylated in N,N-dimethylformamide solution (200 ml) by the addition of barium oxide (18 g), barium hydroxide (1.8 g) and methyl iodide (21 ml).

N-Alkyl derivatives.

The N-alkylation of chitosan can be carried out in the presence of other bases having pK_a values higher than 6.3, according to Nud'ga, Plisko & Danilov (1973). For instance, the addition of methyliodide to a reaction medium including thiethylamine, the former in the molar ratio of 6:1 to chitosan, gives a methylated chitosan with a quaternary nitrogen degree of 0.64. When trimethylamine is added dropwise, this reaches 0.78; evidence is given in Fig. 3.12. These are the best conditions so far developed and they take into account the tendency of the hydroiodic acid formed by the reaction to produce iodide salts with the free or partially substituted amino groups of chitosan.

The chlorides and the iodides of the N-trimethylchitosan and the N-triethyl-chitosan are readily soluble in water; the corresponding bases are soluble in triethylbenzylammonium hydroxide. Iodides are not stable and after a certain time liberate iodine. The reduced viscosity of the solutions of these

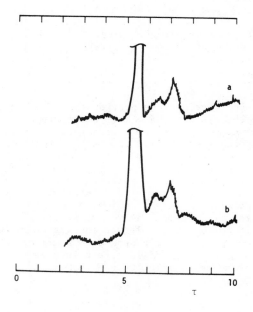

Fig. 3.12. NMR spectra of N-trimethylchitosan with substi-tution degree of: a = 0.58; b = 0.78. Concentration 4 % at room temperature. The presence of a signal at 6.80 τ and its intensity confirm the presence of quaternary nitrogen. Secondary and tertiary nitrogen signals are absent. Nud'ga, Plisko & Danilov, 1973.

derivatives in 2 % acetic acid depend lineraly on the concentration; sodium chloride can depress the polyelectrolyte effect.

The process of N-alkylation introduces a certain amount of polymer chain degradation; it was reported that from a chitosan with average molecular weight 480,000, a N-triethylchitosan was produced with a degree of quaternary nitrogen of 0.36 and average molecular weight of 170.000. As from Fig. 3.6, it was impossible to distinguish the quaternary ammonium groups from the primary amino groups, by potentiometric titration. N-Trimethylchitosanium iodide is an antibacterial substance.

Isomerization of D-glucose to D-fructore by 6-O-(2-hydroxyethyl)chitosan 2-tri-N-methylammonium hydroxide has been performed by Takeda (1970). The reaction rate at 50° in the system catalyzed by the strongly basic chitin derivative was 3 to 10 times that in the NaOH-catalyzed system.

Xanthation.

The sodium chitin can undergo synthetic reactions of the heterogeneous type, such as xanthation, etherification by means of alkylene oxides or active hydrogen compounds and other reactions which alkali metal alcoholates undergo. In the study of Noguchi, Wada, Seo, Tokura & Nishi (1973), alkali chitin was compressed to about one third of the weight of the original material by removing the excess aqueous solution of sodium hydroxide, and then pulverized by a mixer. This alkali chitin was poured into a separating funnel, and after evacuation, it was left at -20° for 10 hr. Carbon disulfide (50 % of laid-in chitin in weight) was poured in utilizing the vacuum, and the xantogenate reaction proceeded by occasional shaking at 30° for 15 hr. Then, the contents were dissolved in an aqueous solution of sodium hydroxide previously cooled to 0°, arranging the mixture to contain 4.5 % alkali and 5 % chitin. After almost all dissolved, it was left at -20° for 5 hr to freeze; then it was heated for about 3 hr to 0 ÷ 5°, and a completely homogeneous viscous chitin solution was obtained. These conditions should be followed strictly; if the immersion in alkali is done at 20° or for a long time even at 11 ÷ 13°, the regenerated fibers tend to be water soluble or the single fibers tend to fuse to each other while spinning, because of some degree of deacetylation (Thor, 1939).

It seems that, due to the freezing operation, the distance between micelles of chitin is widened, accelerating the homogeneous alkali chitin reaction and at the same time making the xanthogenate reaction easy to occur. As the homo-geneous viscose thus obtained shows very high viscosity, the addition of urea is recommended for spinning. The action of urea seems to be due to the disso-ciation of the intermolecular hydrogen bonds between functional groups of chitin (see Figs. 2.2 and 2.3). Cellulose + chitin fibers were obtained with good results (see Chapter 6).

Chitin xanthate can be further reacted with sultones, whose general formula is:

where G is a divalent hydrocarbon radical (alkylene) of three to six carbon atoms containing lower alkyl substituents, to produce polymeric alcohol sulfonic acids containing S-xanthogenate ester substituents and their salts, according to the reaction:

$$\underset{\text{chitin xanthate}}{\underset{\displaystyle \overset{\|}{\underset{S}{}}}{Chit-O-C-S-Na}} \quad + \quad \underset{\text{propanesultone}}{CH_2-CH_2-CH_2-O} \quad \to \quad \underset{\text{chitin xanthate-S-propanesulfonate}}{\underset{\displaystyle \overset{\|}{\underset{S}{}}}{Chit-O-C-S-(CH_2)_3-SO_3Na}}$$

chitin xanthate   +   *propanesultone*   →   *chitin xanthate-S-propanesulfonate*

This class of reactions can be carried out not only on the products obtained from chitin or chitosan with carbon disulfide, but also with carbonyl sulfide; if conditions are properly controlled to exclude competing nucleophiles, xanthates react completely and rapidly with sultones.

Although water is reactive with sultones, it may be used in reaction systems since the sultone + xanthogenate reaction is significantly faster than the sultone + water reaction. Indeed the efficiency with which sultones react with polymeric alcohol xanthogenates to produce the sulfonated polymeric alcohol derivatives is a particularly advantageous feature. Relatively high hydroxyl ion concentrations, however, may give rise to undesirable losses of the relatively expensive sultone reactant. The presence of excess amounts of caustic in water or in other ionizable reaction media should be avoided. When such systems are employed, highly acid conditions at the initiation of the reaction also should be avoided to prevent regeneration of the polymeric alcohol (Bridgeford, Turbak & Burke, 1972).

This example illustrates the preparation of a chitin xanthate-S-propanesulfonate: chitin (26 g) is ground to 60 mesh and held in 50 % sodium hydroxide (300 ml) in a refrigerator. The resultant mixture is centrifuged and a slush obtained which is diluted with water and recentrifuged. The alkali chitin centrifugate obtained is transferred to a closed bottle. Carbon disulfide (100 g) is then added and the resultant contents reacted at 28° for 2 hr while the bottle is slowly rotated. During the reaction, the initially ivory colored chitin changes to yellow-orange. Water (150 ml) is added and the slurry is refrigerated at about 0.5° for about 2 days. The resultant solution is diluted to one l to provide a solution containing 2.6 % chitin; about 550 ml of this solution is diluted with one l of water and glacial acetic acid is added to bring the pH to 6, for the removal of a major portion of the by-products originally present. After sodium hydroxide pellets are added to obtain a 3 % sodium hydroxide concentration in the chitin xanthate solution, propanesultone (11 g) is dispersed in water (300 ml) and added. The resultant mixture is allowed to react at 36° for 16 hr. The resultant product solution is then combined with sodium chloride and 10 % hydrochloric acid in methanol, to yield chitin xanthate-S-propanesulfonic acid having a S-xanthate ester degree of substitution of about 0.21 and a molecular weight of 500.000 daltons. The applications of this class of products are described on page 236.

Chitosan hydroxyl groups react with sulfonium salts in alkaline media (0.1 5 % sodium hydroxide) and liberate the sulfonium cation:

$$\begin{array}{c} CR^1{=}CHR^2 \\ {}^{/} \\ {}^{+}S{-}R^3 \\ {}_{\backslash} \\ R^4 \end{array}$$

where R^1 and R^2 may be hydrogen or a lower alkyl group containing from one to three carbon atoms and may be the same or different, and R^3 and R^4 may be the group $-CR^1=CHR^2$ or an alkyl or certain substituted alkyls (Owen & Sagan, 1967).

Suitable sulfonium salts are:
1) β-haloethyl salts such as $(XCH_3CH_2)_2R^3S^+Y^-$ or $(XCH_2CH_2)_3S^+Y^-$, where X represents halogen and Y^- represents an anion. An example of such salts is tris-(β-chloroethyl)sulfonium chloride;
2) esters of mono-, bis- or tris-hydroxyethyl sulfonium salts such as:

$$
\begin{array}{l}
CH_2CH_2OSO_3^- \\
/ \\
{}^+S-CH_2CH_2OSO_3Na \\
\backslash \\
R^3
\end{array}
$$

The disodium salt of tris-(β-sulfatoethyl)sulfonium inner salt which has the formula:

$$
\begin{array}{l}
{}^+S(CH_2CH_2OSO_3Na)_2 \\
| \\
CH_2CH_2OSO_3^-
\end{array}
$$

is a very suitable sulfonium compound.

Chitin or chitosan articles may be treated simultaneously or consecutively in any order with the alkali, the sulfonium salt and a reactive polymeric substance. These ingredients are added in from aqueous media. Fabrics may be impregnated with the sulfonium salt and the polymeric substance in aqueous solution, for instance cellulose derivatives, polyamides and other polymers.

Other derivatives.

Trimethylsilylchitin has been prepared by Harmon, De & Gupta (1973) by swelling chitin in formamide for 5 ÷ 6 hr at 80°, and after cooling at room temperature, by adding hexamethyldisilazane. After 2 hr at 70° under stirring, the mixture was poured in anhydrous acetone to yield a colorless, amorphous powder whose infrared spectrum exhibited strong peaks at 1250 and 845 cm^{-1} characteristic of C-Si stretching vibrations. The silicon content was 6.19 % and the experimental degree of substitution was 0.6. The trimethylsilyl groups in this derivative are readily cleaved by treatment with water at room temperature. The trimethylsilylchitin was prepared along with other derivatives of cellulose and dextran in view of the production of lipid-soluble derivatives of polysaccharides for industrial applications.

Glycosylated chitin has been obtained by Neely (1964) by an enzymic process which cleaves sucrose and grafts the resulting glucose onto chitin or cellulose. A dextransucrase solution was prepared by culturing *Leuconostoc mesenteroides* and was then used to form a dextransucrase-chitin complex. Glycosylated chitin is characterized by physico-chemical properties which differ from those of chitin.

Danilov & Plisko (1954) reacted alkali chitin with monochloropropanediol in acetone for 40 hr at 60° to obtain the glyceryl ether $C_{11}H_{20}O_{7.25}N_{0.75}$ containing 49 % glyceryl residues; products with lower substitution are soluble in dilute sulfuric, hydrochloric and phosphoric acids, while the above product was insoluble and was not swelled in sodium hydroxide.

Gyorgy, Kuhn & Zilliken (1957) described alcoholized chitin: a suspension of chitin (200 g) in anhydrous methanol (2.5 l) was kept at a temperature just above 0° and was saturated with hydrochloric acid gas for about 2 hr until practically all of the chitin was dissolved. The solution was then evaporated to dryness under reduced pressure. The residue was taken up in a small amount of methanol, treated with decolorizing charcoal, and the solution again evaporated to dryness. The resulting methanolized chitin is a white powder which is readily soluble in water, does not reduce the Fehling solution, does not give a positive reaction when treated with ninhydrin and has a growth-promoting ability for *Lactobacillus bifidus*. In an infant feeding formula based on cow milk, methanolized chitin is added in amounts of 0.5 ÷ 5 % by weight based on the total weight of solids in the formula, to promote the growth of *Lactobacillus bifidus* in the digestive tract of infants.

Other reactions that have been experimented on chitin or chitosan oligomers or monomers and which could be applicable to the polymers, include those described by Kochetkov, Derevitskaya & Molodtsov (1962): glucosamine hydrochloride and N-carbobenzyloxyglycine in aqueous sodium hydroxide and pyridine were treated with dicyclohexylcarbodiimide and after two days gave N-(N-carbobenzyloxy-glycyl)glucosamine. Similarly, N-(N-carbobenzyloxy-DL-alanyl)glucosamine was prepared together with N-hyppuroyl-, N-acetyl- and N-benzoylglucosamine.

6-Mercapto-6-deoxy-D-glucosamine has been prepared by Ogawa & Ito (1965) by a process which entails reacting anisaldehyde with glucosamine (Fig. 3.13); the obtained N-anisylidene-D-glucosamine can be tosylated at O^6 and acetylated to

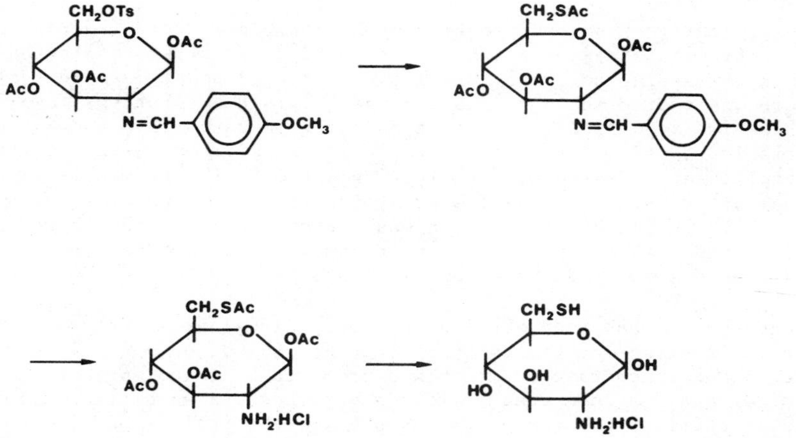

Fig. 3.13. Introduction of -SH groups into glucosamine. Ts is $CH_3C_6H_4SO_2-$ and Ac is CH_3CO-.

N-anisylidene-1,3,4-tri-O-acetyl-6-O-tosyl-β-D-glucosamine; the latter, with
sodium thioacetate, gives N-anisylidene-1,3,4-tri-O-acetyl-6-deoxy-S-acetyl-
-6-thio-β-D-glucosamine. The Schiff base may be decomposed with acids and the
acetyl groups removed to yield 6-mercapto-6-deoxy-β-D-glucosamine. These
reactions should be feasible on chitosan, to afford polymers with novel
characteristics.

Chitosan might undergo mercaptoethylation according to a process developed by
Fields & Reynolds (1965) for primary, secondary, aliphatic, aromatic and
cyclic amines. Treatment should be carried out with a compound of the formula
$ROCOACH_2CH_2B(Q)$ where A and B are dissimilar and may be selected from the
class consisting of sulfur and oxygen. R may be alkyl (methyl through octyl),
aryl (phenyl and substituted phenyl) or aralkyl. Q may be hydrogen, -COR,
-COOR, -CONHR or -CONR$_2$. Ethyl-2-mercaptoethylcarbonate is suitable to carry
out this reaction. Carbamates have been used as well (Reynolds, 1966).

REACTIONS WITH CARBONYL COMPOUNDS.

Anhydrides.

Reactions of the amino groups of chitin and chitosan with carbonyl compounds
leading to formation of Schiff bases have been occasionally reported in the
older literature, and their importance has not yet been fully realized. Only
a few recent papers deal with this important class of reactions. Interest was
first given to acetylation reactions to recover chitin from chitosan,
especially after the chitosan was used to form shaped articles, like membranes
and threads.

Du Pont de Nemours & Co. (1936) reported that chitosan can be used as an
adhesive and rendered insoluble by exposure to ammonia vapors, aldehydes and
acid halides; formamide was indicated as a suitable aldehyde. Zechmeister &
Toth (1934) degraded chitin with acetic anhydride to octaacetylchitobiose;
Shoruigin & Hait (1935) found that acetic anhydride in acetic acid and
sulfuric acid solutions is not as good as for cellulose to obtain chitin
acetate esters. Hopff, Spanig & Steimmig (1954) reacted a number of anhydrides
with D-glucosamine. Hara & Matsushima (1972) carried out the simultaneous
N-acetylation and O-trimethylsilylation of hexosamine. Danilov & Plisko (1954)
mentioned that re-acetylated chitin shows less porosity of structure than the
original native chitin.

More drastic procedures for the N-acylation of chitosan with carboxylic
anhydrides at high temperatures have been published by Karrer & White (1930).
Acetic anhydride in methanol has been used on oligosaccharides of chitosan by
Barker, Foster, Stacey & Webber (1957) and by other authors.

Inoue, Onodera, Kitaoka & Irano (1960) studied the reactions of 2-amino-2-
deoxy-D-glucose in methanol with mixed carboxylic acid anhydrides to give
N-acylated derivatives.

Glyceraldehyde or N,N'-ethylenedimethylolurea were reported by the Oxford
Paper Co. (1963) as cross-linking agents for use in the preparation of lithographic printing plates from chitosan in the presence of zinc salts.

Wieckowska (1968) used acetylacetone and *p*-dimethylaminobenzaldehyde in concentrated hydrochloric acid to determine chitin on the basis of its glucosamine content by spectrophotometric readings at 526 nm. This reaction is based on the condensation of glucosamine with acetylacetone, followed by reaction of the substituted pyrrole thus produced with *p*-dimethylaminobenzaldehyde to form a red compound. Golubchuk, Cuendet & Geddes (1960) used this method to determine the chitin content of wheat as an index of mold contamination and wheat deterioration. They also studied the determination of glucosamine on the basis of the nitrogen liberated by heating with trisodium orthophosphate. Wheats with low mold count and low fat acidity contain 18 ÷ 55 ppm glucosamine while deteriorated wheats have 214 ÷ 393 ppm.

Plisko, Nud'ga & Danilov (1972) used salicylaldehyde to treat chitosan as a preliminary step toward the formation of polyampholytes (see page 117).

Lieflaender (1962) and Kochetkov, Derevitskaya & Molodstov (1962) studied several reactions on glucosamine: they condensed amino sugars with N-carbobenzoxyamino acids in the presence of carbodiimide. The specificity of carbodiimide in the formation of the amide linkage enabled them to use unsubstituted amino sugars directly in the condensation in aqueous pyridine.

Hirano & Ohe (1975) reported on a novel, straightforward procedure for the N-acylation of chitosan. N-Acylchitosans are of interest as they effect the selective aggregation of some cancer cells. Chitosan (100 mg) was dissolved in 5.0 ml of one of the acidic solvents listed in Tab. 3.11. An excess of acid anhydride was added and the mixture was left at room temperature 12 hr to afford a viscous, homogeneous solution in the acid solutions, or a solidified gel in the aqueous, acidic solutions. As an alternative, chitosan (0.5 g) was dissolved in 10 % aqueous acetic acid (10 ml): the solution was diluted with methanol (40 ml) with stirring and the anhydride (2 ÷ 3 mol per hexosamine residue) was added (Hirano, Ohe & Ono, 1976).

For the reaction with higher fatty acid anhydrides (C_{10} to C_{18}), an additional amount (∿1 ml) of pyridine was added and the reaction was accelerated by heating in a boiling-water bath for a few seconds. The product was poured into 100 ml of acetone, and the suspension was stirred for a few hours at room temperature to afford the corresponding acylchitosan.

The presence of both N- and O-acyl groups in the acylchitosans was shown by absorption at 1650 (C=O in N-acyl) and 1750 cm^{-1} (C=O in O-acyl) in the infrared spectra and by the degree of substitution (2.6 for the acetylchitosan, 1.3 for the propionylchitosan, and 1.1 for the butyrylchitosan) per monosaccharide unit; the degree of substitution values were based on the ratio of (N-acyl protons) / (methine and methylene protons of sugar) in the nuclear magnetic resonance spectra. Acylchitosans from the higher fatty acids, exhibited strong infrared absorption for NAc but very weak or almost no absorption for OAc.

O-Deacylation of the foregoing products was carried out by stirring with 0.5 M potassium hydroxide in 95 % ethanol at room temperature overnight. Nuclear magnetic resonance and infrared spectra, together with polarimetry and elemental analyses confirmed the reactions.

Myristoyl, stearoyl, octanoyl, hexanoyl and decanoyl derivatives were described by Hirano, Ohe & Ono (1976).

Table 3.11. Reaction conditions for the acylation of chitosan. Yields between 50 and 78 %. Hirano & Ohe, 1975.

Acylchitosan	Solvent	Anhydride
acetyl	10 % acetic acid, 5 ml	acetic, 2.5 ml
acetyl	50 % formic acid, 5 ml	acetic, 2.5 ml
acetyl	formic acid, 5 ml	acetic, 0.4 ml
propionyl	10 % propionic acid	propionic, 3.4 ml
propionyl	10 % acetic acid, 5 ml	propionic, 2.5 ml
butyryl	acetic acid, 5 ml	butyric, 2.5 ml
butyryl	10 % butyric acid, 5 ml	butyric, 2.5 ml
lauroyl	acetic acid, 5 ml + pyridine, 1 ml	lauric, 1.3 g
palmitoyl	acetic acid, 5 ml + pyridine, 1 ml	palmitic, 2.0 g
benzoyl	acetic acid, 5 ml + pyridine, 1 ml	benzoic, 5.0 g

Vanlerberghe & Sebag (1972) have produced new derivatives of chitosan by using anhydrides of saturated and unsaturated dicarboxylic acids and by carrying out addition reactions to the double bond. The general formula of the compounds used is: $-NH-CO-G-COOH$, where N is attached to the C^2 of chitosan and G can be distinguished as the group:

$$R^1-C=C-H$$

where R^1 is hydrogen, a methyl or the group:

$$R^2R^3C-CHR^4$$

where R^2, R^3 and R^4 are similar or different and are hydrogen, methyl, hydroxy acetoxy, amino, monoalkylamino, dialkylamino groups, eventually substituted with amino, hydroxy, mercapto, carboxy, alkylthio, sulfonic and other groups, or the group:

$$H_2C-O-CH_2$$

The acylation reactions are carried out at room temperature with mole ratios of anhydride to chitosan from 0.3:1 up to 2:1 for saturated anhydrides, in inert solvents, like dioxane or tetrahydrofurane in heterogeneous medium, because chitosan is in suspension as a free base. Chitosan can be acylated with yields of 70 %.

Succinic anhydride, acetoxysuccinic anhydride, methylsuccinic anhydride and diglycolic anhydride are saturated anhydrides useful in this respect. Examples of unsaturated anhydrides are maleic anhydride, itaconic anhydride and citraconic anhydride. With the latter, addition reactions can be further carried out to yield a variety of products, at pH above 7 and at a temperature between 20 and 60°, under nitrogen. Additions can be performed with ammonia, aliphatic amines from methylamine to octylamine, diamines like ethylenediamine hexamethylenediamine; alkanolamines like monoethanolamine, isopropanolamine; mercaptoamines, like β-mercaptoethylamine; amines carrying acidic functions, like glycocoll, alanine, leucine, cysteine, penicillamine, glutathione and protein hydrolysis products. When using a mercaptoamine, the addition may involve the amino or the mercapto function.

The products may be recovered from the reaction medium by pH variation, by precipitation with acetone or alcohol, dialysis or ion-exchange. These products are soluble in alkaline media.

The functional groups carried by the derivatives are ionized in water and give salts with other reagents, or internal salts. Particularly, basic amino acids like lysine and arginine can react to give salts.

These derivatives of chitosan can be used for the collection of trace transition metal ions, to form films or membranes and to prepare cosmetic products.

For instance, an aqueous suspension of chitosan (20 g) is dissolved with a stoichiometric amount of hydrochloric acid. Sodium bicarbonate solution is added in small amounts and maleic anhydride (9.4 g; mole ratio 1:1) is added very slowly in four portions over a total period of 150 min: 87 % of the anhydride is condensed with chitosan. The reaction product is precipitated by the addition of hydrochloric acid, washed with water and acetone. The dry product is a white powder.

The addition reaction of chitosan maleic anhydride with glycocoll is performed as follows: the chitosan maleic anhydride product (7 g) is dispersed in water (40 ml) and sodium hydroxide is added in order to neutralize and dissolve it. Glycocoll (1.5 g) is then added as a solution in 10 ml of 2 M sodium hydroxide and the reaction mixture is warmed at 50° under nitrogen for one hr. The product so obtained is soluble in both alkaline and acidic media. It can be isolated as a sodium salt by precipitation at the isoelectric point, with the help of some hydrochloric acid and ethanol.

Most of the reported addition products are soluble and require ethanol for the precipitation. The addition product of glutamic acid to the chitosan maleic anhydride compound is reported as insoluble in hydrochloric acid medium (Vanlerberghe & Sebag, 1972).

The amounts of carboxylic anhydride and water are important for gel formation. The excess carboxylic anhydride is consumed not only in the acylation of amino groups and some of hydroxyl groups in chitosan, but also in the reaction with water. The mixture should be kept at room temperature a few hours to ensure that the gel rigidly solidifies the whole solution in the flask. The gels are generally dialyzed free of carboxylic acid; they are colorless, transparent and 1̣gid and have no smell of carboxylic acid and almost no taste. The gels are soluble only in formic acid but insoluble in 50 % formic acid, acetic acid, 10 % acetic acid, alkaline solutions, water, methanol, ethanol and acetone. The solidifed gels are stable to heat and a dehydration reaction occurs without melting at higher temperatures over 100° (Hirano & Ohe, 1975).

Aldehydes.

Salicylaldehyde reacts with chitosan to produce a Schiff base; Nud'ga, Plisko

& Danilov (1973) took advantage of this reaction to perform the O-alkylation of chitosan according to the following scheme:

$$|C_6H_7O_2(OH)_2NH_2|_n + n\ R^1C_6H_4CHO \rightarrow |C_6H_7O_2(OH)_2N=CHC_6H_4R^1|_n$$

$$|C_6H_7O_2(OH)_2N=CHC_6H_4R^1|_n + n\ ClR^2ONa \rightarrow |C_6H_7O_2(OH)_{2-x}(OR^2ONa)_xN=CHC_6H_4R^1|_n$$

$$\downarrow + \text{acetic acid}$$

where:
R^1 = H, OH, OCH$_3$, NO$_2$

R^2 = C$_2$H$_4$SO$_2$, CH$_2$CO

x = substitution degree

$$|C_6H_7O_2(OH)_{2-x}(OR^2OH)_xNH_2 \cdot CH_3COOH|_n$$

$$\downarrow + \text{sodium hydroxide}$$

$$|C_6H_7O_2(OH)_{2-x}(OR^2ONa)_xNH_2|_n$$

Benzoic aldehyde, anisic and o-nitrobenzoic aldehydes do not react completely; only a small number of the amino groups form Schiff bases. The Schiff bases obtained from chitosan do not melt, decompose above 220° and are insoluble in organic solvents.

Chitosan can be regenerated from salicylidenchitosan with acetic acid; the fastest and most complete decompostition is made with 50 % acetic acid, 20 ml for 0.5 g of base; to obtain pure chitosan the products can be dialyzed against water, to eliminate the aldehyde and acetic acid.

The reaction products of chitosan with amidox, a dialdehyde starch, have been reported by Lee (1974) and by Muzzarelli, Barontini & Rocchetti (1976); they are yellowish-brown products, whose characteristics depend on the mole ratios; they are very easy to obtain because the reaction between aldehyde and amino groups occurs immediately when a suspension of amidox is made in a chitosan acetate solution.

Moshy & Germino (1969) have reported that a periodate oxidized dialdehyde can be prepared from chitin, since the periodate oxidation is applicable to compounds having two hydoxyl groups or an amino and hydroxyl group attached to adjacent carbon atoms. Several patents are available on the subject of film-forming materials with dialdehyde polysaccharides obtained from starches, cellulose, galactomannans and dextrans: chitin does not appear to have been reacted with these oxidation products.

Aldehyde and dialdehydes other than those obtained from polymers have been occasionally used to cross-link chitosan with other substances, including polymeric substances. Moshy & Germino (1969) used limited amounts of glyoxal to cross-link a chitosan tobacco sheet (see page 247).

Stanley & Watters (1975) and Stanely, Watters, Chan & Mercer (1975) used glutaraldehyde to covalently link chitin with enzymes. They added moist chitin to an aqueous dispersion of the enzyme, and after adsorbing the enzyme on chitin, an aqueous solution of glutaraldehyde was added: usually a large excess, 10 ÷ 50 parts of glutaraldehyde per part of enzyme, where the enzyme is present as 10 % of the solid chitin + enzyme adduct.

Muzzarelli, Barontini & Rocchetti (1976) have pointed out that the reaction of
glutaraldehyde with chitin and chitosan might not be so simple as normally
supposed. In fact this reaction is very fast at room temperature even in
heterogeneous media, in aqueous solutions and over a wide range of pH values.
The modification is irreversible and survives many treatments.

By analogy with the formation of Schiff base in the reaction of aniline with
aryl aldehydes, there is a tendency to assume that the reaction of glutaraldehyde involves the same pattern. However, Schiff base formation is reversible.
Richards & Knowles (1968) have put forward other considerations against the
formation of Schiff bases: they have proposed a mechanism for the reaction of
glutaraldehyde with proteins, which can be reasonably applied to chitin and
to chitosan. α,β-Unsaturated aldehydes in form of oligomers exist in glutaraldehyde and these reactive derivatives of glutaraldehyde would give a Michael
type of adduct with amines. Therefore, the end products would be secondary
amines with unreactive aldehyde functions. Schiff base formation is included
as an intermediate step.

$$OHC-CH_2-CH_2-CH_2-CHO \rightarrow OHC-CH_2-CH_2-CH_2CH{=}C-CH_2-CH_2CHO \rightarrow$$
$$\underset{CHO}{|}$$

$$OHC-CH_2-CH_2-CH_2-CH{=}C-CH_2-C{=}CH-CH_2-CH_2-CH_2-CHO \rightarrow$$
$$\underset{CHO}{|} \quad \underset{CHO}{|}$$

$$OHC-CH_2-CH_2-CH_2-CH{=}C-CH_2-C-CH{=}C-CHO \rightarrow \quad etcetera$$
$$\underset{CH \quad CH_2}{|} \quad |$$
$$\underset{CH_2}{\diagdown \diagup}$$

The amine $R-NH_2$ would react to form species like the following:

$$R-NH-CH-CH-CH_2-C{=}CH-CH_2- \rightarrow R-NH-CH-CH-CH_2-CH-CH-NH-R$$
$$\underset{CHO}{|} \quad \underset{CH{=}N-R}{|} \qquad \underset{CHO}{|} \quad \underset{CHO}{|}$$

The preparation and some aspects of the physico-chemical properties of the
chitosan + glutaraldehyde reaction products have been studied by Muzzarelli,
Barontini & Rocchetti (1976). Generally, rigid gels can be obtained by mixing
glutaraldehyde solutions with chitosan acetate solutions; they do not lose
water easily and occupy the total volume of the solutions from which they
originate. Their color is yellow to brown, depending on the amount of glutaraldehyde. Soluble derivatives can be obtained by a careful selection of
conditions.

Chitosan (450 mg) of average molecular weight $1.2{\times}10^5$ daltons was dissolved in
4 % acetic acid (140 ml) and 25 % glutaraldehyde solution (10 ml) was added,
under vigorous stirring. These amounts are such as to give a clear solution;
with higher concentrations of chitosan, gel formation immediately occurs upon
addition of glutaraldehyde. The resulting clear solution was diluted with 2 l
of acetone; the precipitate was filtered on paper and immediately redissolved
in 10 % acetic acid to a final volume of 100 ml. The precipitate should not be
allowed to desiccate, because it would no longer be soluble: the sample was
dialyzed against 10 % acetic acid and then filtered through a Millipore

membrane. The polymer concentration was checked by weighing a dry aliquot. The actual concentration was in the range of 60 mg×ml^{-1}. The glutaraldehyde : chitosan mole ratio was 1:1. The refractive index increment used for the light scattering measurements was dn/dc = 0.166 ml×g^{-1}; the average molecular weight as measured within two days from the preparation, was 5.2×10^6, but after three months rose to 1.1×10^7 daltons.

Polyelectrolytes.

Many examples can be found in the literature about the interactions of poly-electrolytes: for instance, poly-(vinylbenzyltrimethylammonium chloride) reacts with sodium poly-(styrenesulphonate); methylene blue reacts with heparin.

Chitosan reacts with negatively charged polyelectrolytes like heparin, alginic acid and sodium dextran sulfate. Glycolchitosan and alginic acid have been used to produce anisotropic gels by Thiele & Langmaack (1957). Kikuchi (1973) and Shinoda & Nakashima (1975) have reported on the polyelectrolyte reaction products of chitosan with heparin, obtained by mixing a dilute aqueous solution of heparin with glycolchitosan or with dilute chitosan acetate solution. The products are water insoluble hydrous precipitates or colloidal precipitates, which originate by strong electrostatic attraction between oppositely charged macromolecules which react ionically, giving off their associated counter ions. The polyelectrolyte complex of chitosan with heparin is soluble by heating in water + hydrochloric acid + methanol, or in water + potassium bromide + acetone, or in formic acid alone. Esterification of such polyelectrolyte with various organic acids produced derivatives soluble in formic acid and in dimethylsulfoxide. From the sulfur content of the poly-electrolyte product, the mole ratios of chitosan to heparin repeating units were in the range 6:1 to 3:1 for several different samples.

Kikuchi & Fukuda (1974) determined the mole ratios of nitrogen to sulfur in the polyelectrolyte product obtained from chitosan and sodium dextran sulfate; these were in the range 1.6:1 to 1.3:1 for various different samples. No solvent was found for such a product which is obtained as a fine colloidal precipitate whose composition is not easily reproducible. Its infrared spectrum differs substantially from the spectra of the polyelectrolytes from which it originates.

Sulfuric esters from cartilage extracts have been precipitated with chitosan salts at pH 5 ÷ 7, after prolonged action of proteinase at 37°. Other substances remaining in solution can be easily separated. Strong acids, alkalis and inorganic salts are suitable for recovery of sulfuric esters from the chitosan combination as they are solubilized while chitosan remains as a solid (Sooda & Furuhasi, 1962).

Sodium cellulose acetate sulfate and ammonium cellulose sulfate were treated by Hata & Sakai (1975) with chitosan to give polyelectrolyte complexes forming transparent membranes.

Evans & Kent (1962) demonstrated epithelial and connective tissue mucins by using fluorescein-labeled chitosan. Labeling some amino groups of chitosan with fluorescein, while leaving some free to react with acidic polysaccharides is the basis of the staining. The complete inhibition of fluorescein-labeled

chitosan staining by methylation and the failure of the fluorescein-labeled
chitosan to stain after neuraminidase digestion of bovine salivary glands
sections supported this view.

Chitosan has been shown to precipitate a number of compounds, such as hyaluronic acid, chondroitin sulfate, heparin, bovine submaxillary gland mucin and
nucleic acids from solutions (Evans & Kent, 1962). The interactions between
glycolchitosan and chondroitin sulfate or hyaluronic acid were studied by
turbidimetric, potentiometric and metachromasy measurements by Shinoda &
Nakajima (1975) who indicated a form of pyranose-pairing aggregate structure
for the complexes. Chitosan agglutinates a variety of cells in suspension and
alters the staining characteristics of acidic polysaccharides in tissue
sections. Pre-treatment of sections with chitosan followed by exposure to
acid dyes such as eosin or Congo Red results in intense staining of the acidic
polysaccharides by the acid dyes.

Chitosan blocks the staining of acidic polysaccharides by basic dyes such as
alcian blue and methylene blue. Chitosan agglutinates all mammalian cells
tested by Evans & Kent (1962) as shown in Tab. 3.12. There was some variation
among cell types with regard to the limiting dilution of chitosan for
agglutination: agglutination was seen below 1 µg×ml^{-1}.

Table 3.12. Agglutination of mammalian cells by *Aspergillus* polysaccharide and
chitosan. Evans & Kent, 1962.

Cell type	Lowest concentration for agglutination, µg×ml^{-1}	
	Aspergillus	chitosan
human red blood cells A−	10	1
human red blood cells B+	10	1
human red blood cells O+	10	1
human sperm	10	10
mouse liver	100	10
mouse kidney	100	10
mouse spleen	100	10
mouse sarcoma 180	100	100
mouse adenocarcinoma 755	100	10
mouse leukemia 1210	100	10
monkey liver	100	100
monkey kidney	100	100

Precipitation tests with chitosan were positive with all acidic polysaccharides
tested, as well as with both nucleic acids and certain proteins. Practically
all cells were agglutinated since any cell would be expected to have one or
more of these ingredients on its surface. It is also possible that various

lipids and fatty acids would react with chitosan although these were not
tested. All the compounds that precipitate to some extent with chitosan
contain acidic groups. Only the non-polar dextran and glycogen fail to
precipitate. It is significant that introducing sulfate groups into dextran
causes it to react with chitosan.

Chitosan alters the reactivity of acidic dyes with acidic polysaccharides.
Normally these two classes of anionic substances do not react appreciably with
each other and hence there is no staining of the polysaccharide. If, however,
tissue sections are pre-treated with chitosan, areas known to contain acidic
polysaccharides will stain intensely with acidic dyes like Congo Red, because
chitosan is sandwiched between the two anionic compounds to provide an acid-
base linkage.

CHELATION OF METAL IONS.

An early report that mentions the interactions of metal ions with chitin or
chitosan, without, however, describing them in terms of complexation, is that
by Sadov (1941) who used copper sulfate together with chitosan for textile
finishing. Manskaya, Drozdova & Emel'yanova (1956) wrote that chitin from
mushrooms collects more uranyl ions than crayfish chitin from aqueous
solutions, in the context of the study of metal distribution in geological
samples. Andreev, Plisko & Rogozina (1962) and Glowacka & Popowicz (1965)
mentioned the possibility of collecting uranium on chitin and derivatives
because of the ion-exchange groups of chitin. These authors remarked that
chitin was more effective than cellulose or certain cellulose derivatives,
however they failed to see the complexing ability of chitin and chitosan.

Some information about the interactions of transition metal ions with the
monosaccharide glucosamine, 2-amino-2-deoxy-D-glucopyranoside, is available
since Tamura & Miyazaki (1965) determined the stability and equilibrium
constant of complexes of copper, lead, zinc, cobalt, cadmium nickel and
manganese ions. Complexes having 1:1 and 2:1 molar ratios of glucosamine to
metal ion were found in solution. According to the stability and hydrolysis
constants calculated for these complexes, the copper complexes were the most
stable, but the most susceptible to hydrolysis. Dimerization of the complexes
occurred only with copper complex. A possibility of the presence of a hydroxo
complex in the complex formation between the glucosamine and the metal ions
was ascertained by analysis of the spectra at various pH values. A complex of
the formula |glucosamine-Cu-OH|Cl, very hygroscopic, was isolated after mixing
an áqueous solution of glucosamine with freshly precipitated cupric hydroxide,
by Tamura & Miyazaki (1965). Glucosamine and nickel 2:1 adducts have been
isolated from metahnol solutions by Sahu & Mitra (1971) while ferrous and
cobaltous ions gave evidence of formation of adducts through spectrophotometric studies.

Glucosamine in relation to iron was studied by Ellenbogen & Highley (1961) as
a compound that enhances iron absorption in rats. It is also known that Hale
and Alcian Blue reactions used in the histochemical determination of chitin
are based on reagents containing iron and copper respectively: Runham (1961)
observed that these reactions are blocked by methylation or acetylation,
"presumably because -OH groups were unavailable" thus failing to mention the
role of amino groups. Ikeda & Hamaguchi (1973) studied the interactions of

manganese, cobalt and nickel with N-acetylchitooligosaccharides and lysozyme.

Chelating ability of chitin and chitosan.

Chitosan was first described in terms of a natural chelating polymer by Muzzarelli (1968) and by Muzzarelli & Tubertini (1969). A book has been published recently by Muzzarelli (1973) on natural chelating polymers including substituted celluloses, alginic acid, chitin, chitosan and other polymers of biological interest. Further papers have been published by Muzzarelli & Rocchetti (1973 and 1974), Muzzarelli, Isolati & Ferrero (1974), Muzzarelli, Nicoletti & Rocchetti (1974) and Muzzarelli (1976). The contributions of several other authors are also recalled below.

When chitin and chitosan powders are brought into contact with alkali and alkali-earth metal ions, no swelling, contraction, shrinking or color appearance can be observed. It has been demonstrated that large amounts of salts of lithium, sodium, potassium, cesium, thallium, ammonium, magnesium, calcium and barium do not produce appreciable alterations on chitosan, are not collected to any extent and do not prevent collection of transition metal ions when simultaneously present. The same can be said for rubidium and strontium that were tested at lower concentrations.

Batch measurments of the collection percentages of transtition metal ions on polymers are preliminary information easily accessible and quite useful for predicting the behavior of the metal ions towards the polymer in column chromatography.

The collection percentages, however, do not say enough about the strength of the interaction: for example, certain substituted celluloses show good collection percentages for a few metal ions, but they release the collected metal ions upon washing. This is a point of difference between substituted celluloses and chitosan, in favor of the latter.

Muzzarelli & Rocchetti (1974) have studied samples of diethylaminohydroxy-propylcellulose having various nitrogen contents. The general trend for these celluloses is that the higher the nitrogen content, the higher the degree of collection: the good collection efficiency of diethylaminohydroxypropyl-cellulose was expected on the basis of its nitrogen content, which is higher than in all the other modified celluloses so far produced. These celluloses become violet upon collection of copper; however the presence of large amounts of other salts, and 0.1 M ammonium sulfate prevent the collection of copper.

The nitrogen electrons present in the amino and substituted amino groups can establish dative bonds with transition metal ions, especially in the case of chitosan, where free amino groups are particularly abundant.

These polymers are bases and therefore they can also act to form salts of transition metal anions. In certain cases it is not excluded that several types of interactions may occur simultaneously. Muzzarelli (1973) speculated that the chitosan combines with metals through ion-exchange, sorption and chelation. The data presented by Yoshinari & Subramanian (1976) confirm that all the three processes are important in varying degrees for each individual metal ion. For calcium, the ion-exchange is the dominant process, whereas for

the other metals considered, adsorption with some chelation (indicated by
small shifts in infrared spectra) may be important.

The interactions of the first row transition metal ions with chitin and chitosan are accompanied by color appearance in many instances, namely red for
titanium, orange with metavanadate, green for trivalent chromium and orange
for hexavalent chromium, yellowish-brown with divalent iron, yellowish-green
with trivalent iron, pink with cobalt, green with nickel and blue with copper.
These colors are more intense with chitosan than with chitin. In certain cases
as with copper, the color formation is quite detectable even at very low
concentrations.

The sequence of the curves of the collection rates for the first row transition metal ions is in Fig. 3.14; it can be seen that it is the same as the
Irving and Williams series, and there is a relationship between this order and
the second ionization potentials. Moreover, complexing agents are generally
necessary to perform elution from chitin and chitosan chromatographic columns.
This experimental evidence is in favor of the chelating ability of chitin and
chitosan, whose amino groups play a major role, as confirmed also by measurements carried out on dissolved chitosan.

The curves of the collection rates depend on several factors, the most
important of which are the polymer grain size, the temperature, speed and mode

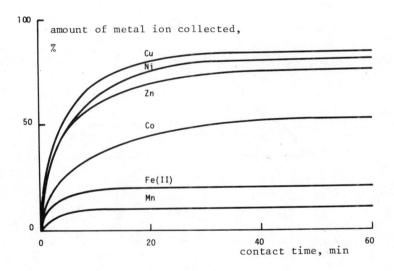

Fig. 3.14. Rates of uptake of manganese, ferrous, cobalt,
zinc, nickel and copper ions on 200 mg of chitosan, when
taken separately in 100 ml of 0.4 mM solutions (0.1 M
potassium chloride). Muzzarelli & Tubertini, 1970.

of stirring, the presence of other ions suitable for collection, oxidation
number of ion and pH of the solution.

The collection rates of zinc on chitosan flakes (2 mm) and powder (100 ÷ 200
mesh) have been measured under similar conditions. The powder collects zinc
ions better, more rapidly and to higher extent than flakes, due mostly to the
larger surface area immediately available for collection.

Temperature plays an important role in collection. There is no general rule
for predicting the influence of the temperature on metal ion collection, and
therefore one should expect variations in both directions. Better collection
at 4° than at room temperature has been observed.

In order to get reproducible results, when measuring distribution coefficients
stirring should be quite energetic, and in any case applied in the same way.
Normally a good magnetic stirrer is sufficient to provide constant and effective stirring. However, ultrasonic stirring can considerably alter the results
as it is a completely different stirring mode. From Fig. 3.15 one can see that
metal ion collection is effectively improved under ultrasonic stirring.

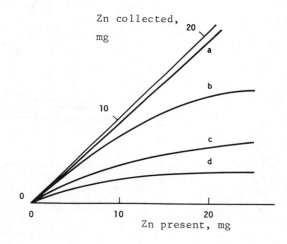

Fig. 3.15. Collection of zinc under various conditions, on
200 mg of chitosan at pH 6.6 and 20° (mg Zn collected vs
mg Zn present in 50 ml); a = precipitated chitosan under
ultrasonic stirring; b = precipitated chitosan under mechanical stirring; c = chitosan powder under ultrasonic
stirring; d = chitosan powder under mechanical stirring.
Muzzarelli, 1971.

The presence of Group I A and II A metal ions, and of ammonium and monovalent
thallium ions have no appreciable effect on the collection of transition metal

ions. The complexes formed between chitin and manganese, ferrous and magnesium are less stable than those of zinc, nickel, cobalt, copper, lead and other transition metal ions. The release of calcium from chitin was more extensive than the release of manganese, ferrous and magnesium ions; after 20 hr, most of the calcium initially adsorbed was released back into the solution (Yoshinari & Subramanian, 1976).

Large amounts of ammonium sulfate allow transition metals present in solution at the trace level to be collected on chitosan columns. For instance, 100 ml of 0.44 M ammonium sulfate solution containing nanograms of radioactive zinc was passed through a 7 × 1 cm column of chitosan powder (100 ÷ 200 mesh). The radioactive profile of the column showed that most of the radioactivity is fixed in the upper part of the column, and that zinc is completely collected. This is a very satisfactory trial of the effectiveness of chitosan in separating trace transition metal ions from very high ionic strength solutions. An even sharper band at the top of the column was observed for copper.

When two or more transition metal ions are present together in the solution under examination, with a quantity of polymer insufficient for the complete collection of both, the cation which forms the most stable complex with the polymer will be collected leaving most of the other cation in solution. Some variations are introduced into these curves by the simultaneous presence of other cations. In any case these variations are less than expected on the basis of a stoichiometric capacity. The rates of nickel and cadmium uptake on chitosan in single and mixed metal solutions were compared and it was found that cadmium suppressed the uptake of nickel but the reverse was not observed. The adsorption of metals by chitin and chitosan from mixed metal ions solution does not follow a stoichiometric path, suggesting that there may be more than one active site for the different metals on the surfaces of these polymers (Muzzarelli, 1973 and Yoshinari & Subramanian, 1976).

When a cation is slightly collected by chitosan, like ferrous ion, and it is in a solution of nickel, which is well collected by chitosan, nickel depresses most appreciably the collection of iron.

Ferrous and ferric ions taken separately, have different behaviors. This indication, as those above, is consistent with chromatographic results as ferrous ions can be easily eluted from chitosan columns, while ferric ions must be reduced to make the elution feasible.

The nature of the anions present in the solution carrying the transition metal ions to be collected on chitosan powder, is very important: chloride can depress the collection extent, while sulfate can enhance it, due to the modifications of chitosan introduced by the presence of anions. Acetate alters the surface of the grains because chitosan acetate is soluble. Thiocyanate depresses collection on chitosan because it is a complexing agent.

Enhanced performances.

While chitosan has been used as such for the early applications, it was remarked by Muzzarelli & Rocchetti (1974) and by Muzzarelli (1974) that chitosan powder treated with copper and cadmium ions shows more numerous and sharper x-ray diffraction bands than untreated chitosan. A comparison of the x-ray patterns for chitosan treated with chloride and sulfate gives the impression that the sulfate anion is partially responsible for the increased degree of crystallinity of the treated polymer. On the other hand, it is known that chitosan forms an insoluble sulfate.

Chitosan has seldom been used in sulfuric acid or in sulfate media. Chitosan bonded with silver and zinc ions was used to collect traces of cyanide from 0.44 M ammonium sulfate solutions (Muzzarelli & Spalla, 1972). At low acidity, (<0.1 M sulfuric acid) chitosan is not dissolved or degraded by sulfuric acid. In accordance with sulfate-induced structure modifications in chitosan, chitosan previously conditioned in sulfate media shows enhanced capacity for transition metal ions.

For conditioning, chitosan (5 g) was introduced into a 0.1 M solution (400 ml) of both sulfuric acid and ammonium sulfate. After 4 hr stirring, the polymer was washed with water until neutral. For some measurements, the polymer was similarly conditioned with ammonium sulfate alone.

The sulfate ions form electrostatic bonds with the protonated amino groups of the polymer: the general reaction mechanisms are those of the protonation of weak bases. The structure parameters of chitosan are not known, but, on the basis of those for the parent polymer chitin (see Chapter 2), the minimum distance between two nitrogen atoms is at least 0.8 nm in the bc plane of the orthorhombic cell. Therefore, assuming that the N-H-O-S-O-H-N group has a 109° angle with interatomic distances N-H 0.1034, O-H 0.1029 and S-O 0.1432 nm, the two nitrogen atoms would lie about 0.55 nm apart. Thus, it seems that an H_2SO_4 molecule cannot bridge two amino groups without altering the form of the macromolecule.

Chitosan conditioned in sulfuric acid would probably carry $-NH_3^{+-}SO_4H$ groups, while chitosan conditioned in a mixture of sulfuric acid and ammonium sulfate, or in ammonium sulfate alone, would probably carry $-NH_3^{+-}SO_4NH_4$ groups.

The metal cations can be regarded as polybasic Lewis acids capable of reaction with several basic entities (including nitrogen and oxygen), the number of which is related to the co-ordination number of the metal cation. The hydrated copper ion is a Lewis acid which competes with sulfate and bisulfate acids for the electron pairs of nitrogen atoms in chitosan:

$$-NH_3^{+-}SO_4NH_4 + Cu^{2+} = -NH_2 \rightarrow Cu^{2+} + NH_4^{+-}HSO_4$$

The complexation of the cations is also due to a contribution of the hydroxyl groups.

The enhancement of the chitosan capacity in sulfate solutions should therefore be regarded as a phenomenon originated by a novel crystallinity induced in the polymer by sulfate, and favored by the availability of sulfate.

Table 3.13 shows that at pH 3.0 in the absence of ammonium sulfate, sulfuric acid conditioned chitosan is less efficient than chitosan as a collecting agent; 0.1 M ammonium sulfate at pH 3.0 enhances the collection of manganese on free chitosan, but sulfate conditioned chitosan does not collect manganese. The chitosan functions more efficiently if conditioned with ammonium sulfate. For instance, while chitosan collects 71 % of the nickel added, the sulfuric acid conditioned chitosan collects only 30 % , but the chitosan conditioned

Table 3.13. Collection of metal ions on 200 mg of chitosan, expressed as fraction (%) extracted from 50 ml of 0.5 mM soln. Muzzarelli & Rocchetti, 1974.

pH	Conditioning	$(NH_4)_2SO_4$	Time, hr	Cr^3	Mn^2	Fe^3	Ni^2	Cu^2	Zn^2	Hg^2
3.0	none	absent	1	53	0	72	70	95	7	100
			12	93	15	74	71	100	22	100
	0.1 M H_2SO_4	absent	1	28	15	18	28	68	9	89
			12	36	10	13	30	79	15	89
	none	0.1 M	1	58	22	94	83	100	57	100
			12	95	23	100	88	100	63	100
	0.1 M H_2SO_4 + 0.1 M $(NH_4)_2SO_4$	0.1 M	1	36	0	25	25	35	23	91
			12	84	0	20	78	76	58	86
	0.1 M $(NH_4)_2SO_4$	0.1 M	1	70	7	96	90	96	63	90
			12	95	0	95	90	100	73	90
5.0	none	absent	1	85	0	–	76	100	30	100
			12	86	13	–	85	100	58	100
	0.1 M H_2SO_4	absent	1	32	10	–	37	81	12	95
			12	76	0	–	55	88	30	93
	none	0.1 M	1	74	12	–	80	100	70	100
			12	93	12	–	80	100	88	100
	0.1 M H_2SO_4 + 0.1 M $(NH_4)_2SO_4$	0.1 M	1	78	10	–	78	100	61	90
			12	90	10	–	86	100	71	90
	0.1 M $(NH_4)_2SO_4$	0.1 M	1	95	5	–	95	100	90	90
			12	90	3	–	95	100	96	90

with both sulfuric acid and ammonium sulfate collects 78 %. Collection is slightly more efficient at pH 5.0. Sulfuric acid at pH 1.0 prevents collection of the transition metal ions on chitosan and conditioned chitosan.

As far as metal oxyanions like metavanadate, molybdate or chromate are concerned, the presence of sulfate depresses somewhat the degree of collection especially at pH 5. The method worked out for the determination of vanadium in sea water (see page 190) is based on these results and takes advantage of the maximum yield at pH 3.0 for collection, and minimum yield at pH 1.0 for elution with sulfuric acid.

The better values for chitosan conditioned with sulfuric acid indicate that the polymer is this case behaves as an anion-exchanger:

$$-NH_2 + H^+ + HSO_4^- = -NH_3^+ SO_4H \quad (conditioning)$$

$$-NH_3^+ SO_4H + oxyanion^{n-} = -NH_3^+ oxyanion^{n-} + SO_4H^- \quad (anion\text{-}exchange)$$

Whenever the sulfate concentration is high, the amount of the oxyanion collected is lowered, in accordance with the second equilibrium.

Data on chitosan salts are reported in Tab. 3.4. It can be seen that separation of metal ions can be performed by taking advantage of the insolubility of certain metal oxysalts of chitosan, when chitosan acetate solutions are used to perform selective and fast precipitations.

In order to make the precipitation as clean as possible and to use it as a way of performing separations, it is convenient to introduce in the mixture a complexing agent like thiocyanate, to prevent other ions present in solution from being carried down for physical reasons. This complexing agent was preferred because there are many data on chromatography in thiocyanate solutions, it can be easily handled in acidic solutions and thiocyanate solutions are suitable for polarographic and spectrographic determinations.

The formation of chitosan polymolybdate is irreversible in the presence of thiocyanate whose concentration in any case should be kept below 2 M to avoid any risk of precipitation of the chitosan thiocyanate. The precipitation of chitosan polymolybdate is quantitative even in the presence of large amounts of thiocyanate. Data on separations of molybdate from other ions are in Tab. 3.14. Thiocyanate was present in excess; a ratio around 1:600 existed between metal ion and molybdenum and a large amount of chitosan was present to ensure the complete precipitation of molybdate.

Alkali-earth metal ions and aluminum can be separated from molybdate even in the absence of thiocyanate. A special mention deserves the phosphate ion which under these conditions forms insoluble chitosan polyphosphomolybdate.

Table 3.14. Separation of molybdenum from phosphate and from metal cations. Chitosan was used to precipitate 275 mg of molybdenum as chitosan polymolybdate in 40 ml of a 0.125 M thiocyanate solution at pH 3.8; washing with 10 ml of 0.125 M ammonium thiocyanate. Muzzarelli & Rocchetti, 1973.

Ion	Amount, μg			Amount, left in solution, %
	added	found in solution	found in washing	
phosphate	1000	0	0	0
Cu^2	318	292	0	92
Pb^2	1036	996	0	96
Ni^2	294	228	6	80
Cd^2	560	518	0	94
Ti^3	240	143	50	80
Co^2	294	268	26	100
Fe^3	225	156	17	77
Hg^2	1003	630	70	68
Zn^2	326	325	0	100
Cr^3	260	52	10	24
Mg^2	121	121	0	100
Ca^2	200	200	0	100
Al^3	125	120	5	100
UO_2^2	1000	870	130	87

Table 3.15. Mercury uptake by chelating polymers, under similar conditions.

Polymer	Initial concn., mgHg×ml^{-1}	Collection	
		mgHg×g^{-1}polymer	%
cellulose, starch	4	12	12
dialdehyde starch + methylenedianiline	40	1050	52
same, reduced	40	830	83
dialdehyde starch + p-phenylenediamine	40	999	83
same, reduced	40	1033	78
diethylaminoethylcellulose	40	290	29
p-aminobenzylcellulose	40	90	9
aminoethylcellulose	40	55	5
chitin	40	175	17
chitosan	4	100	100
chitosan	40	1425	71
poly(p-aminostyrene)	4	100	100
poly(p-aminostyrene)	40	1450	72
wool fiber	40	510	25
diethylaminohydroxypropylcellulose	0.1	25	98
bark, redwood	40	250	10
bark, black oak	40	400	3.5
sulphuric acid lignin	40	150	0.6
Milorganite	40	460	45
cutin	0.088	11	50
polygalacturonic acid	0.088	9	42
alginic acid	0.088	0	0
N-(2-aminoethyl)aminodeoxycellulose	3.7	540	12

Tab. 3.15 lists data from various sources, concerning natural substances having chelating ability. Included are polymers obtained by reacting dialdehyde starch with p,p'-diaminodiphenylmethane or p,p'-phenylenediamine. In the course of the research on these polymers, Masri & Friedman (1972) found out that chitosan is the most effective polymer for mercury in terms of capacity and efficacy. This is recalled just as an example of research intended to prepare artificial chelating polymers, whose outcome has been the appreciation of the exceptional chelating ability of chitosan, a natural polymer.

The same development of research appears from a paper by Koshijima, Tanaka, Muraki, Yamada & Yaku (1973) who synthesized chelating polymers from dialdehyde cellulose with thiosemicarbazide, isonicotinic acid hydrazide and hydroxylamine hydrochloride, which were reportedly as excellent as chitosan that was studied simultaneously. Their results, some of which are presented in Fig. 3.16, show that the presence of magnesium does not depress the collection of copper ions on chitosan. Chitosan is therefore a first-rank chelating polymer which can be favorably compared to Dowex A-1, the only artificial chelating resin so far commercially available.

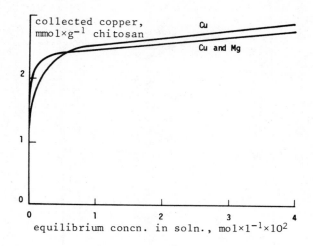

Fig. 3.16. Equilibrium curves for copper ions and chitosan
at 30°, alone and in the presence of magnesium ions. The
collection of copper is not depressed by the presence of
large concentrations of magnesium. Koshijima, Tanaka,
Muraki, Yamada & Yaku, 1973.

Tab. 3.16 shows the uptake of various metal cations by chitosan and by
p-aminostyrene; the results clearly show that chitosan has higher binding
capacity (above 1 $mmol \times g^{-1}$) with all the metals tested (Masri, Reuter &
Friedman, 1974). It was also observed that chitosan forms a liquid gel with
concentrated solutions of ferric chloride. Although chitosan and p-amino-
styrene are both polyamine polymers, they have not equal effectiveness. The
greater effectiveness of chitosan for metal binding may be due in part to the
greater basicity of the aliphatic primary amino group compared to the aromatic
amino group. Nevertheless, p-aminostyrene bound more than 1 $mmol \times g^{-1}$ in the
cases of mercury, copper, silver, gold and platinum. For these data, 1 g of
polymer was equilibrated for 24 hr with 20 mmol of ion dissolved in 50 ÷ 100
ml, then filtered, rinsed, dried and weighed.

While Masri & Friedman (1972) used high metal ion concentrations (20 mmol in
100 ml against 1 g chitosan), Koshijima, Tanaka, Muraki, Yamada & Yaku (1973)
used 50 ml of solutions at the concentrations of 50 ÷ 500 ppm against 100 mg
of chitosan. The first team used extreme conditions to reach saturation values
like those in Tab. 3.16, while the second team used dilute solutions to study
the metal ion collection dependence on the concentration, without reaching
saturation conditions. Some of the latter results are in Tab. 3.17.

It should be also recalled that the pH indicated in determinations involving

Table 3.16. Collection of metal cations on chitosan and *p*-aminostyrene, one g stirred 24 hr with 20 mmol of metal chloride or nitrate in 50-100 ml solution. Masri, Reuter & Friedman, 1974.

Cation	Collection, mmol cation$\times$g^{-1}polymer	
	chitosan	*p*-aminostyrene
Hg^{2+}	5.60	5.70
Cd^{2+}	2.78	0.10
Pb^{2+}	3.97	0.17
Zn^{2+}	3.70	0.52
Co^{2+}	2.47	0.07
Ni^{2+}	3.15	0.02
Cr^{3+}	0.46	0.03
Cu^{2+}	3.12	1.31
Fe^{2+}	1.18	0.05
Mn^{2+}	1.44	–
Ag^{1+}	3.26	1.98
Au^{3+}	5.84	5.75
Pt^{4+}	4.52	3.82
Pd^{2+}	6.28	0.87

chitosan, are normally initial pH values which increase after the introduction of chitosan into the solution. Relatively large amounts of sulfuric acid can be neutralized by chitosan which gains weight because of the sulfate salt formation. The hydrogen ion competes with metal cations for binding sites of chitosan.

The important differences between metal cation uptake on chitosan and on *p*-aminostyrene underline the contribution to the collection efficiency of the regular distribution of amino groups on the chain and of the presence of hydroxyl groups; the different nature of the polymers are also important in this respect.

The chelating ability of chitin has also implications in ecology: the concentrations of the following metals, cadmium, copper, cobalt, iron, magnesium, manganese, nickel, lead, strontium and zinc in surface zooplancton samples have been compared by Martin (1970) with the values determined on samples collected at 100 m or more depth in the Ocean. Average values for copper, iron, manganese, nickel, lead, strontium and zinc were higher in the samples collected at depth because food-dependent molting rates were lower; thus more time was available for metal collection to take place. Molted copepod exoskeletons may be important in biogeochemical cycles. Copepods are the most abundant multicellular animals in the world, and there are eleven cast exoskeletons for every adult; these slow-sinking chitinous exuvia are capable of taking up transition metal elements.

Table 3.17. Collection of metal cations on chitosan, after shaking 100 mg of powder with 50 ml of solution at the concentration of 500 ppm for 74 hr at 30° compared to collection on cellulose derivatives. Koshijima, Tanaka, Muraki, Yamada & Yaku, 1973.

Cation	Thiosemicarbazide cellulose, $mmol \times g^{-1}$	Isoniazid cellulose, $mmol \times g^{-1}$	Dioxime cellulose, $mmol \times g^{-1}$	Chitosan, $mmol \times g^{-1}$
Ag^{1+}	2.38	0.45	0.60	0.88
Hg^{2+}	3.12	0.89	0.51	2.55
Cu^{2+}	0.61	1.97	0.51	2.29
Cd^{2+}	0.20	0.13	0.014	0.78
Cr^{3+}	0.0011	0.0083	0.0095	0.0026
Mn^{2+}	0.015	0.025	0.010	0.191
Fe^{2+}	0.069	0.313	0.381	1.28
Co^{2+}	0.051	0.146	0.050	0.033
Ni^{2+}	0.08	0.31	0.04	1.18
Pb^{2+}	0.57	0.34	0.17	0.088

ANALYTICAL DETERMINATIONS.

Spectrophotometric determinations.

Tracey (1955) has published a review of the current analytical methods developed up to 1953, which include the Van Wisselingh chitosan test modified by Campbell; the chitosan sulfate test modified by Roelofsen and Hoette; the colorimetric determination of glucosamine following hydrolysis in sulfuric or hydrochloric acid according to Elson and Morgan, modified by Blix and Immers, and Vasseur, and the alkaline distillation method for glucosamine. N-Acetyl-glucosamine determination according to Morgan and Elson is also described in detail, while x-ray diffraction and enzymic techniques are of course only mentioned. The review by Tracey (1955) contains valuable details on analytical determinations according to early techniques.

A classical and widely used method employed in the demonstration of chitin, depends on the detection of glucosamine set free on alkali hydrolysis of chitosan, by the iodine + sulfuric acid test (Campbell, 1929). Modifications of the chitosan test have been made in order to suit the nature of the sample and the quantity of chitin in it. It has been widely held that the results of the chitin test are sometimes unreliable.

When chitosan is treated with 0.2 % iodine solution in potassium iodide, followed by 1 % sulfuric acid, a bright red-violet color develops; Richards & Cutkomp (1946) reported that certain chitin samples, when subjected to chitosan test, develop a green color. The same can be said for chitin from *Scrupocellaria berthelotti*, which, however, after prolonged treatment with potassium hydroxide turned the green color into the red-violet expected color. The chitin of the *Palamneus swammerdami* gives orange color in the chitosan test but after treatment with potassium hydroxide the typical violet color can be obtained. The green and orange colors seem to be related to the different sugars present in those chitins (see page 89). Sundara Rajulu & Gowri (1967) have pointed out that the chitin of the cenosteum of Millipora contains galactose and mannose in addition to glucosamine.

Chondroitin sulfate of vertebrate connective tissues also reacts positively to the chitosan test. Therefore, the chitosan test which has been used for a long time, is no longer reliable today. A number of sources of chitinases have been reported by Tracey (1955) and by Richards (1951): the use of chitinases should enable one to explain that the fluctuations in the nitrogen values recorded for chitin from different sources in the range 6.45 ÷ 1.16 % while the theoretical value is 6.9 %.

Prakasam & Azariah (1975) have recently modified the chitosan test: saturated potassium iodide at 160° is used for 45 min to deacetylate chitin and after washing the chitosan in running water, the Lugol's iodine solution (iodine + potassium iodide + water, 1:2:300, w:w:v) is applied in a pH range of 1.4 ÷ 4.0, thus avoiding the use of dilute sulfuric acid. The color observed in this range varies from violet to red-violet and results are more reproducible and reliable. Ravindranath & Ravindranath (1975) also underline the importance of the neutralization step and applied the chitosan test to the delicate chitinous structures: time and temperature for deacetylation in this case may differ from material to material. The incomplete neutralization of the chitosan may give rise to a yellow-green color which turns red-purple after repeated washing.

A method based on the highly specific properties of chitinases consists in the hydrolysis of a chitinous structure treated with 1 N sodium hydroxide at 100° to remove protein and stained with Congo Red. Bussers & Jeuniaux (1970 and 1974) used it to search chitin in the metaplasmatic productions (cysts, loricas and shells) of 22 species of Ciliata, 14 of which were found to be able to produce chitinous structures. This method has not yet been adapted to the study of histological sections.

The method of Wieckowska (1968) based on *p*-aminobenzaldehyde has already been mentioned on page 132 and it is fully described in a paper by Kumar & Hansen (1972). The determination of chitin is also feasible by gas chromatography through the determination of acetic acid (Holan, Beran, Prochazkova & Baldrian 1971) or the trimethylsilyl derivatives (Sweeley, Bentley, Makita & Wells, 1963).

Ride & Drysdale (1971) proposed a colorimetric estimation of N-acetylglucosamine from chitin after enzymic release. Tsuji, Kinoshita & Hoshino (1969) have reported a highly sensitive assay for glucosamine, based on deamination with nitrous acid to yield the aldehyde 2,5-anhydromannose which can be determined by the method of Sawicki, Hauser, Stanley & Elbert (1961). Foster, Martlew & Stacey (1953) showed that the rate of deamination of chitosan with nitrous acid is similar to that of β-methylglucosaminide. Since deamination results in both production of aldehyde and cleavage of the chitosan, it appeared that chitosan can be assayed by a method based on the above information. Chitosan, however, contains residual acetyl groups which protect some of the amino groups and thus only some of the glucosamine residues from the original chitin are converted to aldehyde and can be measured by the proposed method. Nevertheless, if the number of aldehyde groups produced in replicate analyses is reasonably constant, alkaline degradation could still form the basis for an assay, since this method is much more sensitive than those previously used. Ride & Drysdale (1972) applied this method for the estimation of filamentous fungi in plant tissues and found it shorter, more convenient and more sensitive than the previous ones based on the enzymic glucosamine release. The only preliminary necessary is the determination of a glucosamine value for the fungus, preferably grown in vivo. Alternatively, fungus grown in vitro on an appropriate medium may be used if account is taken of the possible variation of glucosamine content with age and cultural conditions.

The method was applied to measure mycelial growth in animal tissues by Lehmann, White & Ride (1974) and by Lehmann & White (1975). The assay is reported here in detail; it includes two parts, the first concerning the deacetylation of chitin to chitosan and the second describing the analytical determination of chitosan.

First part: chitin deacetylation. Thaw deep-frozen homogenate and place in graduate centrifuge tubes; spin at 1500 g for 5 min. Discard supernate and resuspend pellet in 3 % sodium lauryl sulfate (4 ml), heat at 100° for 15 min, cool, spin at 1500 g for 5 min; discard supernate. Resuspend pellet in water (10 ml) and spin again, discard supernate. Suspend pellet in potassium hydroxide solution (exactly 3 ml of a 120 % w:w solution) and heat 1 hr at 130°; cool, add exactly 8 ml of ice-cold ethanol (75 % v:v) and mix to give a single phase, keep in ice-water 15 min; add Celite and spin at 1500 g 5 min at 2°. Discard supernatant, resuspend in ice-cold ethanol (10 ml, 40 % v:v), spin at 1500 g for 5 min at 2°. Wash the pellet in ice-cold distilled water: it contains chitosan to be assayed in the second part of the method.

Second part: assay of chitosan. Make up centrifuge tubes to the 0.5 ml mark with distilled water: these tubes are the tests. Set up 2 tubes containing glucosamine solution (0.2 ml, 10 $\mu g \times ml^{-1}$): these are the glucosamine controls. Set up 2 tubes containing 0.2 ml water: these are the water controls. Add 5 % potassium hydrogen sulfate (0.5 ml) to each test and (0.2 ml) to each control. Add 5 % sodium nitrite (0.5 ml) to each test and (0.2 ml) to each control. Mix contents and leave tubes for 15 min, mix again at the end of 5 and 10 min during this period. Spin tests (1500 g, 2 min at 2°) and place controls in ice. Pour off supernate into tubes in ice; quickly pipette off 2 aliquots (0.6 ml) into 2 sample tubes. Place all tubes at room temperature. Add 12.5 % ammonium sulfamate (0.2 ml) and mix tubes vigorously 5 times over 5 min. Add 0.5 % solution of 3-methylbenzo-2-thiazolone hydrazone hydrochloride monohydrate (0.2 ml) and place covers on tubes. Heat in boiling water bath for exactly 3 min and cool to room temperature in tap water. Add 0.5 % ferric chloride solution (0.2 ml) and read at fixed time (20 ÷ 30 min) after the addition at 650 nm, by using micro-cuvettes.

Sensitive determinations of chitin are also requested to detect early stages of biodegradation of hydrocarbon fuels by fungi and for chemically differentiating a massive fungal contamination from a non biological one (Ernst, Emeric & Levine, 1975) and to measure fungal infiltration of cellulosic materials (Ernst, Emeric & Levine, 1976).

Colloid titrations.

The colloid titration is based on the reaction between positively charged
colloidal particles and negatively charged ones (Senju, 1969). The colloidal
particles in the solution are kept stable by their charges. If their charges
are neutralized, the colloid particles tend to associate and eventually
precipitate. Thus, when the oppositely charged ionic macromolecules are added
to such colloid solution, the neutralization reaction will proceed stoichio-
metrically. Therefore, if the ionic polymers whose chemical structures,
charges and molecular weights are known, are selected as titrants, the colloid
particles in the sample can be determined volumetrically as far as the
equivalent point is visible.

To titrate negative colloid solutions, an excess of 0.005 N glycolchitosan or
methylglycolchitosan solution (positive colloid) is added to the sample
solution to form a precipitate, and the excess is back titrated with 2.5×10^{-3}
N potassium polyvinylsulfate solution.

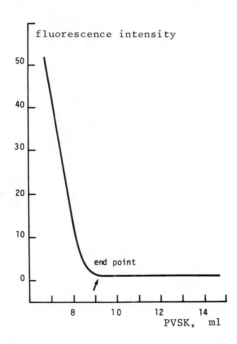

Fig. 3.17. Titration of 40 ml of 5.5×10^{-5} M aqueous methylglycolchitosan with 2.5×10^{-4} M aqueous potassium polyvinylsulfate standard solution. Kina, Tamura & Ishibashi, 1974.

Among the negative colloids that can be titrated with chitosan are clay, polyphosphate, plant mucilages, hyaluronic acid, heparin, tannins, humic acids,
lignin, cellulose phosphate or sulfate or acetylphtalate, carboxymethylcellulose, polyacrylic acid, polymethacrylic acid, polystyrene sulfuric acid and
many others.

Toluidine Blue was preferred as an indicator; Kina, Tamura & Ishibashi (1974)
have developed 8-anilino-1-naphthalenesulfonate as a fluorescent indicator in
colloid titrations. Potassium polyvinylsulfate and methylglycolchitosan have
been proposed as standard substances of the negatively and positively charged
colloids, respectively. The indicator is practically non fluorescent in water
but becomes strongly fluorescent when it is bound to cationic colloids such as
methylglycolchitosan (λ_{ex} 374 nm; λ_{em} 480 nm). The fluorescence of the
indicator-chitosan complex is diminished by the titration with polyvinylsulfate solution.

Fig. 3.17 shows the titration curve. Transition metals quench the fluorescence
and should be removed by dialysis or masked, if necessary.

Toei & Kohara (1976) have quantitatively titrated glycolchitosan and methylglycolchitosan by both the indicator and conductometric methods. They point
out that the amino groups in glycolchitosan are in the ammonium form below pH
5.5, while they gradually rearrange to amino groups above said pH value.

CHAPTER 4

CHITINASES AND
RELATED ENZYMES

CHITINASE AND CHITOBIASE.

Enzymatic hydrolysis of chitin to acetylglucosamine is performed by a system consisting of two hydrolases: chitinase (chitin glycanohydrolase, E.C. 3.2.1. 14) and chitobiase (chitobiose acetylaminodeoxyglucohydrolase, E.C. 3.2.1.29). Chitinases are widely distributed enzymes synthesized by bacteria, fungi and digestive glands of animals whose diet includes chitin.

Reviews have been made on this subject by Jeuniaux (1963, 1966 and 1971). The occurrence of chitinase in sugar maple and oaks is explained by Wargo (1975) in terms of protection against invasion by microorganisms.

In an attempt to separate the enzyme system taking part in the decomposition of glycolchitin, Ohtakara (1963) carried out the purification of chitinase from *Aspergillus niger* and determined both liquefying and saccharifying activities. Using ammonium sulfate fractionation and hydroxyapatite chromatography, the chitinase system of the mold was separated into four different enzyme fractions, which were required for the complete hydrolysis of glycolchitin. One of these fractions caused a rapid decrease of viscosity of glycolchitin solutions, another enzyme possessed N-acetyl-β-glucosaminidase activity upon N,N'-diacetylchitobiose and β-methyl-N-acetylglucosaminide: glycolchitin was decomposed to constituent amino sugar by successive action of the two enzymes (Fig. 4.1).

Ohtakara (1964) concluded that one of the enzymes rapidly cleaves the endo-β-glucosaminidic bonds in the polysaccharide chain, forming chitodextrin and oligosaccharides, while the other one produces monosaccharide.

During studies on the relationship of viscosity reduction of substrate solutions to enzyme concentrations, Ohtakara (1961 and 1962) reported that the black-koji mold enzyme causes a rapid reduction in the viscosity of substrate solutions, as shown in Fig. 4.2, where the viscosity reduction is plotted against reaction time. The time required to halve the viscosity increment (F in Fig. 4.2), is inversely proportional to enzyme concentration. It was confirmed that the activity is proportional to the amount of enzyme used. On the other hand, Elson-Morgan and Morgan-Elson tests on the reaction mixtures were negative, as no amino sugar is liberated. To examine the effect of substrate concentrations and different preparations of substrate on the activity, experiments were carried out in 0.2 % (final concentration) glycolchitin under the same conditions and gave curves similar to those in Fig. 4.2 and consequently the proportional relationship between the activity and the amount of enzyme used was ascertained.

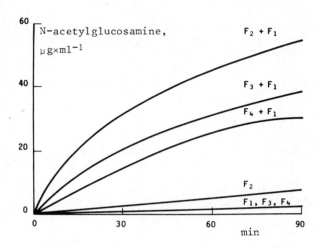

Fig. 4.1. Synergic action of different fractions of raw
chitinase upon glycolchitin at 40°. 1, 2, 3 and 4 refer
to the chromatographic fractions eluted from hydroxyapa-
tite concentrations of 10^{-4}, 5×10^{-3}, 5×10^{-2} and 0.1 M.
Ohtakara, 1963.

As far as chitosan is concerned, Ohtakara (1964) reported that glycolchitosan
and chitosan, tested under the same conditions as chitin and glycolchitin, are
not degraded by the chitinase system of *Aspergillus niger*.

CHITINASE EXTRACTION AND PURIFICATION.

Chitinases contained in the culture media of *Streptomyces* (Jeuniaux, 1963 and
1966; Reynolds, 1954) and *Actinomyces* (Tiunova, Pirieva & Epshtein, 1973) or
present in the *Periplaneta* (Bernier, Landureau, Grellet & Jolles, 1971) or
Capra serum (Lundblad, Hederstedt, Lind & Steby, 1974) have been purified.
Dandrifosse (1975) studied the purification of chitinase contained in pancreas
or gastric mucosa of the frog *Rana temporaria temporaria* L.; these dissected
parts were washed in a saline solution, dried, pooled and stored at -70°.
After thawing, the glands were homogenized; after centrifugation at 4°, the
supernatant was slowly acidified at pH 5. The precipitate obtained was eliminated and colloidal chitin was added to the supernatant in the ratio 2 ml of
chitin per 100 units of chitinase. After half an hour of slow magnetic
stirring at room temperature and centrifugation, the sediment was resuspended
in 25 ml of disodium orthophosphate + citric acid buffer and centrifuged
(10,000 g, 10 min, 4°). The precipitate was incubated in the same buffer at 37°
for 3 days. Every 8 hr of this period, the centrifugation was repeated and the
supernatants were kept at -70°. They contained all the chitinase activity

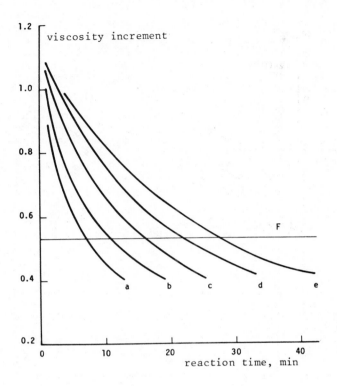

Fig. 4.2. Viscosity reduction of glycolchitin solutions
(0.1 % final concentration) at 30° and pH 3.6. a,b,c,d,e
refer to dilutions of the enzyme to 100, 150, 200, 300 and
400 respectively; F indicates the half-viscosity times.
Ohtakara, 1961.

that could be desorbed from chitin, and were dialyzed against aquacide at 4°
in order to obtain a couple of ml of solution. This was submitted to chromatography at 4° on a 100 ml Biogel P60 column equilibrated with 0.02 ammonium
sulfate solution. Chitinases were eluted with this solution to avoid the
denaturation of chitinases which could result if distilled water was utilized.
The chitinases appeared in the 6 ÷ 20 ml effluent interval.

The activity of the enzyme was measured following the Jeuniaux's method (1961
and 1966) with lobster serum diluted 10 times as chitobiase source. The unit
of chitinolytic activity was defined as the quantity of chitinases required to
catalyze the release of 11 μg of acetylglucosamine originating from native
chitin per hour at 37° and pH 5.2. The electrophoresis results were interpreted as indicating that the frog gastric mucosa and pancreas contain two iso-
enzymes of chitinase. This observation agrees with those of Berger & Reynolds

(1958), Skujins, Pukite & McLaren (1970) and Jeuniaux (1963) demonstrating the existence of two or three forms of this enzyme in the case of other enzymatic extracts.

When the most purified chitinase solutions were submitted to electrophoresis on alkaline urea gel slabs, two bands of glycoproteins were observed exactly located where the chitinase activity could be observed. This result agrees with previous observations on enzyme secretion by the gastric mucosa of Reptiles (Dandrifosse, 1969 and 1974).

Kimura (1965) described the purification of chitinase from digestive juice of *Helix peliomphala* and determined its general physico-chemical properties. He pointed out that the activity of the enzyme depends on the purification treatments given to the substrate, and showed in Fig. 4.3. that commercial chitin needs to be purified from proteins, lime and pigments for the best performances of chitinase. He defined the pH interval where chitinase is active, and reported a maximum activity at the temperature of 37° with the

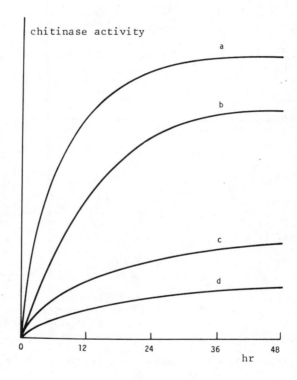

Fig. 4.3. Substrate purification and enzymatic activity of chitinase; a = purification; b = deliming; c = protein removal; d = unsettled. Kimura, 1965.

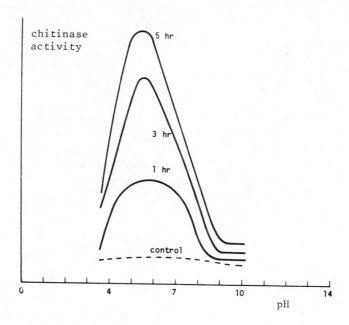

Fig. 4.4. Enzymatic activity of chitinase as a function of
pH. Kimura, 1965.

complete deactivation of chitinase at 60° (see Fig. 4.4).

The purification of chitinase from *Aspergillus niger* was carried out by
Ohtakara (1961) from its wheat-bran culture, using a chromatographic separation
on hydroxyapatite and the enzyme was concentrated 40-fold. The purified
chitinase was nearly free from cellulase and amylase and was powerfully active
in the reduction of viscosity of glycolchitin solutions. According to Koaze,
Kojima & Hara (1975), the enzymatic activity is maximum in 5 ÷ 7 daus. *Aspergillus* is grown on a medium of glucose + starch + corn steep liquor; the
culture filtrate is treated with ammonium sulfate to give a precipitate that
contains 1600 $U \times g^{-1}$.

Kimura (1966) investigated the specificity of chitinase for different substrates including cellulose, mannan, pectic acid, chitin, chondroitin sulfate,
hyaluronic acid and heparin. They found that chitinase from *Helix peliomphala*
does not react with those polysaccharides except chitin, in the pH interval
3.6 ÷ 10.1.

Kimura (1966) examined the effect of metal ions on the chitinase activity and
found that calcium, lead and mercury ions inhibit chitinase activity whilst
manganese, iron, tin and zinc (all as chlorides) enhance the chitinase acti-

vity. Potassium dihydrogen phosphate also depresses the activity, as shown in Figs. 4.5 and 4.6. The inhibition and activation of the enzyme was shown to be reversible by using chelating substances in the class of EDTA.

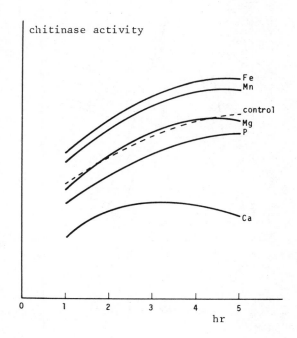

Fig. 4.5. The effect of metal ions at the concentration of 0.1 ng×ml^{-1} on the activity of purified chitinase. Kimura, 1966.

Fowl, particularly poultry, possess considerable recoverable chitinolytic activity in their digestive tract. Chickens in particular have been found to have significant quantities of chitinase activity in their entrails. Poultry processors and meat packing plants have a steady supply of large numbers of these birds. Trials with turkey, goose and duck stomach and gastric tract yielded less enzyme than with chicken.

An extraction method for separating and recovering chitinase activity from the entrails of fowl has been developed by Smirnoff (1975). The method basically comprises contacting the entrails with dilute acid medium until chitinolytic activity is extracted; adding colloidal chitin to the extract and collecting and washing the resulting insoluble deposit; allowing the hydrolysis of chitin in the deposit to proceed in further dilute acid medium and removing any insoluble residue; saturating the solution with a neutral salt and, finally, recovering the precipitated chitinase material.

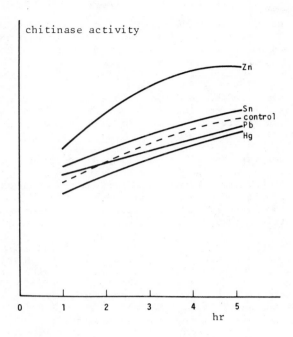

Fig. 4.6. The effect of metal ions at the concentration of
0.1 ng×ml^{-1} on the activity of purified chitinase. Kimura,
1966.

The entrails, particularly gizzards and intestines are submerged in dilute
acid medium to extract the enzyme. The medium may be entirely aqueous or a
mixture of water and ethanol. The dilute acid may have to be replenished
during the extraction to maintain an acidic pH. Preferably the medium is
buffered to a pH of about 5 to 5.5. Various suitable buffers are known with a
mixture of citric acid and dibasic sodium phosphate; an antibacterial agent is
usually incorporated to avoid any bacterial contamination.

After separation of the extracting medium from solids, colloidal chitin is
added to cause preferential adsorption of the enzyme. A colloidal form of
chitin is desired to give good dispersion and high surface area to effect
efficient adsorption without excessive amounts and times being required. The
colloidal form is also desired to enable hydrolysis to proceed to completion
in reasonable times. One suitable technique for preparing a colloidal form of
chitin is described on page 57. Still other techniques could be used.
Technical or crude chitin can be obtained as a by-product where chitin-containing natural foods are being processed, e.g. from fisheries processing
crustacea.

The chitin + chitinase deposit is washed with neutral or dilute acid solution to remove impurities. Preferably the same buffered solution as the extractant is used.

The salt added is normally ammonium sulfate, although some other neutral and soluble salt may be useful. This salting-out effect has been used in recovering other enzymes and high molecular weight proteins.

The steps of adsorption on colloidal chitin, removing unadsorbed material, hydrolyzing chitin, adding a neutral salt to the clear solution and recovering precipitated chitinase, can be repeated a number of times (until the recovered enzyme material has the desired activity). An activity of at least 850 ÷ 900 nephelometric units is preferred for insecticidal applications, and is usually obtained after two adsorptions on chitin.

The nephelometric analysis for enzyme activity is based on measuring the decrease in turbidity by nephelometry or photometry in a colloidal suspension of chitin on enzymatic lysis. The nephelometric units measure the activity of enzyme per mg which will cause the turbidity due to 0.3 mg of colloidal chitin per ml to decrease 50 % after 2 hr incubation at 37.5° and pH 5.3 (buffer solution: 0.1 M citric acid and 0.2 M sodium dihydrogen orthophosphate).

Smirnoff (1975) placed some 60,000 gizzards and intestines of freshly killed chickens in 4000 l of buffer solution, pH 5.2. The buffered solution contained 0.02 M citric acid and 0.02 M Na_2HPO_4 with 10 mg per l of an antibiotic. The solution was filtered through glass-wool and centrifuged at 4500 g for 15 min and the deposit was collected. The deposit was then washed in 100 l of buffer at pH 5.2 and this suspension maintained at 37° with light stirring until nearly complete hydrolysis of the chitin. After centrifugation the supernatant was treated with 55 kg of ammonium sulfate (55 % saturation). The saturated solution was centrifuged at 4500 g for 10 min. The deposit obtained was composed of chitinase having an activity of 100 ÷ 150 nephelometric units.

The enzyme preparation was returned to 20 l of buffer solution at pH 5.2. with 1.2 kg of colloidal chitin added. Hydrolysis at 37° with weak stirring was carried out until total reduction of the turbidity was obtained. The solution was centrifuged at 4500 g for 15 min, the supernatant was saturated by 11 kg of ammonium sulfate, then stirred and centrifuged 10 min at 4500 g. The deposit was 100 g of enzyme material with an activity of 850 ÷ 900 nephelometric units.

Fugono & Yamano (1969) have reported that a large number of basidiomycetes possess intense chitinolytic activity and have published a method for the industrial production of chitinase from the cultures of the Agaricaceae: *Armillaria mellea*, *Lentinus edodes*, *Lampteromyces japonicus*, *Flammulina velutipes*, *Pleurotus ostreatus* and *Schizophyllum commune*; from the Polyporaceae: *Coriolus hirsutus*, *Formitopsis rhodophaeus*, *Irpex lacteus*, *Poria vaporaria*, *Trametes sanguinea* and *Trametes cinnabarina* and from the Hydnaceae: *Corticium centrifugum* and *Stereum hirsutum*. Their enzymatic acitivity has been determined on colloidal chitin at 45° according to Kawasaki, Ito & Kurata (1964); optimum pH has been found close to 5 for most of them. After purification of the raw chitinase on DEAE-Sephadex at pH 4, the pH interval where the enzymes kept most of their activity was found to be between 4 and 10; the optimum temperature interval is 45 ÷ 50°; thus, fungal chitinase is inactivated at acidic pH values and at temperatures around 70°. Chitin hydrolysis leads to

N-acetyl-D-glucosamine and N-acetylchitobiose.

As an example of procedure for the production of chitinase from fungi, Fugono & Yamano (1969) described the culture of *Trametes sanguinea* in 1000 l cultures containing 2 % dextrose, 1 % cellulose, 0.3 % peptone, 0.2 % yeast extract, 0.15 % potassium dihydrogen phosphate, 0.05 % manganese sulfate and vitamin B (0.2 mg), with 500 l of air per min at 28° and applied stirring at 160 g. After 120 hr, the pH was adjusted to 4 with sulfuric acid, and after filtration, 225 kg of ammonium sulfate are added to 450 l of the filtrate; the precipitate is centrifuged and after drying at 35°, chitinase powder is obtained with an overall yield of 3.3. kg.

Insect cells and forest trees are equally able to produce lytic enzymes having antifungal action. Bactericidal or bacteriolytic properties of insect tissues, fluids and secretions have been reported by Landureau & Jolles (1970). Insect cells are able to resist different infections by bacteria or fungi and the lytic enzymes produced in vitro are excellent chitinases and poor muramidases.

The in vivo hyphal lysis of fungi in a pathogenic relationship with the host plants has been observed in tomato plants; production of chitinase was admitted as a defence against many fungi. Chitinase has been detected in seeds of woody and herbaceous plants, in sweet almond extracts and in the leaves of red kidney beans. Wargo (1975) reports chitinase in vegetative tissues of perennial woody plants and remarks that *Armillaria mellea*, an aggressive pathogen attacking healthy trees, colonizes tissues by new growth from hyphal tips. There is evidence that growth of the hyphal tips of fungi depends on a delicate balance between wall synthesis and wall lysis: chitinase of course can alter the balance.

Chitinases in plants are accompanied by β-1,3-glucanases: both enzymes seem to be necessary to cause a high degree of lysis of *Armillaria mellea* for example, since the hyphae contain significant quantities of β-1,3-glucan in addition to chitin. By this mechanism the enzymes are effective as deterrent to infection.

On this basis, associations of β-1,3-glucanase and chitinase look interesting as antimicrobial agents for agricultural applications. Tanaka, Ogasawara, Nakajima & Tamari (1970) and Ogasawara, Tanaka, Yoshinaga & Hiraga (1973) used both enzymes in water with added isopropanol and polyoxyethylenelaurylether for treatment of rice blight; the emulsion was sprayed over rice plants. The plants were inoculated with *Pyricularia orizae* two days later and the number of infected spots were counted one week after inoculation; infection occurrence was about 4 % compared to chitinase-untreated controls.

The insecticidal activity of microorganisms (including viruses) is enhanced by chitinase; Smirnoff (1975) sprayed balsam fir trees infested with spruce budworms with Thuricide alone or with Thuricide plus chitinase. Chitinase contributed in arresting feeding of larvae and in increasing the mortality rate.

A review of microbial decompositon of cellulose, keratin, chitin, lignin and rubber was published by Nickerson (1971).

CHITOSANASES.

A new class of enzymes, chitosanases (E.C. 3.2.1.99) that are active in hydro-
lysing chitosan has been proposed by Monaghan, Eveleigh, Tewari & Reese (1973).
Among 200 microorganisms studied, 25 fungal and 15 bacterial isolates degraded
Rhizopus rhizopodiformis cell walls as well as chitosan. Chitosans occur
widely in Zygomycete walls but not in other fungal classes. As Zygomycetes are
widespread in nature, chitosanases are also probably ubiquitous (Monaghan,
1975).

A strain of *Bacillus* sp., *Bacillus* R-4, has been isolated as a microorganism
which lyses the cell walls of the fungi belonging to *Rhizopus* species and
other related species. A carbohydrolase produced by this organism in the
culture broth has been identified as a chitosanase. Since both the chitosanase
and the lytic activity show quite similar behavior in all purification procedures, as well as in experiments on the enzymatic properties, it has been
concluded that the chitosanase is responsbile for the lysis of *Rhizopus* cell
walls. The organism has been shown to produce a protease which is also responsible for the lysis (Tsujisaka, Tominaga & Iwai, 1975). An abrupt decrease in
the viscosity of the reaction mixture in which chitosan was used as a substrate has indicated that the enzyme hydrolyzes chitosan in an endowise
manner. A rapid lysis of *Rhizopus* cell walls in the early stage of the
reaction, as reported by Monaghan, Eveleigh, Tewari & Reese (1973) may be due
to this mode of action of the chitosanase.

The activity of the purified chitosanase from *Bacillus* R-4 was examined by
Tominaga & Tsujisaka (1975) using soluble starch, dextran, pullulan, laminaran,
carboxymethylcellulose, colloidal chitin, glycolchitin, chitosan and glycolchitosan as substrates. The enzyme hydrolyzed chitosan and glycolchitosan, but
not the other glycans tested.

The molecular weight of the chitosanase was estimated to be about 31,000 by
gel filtration on Sephadex G-100. Its isoelectric point determined by isoelectric focusing was estimated to be pH 8.30. The enzyme exhibited the
maximum ultraviolet absorption at 278 nm and the ratio of the absorbancy at
280 nm to that at 260 nm was 1.95. The molecular extinction coefficient was
calculated as being $4.96 \ 10^4 \ \text{mol}^{-1} \times \text{cm}^{-1}$ at 278 nm.

Chitosanase is stable at temperatures below 50°. If the enzyme is incubated at
70° for 15 min, about 70 % of the original activity remains. Chitosanase is
most active at pH 5.6. The enzyme is stable in the range of pH 4.5 ÷ 7.5 at
40° for 3 hr and is inactivated more rapidly at alkaline pH than at acidic pH
values.

The effect of various metal ions on the chitosanase activity is summarized in
Tab. 4.1. Among the metal ions examined, none of them activated the chitosanase, whereas mercuric, ferrous, nickel and zinc ions showed strong inhibitory effects.

The effect of reduced glutathione and of L-cysteine on the chitosanase inhibited by mercuric ion or sulfhydryl reagents was examined by Tominaga &
Tsujisaka (1975). The enzyme treated with mercuric ion, *p*-chloromercuribenzoate
or monoiodoacetate lost about 97 %, 74 % or 85 % of the original activity,

Table 4.1. Effect of metal cations on chitosanase; the enzyme solution (1 ml) in 0.2 M acetate buffer (pH 5.6) containing 10 µg chitosanase was kept at 40° for 30 min in the presence of each metal cation and the remaining activity was measured. Tominaga & Tsujisaka, 1975.

Concentration of cation, mM	Cation	Relative activity of chitosanase, %
10	Ca^{2+}	99.5
10	Ba^{2+}	90.2
10	Mg^{2+}	92.1
10	Mn^{2+}	92.1
10	Sr^{2+}	89.5
10	Mo^{2+}	77.6
10	Sn^{2+}	60.5
1	Co^{2+}	100.0
1	Ni^{2+}	18.8
1	Zn^{2+}	37.6
1	Fe^{2+}	12.2
1	Cr^{2+}	0.0
1	Hg^{2+}	3.5
1	Pb^{2+}	0.0
1	Cu^{2+}	0.0
–	none	100.0

respectively. However, if either reduced glutathione or L-cysteine was added at a concentration three times higher than those of the inhibitors present in the enzyme solution, the inhibited chitosanase restored its activity almost completely in each case. It was also recognized that the addition of reduced glutathione to the native chitosanase preparation used, elevated the activity about 10 %.

Extracellular chitosanase synthesis was induced in *Streptomyces* by glucosamine (Price & Stork, 1975). The purified enzyme hydrolyzes chitosan but not chitin, nor carboxymethylcellulose. The only hydrolysis products are di- and trigluco-samine, the maximum activity at 25° being between pH 4.5 and 6.5. The rate of the reaction was depressed by concentrations of soluble chitosan higher than 0.5 g×l^{-1}; the apparent K_m was 0.688 g×l^{-1}. The chitosanase, whose composition is reported in Tab. 4.2 , prevented spore germination and caused a sharp decrease in the turbidity of germinated spore suspensions of *Mucor* strains. An enzyme from *Myxobacter* that has both β-1,4-glucanase and chitosanase activi-ties has been characterized by Hedges & Wolfe (1974).

Evidence for homogeneity was obtained from electrophoresis and sedimentation velocity studies; only one N-terminal amino acid, valine, was found. Results of denaturation studies showed that β-1,4-glucanase and chitosanase activities decreased at equal rates. With carboxymethylcellulose as the substrate, a K_m of 1.68 g of carboxymethylcellulose per l of solution and a V_{max} of 2.20× 10^{-9} mol×min^{-1} were found. With chitosan as the substrate, a K_m of 0.30 g of

Table 4.2. Aminoacid analysis of chitosanase, M_w 29,000, from *Streptomyces*. Price & Stork, 1975.

Aminoacid	Protein, % weight	Number of residues per mol
lysine	8.99	20
histidine	2.34	5
arginine	5.05	9
tryptophan	1.93	3
aspartic acid	12.56	32
threonine	4.31	12
serine	4.57	15
glutamic acid	9.84	22
proline	3.14	9
glycine	4.52	23
alanine	6.85	28
valine	4.11	12
methionine	2.24	5
isoleucine	2.83	7
leucine	7.38	19
tyrosine	6.47	11
phenylalanine	4.76	9
cysteine	0.00	0

chitosan per 1 of solution and a V_{max} of 0.75×10^{-9} mol$\times$min^{-1} were found. A pH optimum of 5.0 was found for β-1,4-glucanase activity, and pH optima of 5.0 and 6.8 were found for chitosanase activity. β-1,4-Glucanase activity had a temperature optimum of 38° and chitosanase activity has a temperature optimum of 70°. Chitosan stabilized both enzyme activities at 70°. Cellotriose was the smallest oligomer capable of hydrolysis. Glucosamine was released by action of the enzyme upon cell wall preparations of several fungi.

It appears, therefore, that the chitosanases from *Bacillus* R-4 and *Streptomyces*, which do not hydrolyze carboxymethylcellulose, differ from that of *Myxobacter*. An important point of difference is the optimum temperature: 60° for the chitosanase studied by Price & Stork (1975), 70° for the chitosanase described by Hedges & Wolfe (1974), 40° for the chitosanase reported by Tominaga & Tsujisaka (1975).

In addition, unlike for the *Myxobacter* enzyme, the *Streptomyces* chitosanase produces di- and trisaccharides from chitosan. As a further point of difference, the chitosanase from *Myxobacter* contained 1.3 % carbohydrase and an estimated 3 cysteine residues per mole of enzyme; other constituents were also present in different proportions.

Many reports have already pointed out that the cell walls of *Aspergillus* and *Penicillium* which belong to Euascomycetes consist mainly of some kinds of glucan and chitin (Applegarth, 1967), and therefore the carbohydrolases participating in the lysis of these cell walls are principally some kinds of glucanase and chitinase (Horikoshi & Ida, 1959; Skujins, Potgieter & Alexander, 1965; Troy & Koffler, 1968). It has been known, however, that the cell walls of *Rhizopus* and *Mucor*, which belong to Zygomycetes, are distinguishable from those of Euascomycetes in that they contain chitosan as a main component besides glucan and chitin (Bartnicki-Garcia & Nickerson, 1962). Almost all lytic enzymes reported so far are not able to degrade these cell walls. *Rhizopus* cell walls were decomposed by the chitosanase described by Tominaga & Tsujisaka (1975), which has no glucanase and chitinase activities. This fact may indicate that chitosan is present in *Rhizopus* cell walls, playing an important role in maintaining their structure.

Chitosanase are stable substances readily soluble in water and in a variety of buffers. They could be used for taxonomic studies of fungi and for the study of some mycoses of plants and animals. They lend themselves to the investigation of the structure and biosynthesis of the cell wall of *Mucorales*, also in combination with chitinases.

LYSOZYMES.

Endo-β-N-acetylmuramidases or lysozymes (E.C. 3.2.1) have been isolated from a great variety of sources: mammalian tissues, insects, plants, birds and micro-organisms. They degrade the glycan of the bacterial cell walls to oligosaccharides of the N-acetylglucosaminyl-N-acetylmuramic acid type, while endo-β-N-acetylglucosaminidases give saccharides with N-acetylglucosamine at the reducing end. An enzyme of the latter type is produced by strains of *Staphylococcus aureus*, *albus* and *pyogenes* and of *Clostridium perfringens*: it degrades the β-N-acetylmuramyl-(1→4)-N-acetylglucosamine polymer of the bacterial cell walls to yield disaccharides of the N-acetylmuramic acid - N-acetylglucosamine type (Nord & Wadstrom, 1972).

The ability of hen's egg white lysozyme to degrade chitin and hydroxyethyl-chitin besides bacterial cell walls has been reported by many authors; purified lysozyme from fig and papaya plants, human urine and hen's egg white degrade both chitin and cell wall glycans $(NAG-NAM)_n$. The plant enzymes should be regarded primarily as chitinases. *Streptomyces* chitinase has no action on the N-acetylglucosaminyl-N-acetylmuramic polymer of the bacterial cell walls. On the other hand, the staphylococcal glucosaminidase and other enzymes from animals are more active on the bacterial substrates. While some muramidases and some chitinases tend to overlap in their specificities, the animal enzymes are probably involved in the host defence against bacterial invasion or in the metabolism of the bacterial cell wall, while the plant enzymes have other physiological significance.

Hydrolysis.

The study of the kinetics of lysozyme and chitinase adsorption on chitin is complicated by the fact that a decreasing chain length among the depolymerized substrate molecules lowers the rate of the digestion by lysozyme as chitin oligosaccharides are digested at different rates and therefore few attempts have been made to estimate affinity constants for chitinases.

The kinetics of the interaction between lysozyme and ethyleneglycolchitin have

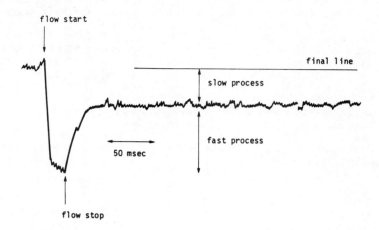

Fig. 4.7. The trace of the fluorescence change observed
after mixing lysozyme and ethyleneglycolchitin solutions
by the stopped-flow method at pH 3.6 and 25°. Lysozyme
was 5×10^{-5} M and ethyleneglycolchitin was 0.2 %.
Nakatani & Hiromi, 1974.

been studied by the fluorescence stopped-flow method by Nakatani & Hiromi
(1974). As from Fig. 4.7, they have found that there is a faster concentration
dependent process, which is a typical feature of bimolecular association
processes, corresponding to the association of the enzymes with the substrate;
this precedes the slow process which represents the steady-state hydrolysis of
the polymer substrate. On chitin, lysozyme at 25° is linearly adsorbed from pH
2.5 to pH 9.5 with a sudden decrease to zero at pH 11. The amount adsorbed at
pH 5.5 was 100 $\mu g \times g^{-1}$ and at pH 9.0 was 180 $\mu g \times g^{-1}$ for initial concentrations
of substrate and enzyme of 250 $\mu g \times ml^{-1}$. Data showing the enzyme activity are
in Fig. 4.8.

Among the lysozyme substrates of known structure, chitohexose is hydrolyzed
with greatest efficiency, giving chitotetraose and chitobiose as the only
products (Rupley & Gates, 1967). In fact the active site of lysozyme can
acco. date 6 units of a polymeric substrate. Only those saccharide molecules
whose pyranose ring makes contact with sites D and E are catalytically hydrolyzed.

The free energy requirement for distortion, gives rise to partial binding of
the substrate to sites A, B and C of the enzyme, i.e., in a non-productive
binding mode. The binding modes of chitohexose are shown in Fig. 4.9. Unproductive complexes are produced when saccharide binds at the A ÷ C sites of the
enzyme, with the reducing ends at site C. In this binding mode the units at
the non-reducing end of chitotetraose and higher oligomers protrude into the

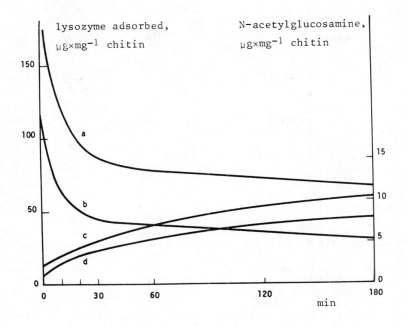

Fig. 4.8. Sorption and activity of lysozyme on chitin at 37° and pH 5.5. Phosphate + acetate buffer was 0.05 M and chitin was 1.0 mg×ml^{-1}. Skujins, Pukite & McLaren, 1973.

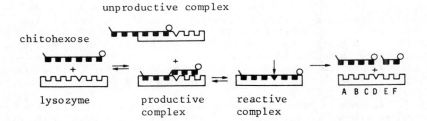

Fig. 4.9. Graphical representation of formation of the two main lysozyme-chitosaccharide complexes. Holler, Rupley & Hess, 1975.

solvent. Also shown in Fig. 4.9 are stable, productive complexes in which the saccharides also fill the A ÷ C sites. However, in this binding mode, the non-reducing end of the saccharide occupies site A and the reducing end units protrude into the D ÷ F region of the enzyme.

Holler, Rupley & Hess (1975) used temperature-jump relaxation, spectrophoto-metric and spectrofluorimetric stopped-flow techniques to carry out kinetic investigations on productive and unproductive lysozyme-oligochitosaccharide complexes.

When chitohexose binds to lysozyme, it perturbs the hydrogen ion equilibria of the enzyme. At least two different processes occur: one process occurs in the 5-msec time region and it is observed with chitotriose and larger oligomers under conditions where chitotriose is known to bind exclusively to the A ÷ C sites of the enzyme. This process is considered to be associated with the formation of unproductive lysozyme-substrate complexes. The other process occurs in the 100-msec time region and is observed only with oligomers larger than chitotriose; it is considered to be associated with formation of produc-tive complexes.

x-Ray analysis studies on the association of various inhibitors and substrates with lysozyme have led to a proposed mechanism of catalysis involving the pro-duction of a catalytic carbonium ion electrostatically and sterically stabi-lized by groups of the enzyme. Tritium kinetic isotope effects have been utilized to probe the nature of the transition state in the lysozyme-catalyzed hydrolysis of chitotriose. To reach this information, the variation of the tritium isotope effect was studied on the basis of the $^3H/^{14}C$ ratios in tritiated chitotriose before hydrolysis and in the monosaccharide product altered by the isotope effect of enzyme hydrolysis.

The hydrolysis of the chitotriose yields chitobiose and N-acetylglucosamine in the ratio of 2:1. The ratio of chitotriose to lysozyme of 2:1 allowed binding of one mole of trisaccharide to the unproductive sites of the enzyme, thus leaving an enzyme : substrate ratio of 1:1 for catalysis. According to Smith, Mohr & Raftery (1973), for a carbonium ion mechanism the reacting carbon changes hybridization from sp^3 to sp^2 and the out-of-plane bending vibrations of a carbon-hydrogen bond to an sp^2 carbon should be at lower frequencies than the corresponding bending vibration of a similar bond to an sp^3 carbon. The difference in frequency accounts for the isotope effect 10 ÷ 15 % for S_N1 reactions. Thus the transition state of a reaction at the reacting carbon governs the isotopic effect through a weakening of the C-H bond during the transition state of the reacting carbon. The 19 % isotopic effect reported for lysozyme catalyzed hydrolysis of chitotriose suggests a transitional state with considerable carbonium ion character. An isotopic effect of this size would rule out both single and double displacement mechanisms involving attack by the solvent or by the enzyme itself at least in any rate-determining step.

The main differences in behavior between the longer and smaller substrates may arise presumably from the differences in their diffusion constants and their ability to assume the highly restricted forms necessary for specific binding in the active center cleft of the enzyme.

There is ample evidence of the occurrence of tryptophanyl groups at the binding sites of the lysozyme. The structural determinations of the enzyme in the crystalline state and the identification of the regions of the enzyme to which inhibitors bind have shown that three tryptophans occur at the binding site, as also confirmed by chemical evidence. It was therefore assumed that the tryptophan is responsible for the red shift on interaction with lysozyme with substrates and inhibitors, in its ultraviolet spectrum (Dahlquist, Jao & Raftery, 1966) as well as for the shift in the fluorescence spectra reported by Nakatani & Hiromi (1974). In fact the spectrum of lysozyme and chitotriose at pH 5 exhibits an important alteration at 293.5 nm. Assuming that the alteration of peak height is proportional to the ratio of the enzyme-substrate complex to the total enzyme concentration, ES/E°, the following calculation of the binding constant K_S can be made:

$$E \; + \; S \; = \; ES$$

$$K_S \; = \; E \times S \, / \, ES$$

$$\log K_S \; = \; \log(E \, / \, ES) \; + \; \log S$$

The plot of $\log S$ vs $\log(ES/E)$ gives a line with intercept of $-\log K_S$ or pK_S. Such a plot for the binding of chitoriose to lysozyme at pH 5.5 gives a slope equal to one, indicating one molecule of chitotriose binding to the enzyme. Chitotetraose and chitobiose represent the greatest amounts of the products of the digestion of chitohexose and are similar to the products obtained from the lysozyme digestion of chitin and chitin precipitated with sulfuric acid.

Transglycosylation.

It has been well recognized by a number of authors that most carbohydrolases catalyze transglycosylation as well as hydrolysis of glycosidic linkages. Transglycosylation follows the scheme in Fig. 4.10.

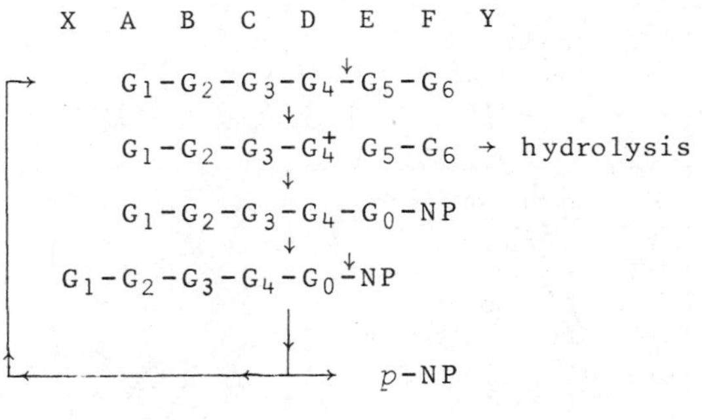

Fig. 4.10. Release of p-nitrophenol through transglycosylation. X-Y = binding sites of lysozyme molecule; $G_1 \div G_6$ = hexasaccharide of N-acetylglucosamine; p-NP = p-nitrophenol; the acceptor is p-nitrophenyl N-acetylglucosamine. Hayashi, Inohara, Taira, Doi & Funatsu, 1974

It is also known that the efficiency of lysozyme catalyzed transglycosylations is much greater than that of hydrolysis. There may be a special mechanism or molecule providing the high efficiency of the transglycosylation, because it takes a more complicated pathway, such as release of fragments of the substrate from the sites E and F and the binding of the acceptor molecule at the same sites, than in the case of hydrolysis with the simple attack of the water molecule.

To obtain information about transglycosylation, Hayashi, Inohara, Taira, Doi & Funatsu (1974), used a reaction mixture containing lysozyme (2 ÷ 10×10^{-4} M), glycolchitin (1 ÷ 6×10^{-3} M) and p-nitrophenyl N-acetylglucosamine (0.1 ÷ 5× 10^{-3} M) in 0.05 M phosphate buffer, pH 6.0. It was incubated at 40° for given periods, and the amount of released p-nitrophenol through transglycosylation and following hydrolysis (see Fig. 4.10) was determined by spectrophotometry at 400 nm after the pH value of the reaction mixture was adjusted to 9 by addition of 0.2 M sodium dihydrogen phosphate.

The maximum rates of the hydrolysis and the transglycosylation were found to be nearly in the same pH region, as shown in Fig. 4.11. Contrary to the pH vs profile of the hydrolysis, the transglycosylation exhibited great efficiency in a pH region from 6 to 8. When the large excess of the substrate was added to the enzyme solution, it was found that the lysozyme-substrate complex was

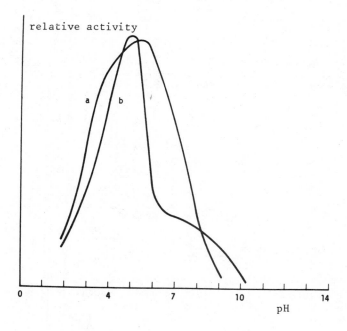

Fig. 4.11. pH-Dependence of release of p-nitrophenol from p-nitrophenyl N-acetylglucosamine in the presence of glycolchitin, by the action of lysozyme; a = release of p-nitrophenol; b = hydrolysis. Hayashi, Inohara, Taira, Doi & Funatsu, 1974.

formed fully in the same region, pH 6 ÷ 8; there, the cleavage of β-1,4-glyco-
sidic linkage may conduct preferentially the transglycosylation with an
appropriate acceptor. The maximum transglycosylation occurs at 70°, i.e., at a
much higher temperature than hydrolysis.

At 0.01 % substrate concentration, only the pentamer and glycolchitin were
hydrolyzed by lysozyme. However, the hydrolysis of the pentamer ceased within
1.5 hr incubation. The kinetic parameters were: V_{max} = 1.51×10^{-6} mol×min^{-1} and
K_m = 9.1×10^{-5} M. The V_{max} of the hydrolysis was twice as large as that of the
transglycosylation and K_m was about half of the transglycosylation value.
The distinct difference between the hydrolysis and the transglycosylation of
oligosaccharide was the time-course of the reactions, which completely resem-
bled those observed when glycolchitin was used as a substrate. Thus, it was
confirmed again that when the oligosaccharides were used as the substrate,
lysozyme catalyzed only the transglycosylation. The efficiencies of several
substrates on the release of p-nitrophenol are shown in Tab. 4.3, in comparison with their relative values of digestibility and capability to form the
lysozyme-substrate complex.

Table 4.3. Relative rates of release of p-nitrophenol and of hydrolysis, at
40° and pH 5.5 for 4 hr. Hayashi, Inohara, Taira, Doi & Funatsu, 1974.

Substrate	Release of p-nitrophenol	Hydrolysis	Amount of complex
glycolchitin	100	100	100
glycolchitin (pentamer)	30	0	100
carboxymethylchitin	75	–	100
carboxymethylchitin, low M_w	42	0	100
carboxymethylchitin, high M_w	40	0	100
chitosan, 50 % deacetylated	81	35	95
chitosan, 75 % deacetylated	55.	10	70

OLIGOMER PRODUCTION.

N-Acetylglucosamine and acetylated oligomers.

N-Acetylglucosamine has been produced by Pope & Zilliken (1959) by a fermentation process where a mold of the *Aspergillus* genus is propagated in a medium
containing chitin. They found that if the aerobic fermentation conditions
necessary during the propagation of the mold are changed to anaerobic after
the fermentation has proceeded for a period of time, the N-acetylglucosamine,
which is held by the mycelium in a form in which it is not easily recovered
from the medium, is converted to a form suitable for recovery. For example,

a fermentation medium of 0.1 % suspended chitin (reprecipitated chitin after dissolution in concentrated hydrochloric acid), 0.15 % potassium dihydrogen phosphate, 0.05 % magnesium sulfate and 0.05 % ammonium nitrate, with initial pH 3.5, sterilized with steam in a 200 l tank under pressure, was seeded with a 10 % inoculum of *Aspergillus terreus* and allowed to ferment at 32° for 24 hr under stirring and under aeration. At the end of this period of time, aeration was stopped and carbon dioxide was supplied for 16 hr. The broth was then harvested and the N-acetylglucosamine recovered after preliminary treatment of the broth with ion-exchange resins to remove salts, with the help of activated carbon. Elution from the carbon filter was performed with ethanol aqueous solution. Degradation may be performed in acetic acid as well; chito-biose octaacetate is one of the main products of the acetolysis of chitin.

Barker, Foster, Stacey & Webber (1958) described the acetolysis of chitin in a mixture of acetic anhydride and sulfuric acid at room temperature for 40 hr and at 55° for 11 hr. After dilution with sodium acetate, the mixture was extracted in chloroform from which chitobiose octaacetate was crystallized. Thomas (1973) has pointed out however that chitobiose octaacetate is often contaminated by an acetylated amino sugar having an α,β-unsaturated aldehyde function.

A variety of solvent mixtures for use in thin-layer chromatography have been examined by Powning & Irzykiewicz (1967) who have proposed single run separations of chitin oligosaccharides up to chitohexose. The best results were obtained on silica gel plates. A linear relationship exists between R_M values and degree of polymerization, as shown in Fig. 4.12; ammonia added to the solvent mixtures increased resolution.

Capon & Foster (1970) examined two methods for the preparation of chitin oligosaccharides; acetolysis followed by de-O-acetylation, and hydrolysis in 2 M hydrochloric acid followed by neutralization and desalting on Sephadex G-25: the latter method was found more convenient. Chitin (100 g) was added slowly with stirring to cold hydrochloric acid (200 ml). The mixture was left at room temperature for 2 hr, then heated at 40° for 1 hr, and finally cooled in ice-water and neutralized with cold sodium hydroxide. The fractions free of sodium chloride were pooled and liophilized to yield a mixture of acetylated oligosaccharides, whose separation was then performed with Sephadex LH-20 columns.

The introduction of an electrolytic desalting step, prior to separation by gel exclusion chromatography, proposed by Berkeley, Brewer & Ortiz (1972), allows the use of small Sepahdex columns with satisfactory yields. The desalter is a glass cell separated from the anode and cathode compartments by ion-exchange membranes. The platinum mesh electrodes were cooled by a continual flow of tap water serving also to remove the ions transported. The potential difference was 50 V with initial current of 4.2 A and final current 0.5 A at 24°. The final products, obtained according to the above method applied to a preparation as described by Rupley (1964) were analyzed by thin-layer chromatography following Powning & Irzykievicz (1965) and by gas chromatography according to Bhatti, Chambers & Clamp (1970).

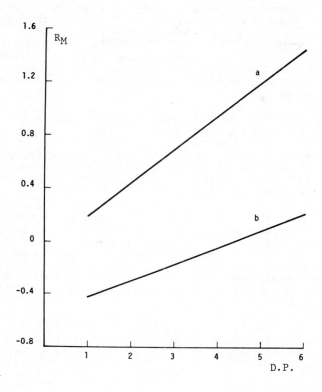

Fig. 4.12. Relationship between R_M value $\equiv \log (1/R_f - 1)$
and degree of polymerization, D.P., of chitin oligosaccharides on silica gel thin layers; a = 50 isoamyl alcohol +
60 ethanol + 30 water + 1 ammonia; b = 50 ter-butanol +
70 ethanol + 40 water. Powning & Irzykievicz, 1967.

Glucosamine and deacetylated oligomers.

Deacetylated oligosaccharides can be obtained from the degradation of chitosan.
While controlled acidic treatments like those described above lead to chitin
degradation to produce N-acetylated oligosaccharides, it should be pointed out
that harsh acidic treatment of chitin can lead to both degradation and deacetylation. Many early papers are unfortunately not very clear in describing the
degradation products of chitin.

Horowitz, Roseman & Blumenthal (1957) dissolved chitosan hydrochloride (5 g)
in water (100 ml) heated at 53° and added concentrated hydrochloric acid (900
ml). After 24 hr at 53°, glucosamine was obtained together with oligomers,
while, after 72 hr, the conversion was to the monosaccharide. They carried out

the hydrolysis for 48 hr and studied the separation of glucosamine hydrochloride from the dimer and traces of trimer and tetramer. Lenk, Wenzel &
Schuette (1961) produced further results by chromatography on Dowex 50.

Barker, Foster, Stacey & Webber (1958) also studied the acidic hydrolysis of
chitosan at 100° for 34 hr: they obtained deacetylated oligosaccharides in the
range mono- to hexa-. For this they used concentrated hydrochloric acid with
chitosan hydrochloride in water, and mentioned the difficulty of the chitosan
degradation in acids.

Many reports deal with the acid hydrolysis of chitin to prepare glucosamine.
Ikan (1969) accomplished the degradation of chitin by heating with concentrated hydrochloric acid. Crude glucosamine hydrochloride was purified by
filtration through Celite and activated carbon. Final purification was
effected by dissolving the product in hot water and adding ethanol, whereupon
the α-anomer crystallizes, while the β-anomer remains in solution and may be
recovered by precipitation with ether.

Zimmerman (1958) has also described the preparation of glucosammonium chloride
and its ultraviolet absorption behavior in borate buffers. Purchase & Braun
(1946) described the preparation of this compound based on the use of filter
aids, concentration under diminished pressure and crystallization. Their
method was slightly simplified by Navratil, Murgia & Walton (1975) who used
12 M hydrochloric acid at 95° for 30 ÷ 60 min. Oeriu, Lupu, Dimitriu & Craesco
(1962) suggested the use of 30 % hydrochloric acid at 90 ÷ 95° for best
results, during 5 hr.

Matsushima (1948) boiled 1.1 kg of raw chitin with 2 1 of concentrated hydrochloric acid for 2 hr: glucosamine hydrochloride crystallizes out on cooling
and can be purified by recrystallization and decoloration.

CHITINASE ACTIVITY DETERMINATION.

The substrate used by Jeuniaux (1963) is native chitin from *Sepia officinalis*.
The chitinolytic activity is measured by Jeuniaux's method which had been
worked out for enzyme solutions deprived of chitobiase, for instance vertebrate organ extracts: it consists of the hydrolysis of chitobiose originating
from chitin, by adding chitobiase. To 1 ml of chitin suspension (5 mg×ml^{-1}),
1 ml of citric acid + disodium hydrogen orthophosphate buffer at pH 5.2 is
added with 1 ml of lobster serum diluted 1:10 as a source of chitobiase, and 1
ml of enzyme extract. N-Acetylglucosamine is measured at different reaction
times after inactivation of the enzymes at 100°. The activity of chitinase is
expressed as μg of N-acetylglucosamine per hr and per g of fresh tissue.

When the effect of pH on the chitinase activity is under study, one first step
includes the hydrolysis of chitin at the desired pH values; after incubation,
the chitinase is inactivated and the soluble oligomers are separated. During
the second step, the pH value of the oligomer solution is adjusted to 5.3, the
optimum value for chitobiase which is added and incubated 1 hr to complete
hydrolysis of chitobiose and chitotriose (Micha, Dandrifosse & Jeuniaux, 1973).

By this method, the activity of gastric chitinase of two species of Reptiles
was measured as a function of pH. These chitinases have optimum pH 3.0, but

are still active at pH 1.5, while they are inhibited at pH 7.0 and are
adequately adapted to the acidic medium of the gastric fluid. Optical methods
were used to determine N-acetylglucosamine, for instance the colorimetric
method by Reissig, Strominger & Leloir (1955) which is a modification of the
method of Elson & Morgan (1933).

Powning & Irzykievicz (1965) assayed chitinase in extracts of seeds by adding
them to colloidal chitin in 0.125 M citrate buffer at pH 4.5. After 2 hr at
35°, sodium carbonate is added to stop the reaction and the optical density is
measured at 420 nm. Blanks were prepared by incubating separately extracts and
chitin, and combining after addition of carbonate. Lunt & Kent (1960) observed
that the decrease in turbidity was still taking place after the release of
N-acetylglucosamine stopped. The turbidimetric method has been described by
Jeuniaux (1966).

Other methods for the determination of chitinase activity include the viscosi-
metric method on glycolchitin by Lundblad & Hultin (1966) and by Solder, Nord,
Lundblad & Kjellman (1970) and the determination of the release of reducing
sugar by Momose, Inaba, Mukai & Watanabe (1960). Chitobiase, β-N-acetylglucosaminidase, is assayed colorimetrically on the basis of nitrophenol liberated
from nitrophenyl-N-acetyl-β-D-glucosaminide by incubating for 15 min at 35°,
pH 5.7 adjusted with 0.2 M acetate buffer (Kimura, 1973).

FLUORESCENT CHITINASES.

Benjaminson, Bazil, D'Orai & Carito (1966) coupled chitinase with horse spleen
ferritin and mentioned the possibility of coupling chitinase with fluorescein
isothiocyanate. Benjaminson (1969) conjugated chitinase with lissamine
rhodamine B-200 chloride and with fluorescein isothiocyanate to obtain
specific stains for chitin in situ. In fact, staining through the activity of
the specific enzyme avoids destructive prestaining treatment and allows
cytologic relationships to be preserved. The method is based on the affinity
of the chitinase for this substrate in the fungal wall and in the insect
exoskeleton: no staining is observed with other polysaccharides in bacterial
and algal cell walls. Only those structural sites where chitin is present
fluoresce when the stained preparation is excited with ultraviolet light. The
preparation of the dye conjugate is as follows: chitinase (120 mg) is
dissolved in 0.02 M phosphate buffer (2 ml) at pH 7.4; lissamine rhodamine
B-200 chloride (6 mg) or fluorescein isothiocyanate is dissolved in 0.05 M
carbonate + bicarbonate buffer (2 ml) at pH 9.0; the solutions are combined
at 5° for 24 hr and centrifuged at 32,000 g; the supernatant is removed and
dialyzed against 0.02 M phosphate buffer; residual dye is removed by passage
on Sephadex. Upon elution with water, the fluorescent chitinase is collected
and liophylized.

CHITINASE AND CHITOBIASE OCCURRENCE.

The enzymatic system that degrades chitin is very important because it allows
many plants and animals to perform various vital functions, including
digestion, growth, chemical attack and other metabolic activities. On the
basis of exhaustive studies on the distribution, tissular location and

properties of chitinase and chitobiase in animals, the evolution of the biosynthesis of these enzymes as been drawn as follows by Jeuniaux (1971).

The widespread distribution of chitinases and chitobiases in bacteria, molds
and Protozoa suggests that the synthesis of both chitinolytic enzymes is a
property appertaining already to the primitive unicellular animals. In
diblastic Metazoa particularly in the sea-anemones, chitinases and chitobiases
are synthesized by both ectodermal and endodermal tissues. From this situation
some evolutionary tendencies can be observed, along which the synthesis of one
or both enzymes of the chitinolytic system is apparently lost by one or both
the embryonic layers.

The synthesis of chitinase is generally lost by the ectodermis of Protostomia,
while the synthesis of chitobiase is always retained by these cells. Two
remarkable exceptions must be emphasized: the synthesis of chitinase in the
epidermis is retained by the embryos of Nematoda and during the whole life in
Arthropoda (except in the adults of insects). The preservation of these biosynthetic abilities by these animals is related to the molting phenomenon in
Arthropoda and to the hatching process in Nematoda.

On the other hand, the biosynthesis of chitinase is generally retained by
tissues and organs of endodermic origin (digestive tract and glands) in most
invertebrates, except in some species which are strongly adapted to a highly
specialized diet entirely devoided of chitin, such as in the case of strictly
phytophagous or xylophagous insects: for example the stick-insect *Carassius
morosus*, the silk-worm *Bombyx mori* and the wood-boring larvae of Longhorn
beetles *Ergates faber*.

In deuterostomian animals, and mainly in vertebrates, a marked tendency towards
deletion of the biosynthesis of chitinase and chitobiase is evident. The
distribution of chitinases in the digestive system of vertebrates appears as
being a typical example of regressive evolution by the loss of a biosynthetic
property obviously present in the ancestral lineage.

In fishes, which are generally chitin-eating animals, the biosynthesis of
chitinase is very widely distributed, with only a few exceptions. In more
evolved vertebrates, besides a more restricted location of chitinase secretion
in the digestive tract, one can observe the absence of chitinase secretion in
those species which are adapted to a diet devoid of chitin, such as
terrestrial tortoise, pigeon, rabbit, sheep, guinea-pig, sloth and cat. This
is also the case of man, although gastric chitinases have been found in some
monkeys.

For the interpretation of these facts, one should take into account the
results of experiments concerning the effects of change of diet on the chitinase secretion. It has been shown that animals, which are able to secrete
chitinase when they eat their normal food, continue to secrete this enzyme
when reared on a diet devoid of chitin (mice or rats for instance). On the
other hand, animals which do not secrete chitinase when they are normally fed,
do not acquire this ability when chitin is added to their food or when they
receive a chitin-containing diet, whatever the duration of the experiment
(with guinea-pigs for instance). It thus appears that the chitinase secretion
is the consequence of the genetic adaptation of the species (or the genus, or

the family) to that diet.

Cornelius, Dandrifosse & Jeuniaux (1975) have examined the secretions of chitinases in six species of mammals belonging to the order Carnivora; chitinases were found only in the extracts of the gastric mucosa of two species not strictly relying on a meat diet (dog and fox), while those with exclusive carnivorous habits (stoat, ferret, marten and cat) do not secrete the enzyme.

Micha, Dandrifosse & Jeuniaux (1973) in a study on the localization of chitinase synthesis in reptiles, batrachians and fishes have found that species with chitin-rich diet secrete chitinase at the level of gastric mucosa and in some cases in pancreas, pyloric caeca, intestinal mucosa and liver. In many fishes, the extracts of kidneys and spleen exhibit some chitinolytic activity.

An evident correlation exists between the composition of the diet and the ability to secrete chitinase even though this correlation is less clear-cut among fishes (Colin, 1972; Saxena & Sarin, 1972; Spencer, 1961; Tracey, 1961). Chitinase is not an inducible enzyme in vertebrates; the lack of chitinase secretion in an animal is not the consequence of the nature of the usual food of the individuals, but the consequence of the genetic adaptation by loss of the gene corresponding to a certain enzyme. The ancestors of the lower vertebrates were necessarily chitin-eating animals or at least secondary consumers, eating chitin-covered animals like zooplankton, but not plants (Jeuniaux, 1971).

Dandrifosse (1970) has studied the action of a number of inorganic ions on the chitinase secretion by gastric mucosa of *Lacerta viridis*. Potassium, ammonium, cesium, chloride and bicarbonate ions increase the cellular permeability to the chitinase, while iodide limits it.

Chitinase is not only present in digestive systems but occurs as well in the integument of those animals which carry out molting. Chitinase synthesis has been found to be stimulated periodically by the molting hormone ecdysone. Arthropods exhibit the ancestral property of synthesizing, using and renewing chitin, as they possess an enzyme system which enables them to degrade their cuticle periodically. The enzymes liberate N-acetylglucosamine and amino acids in the exuvial fluid; these products are rapidly readsorbed in the epidermis. Chitinases are no longer secreted just before exuvia casting. The aqueous extracts of exuvia are rich in chitinase, chitobiase and proteolytic enzymes; these hydrolases are therefore strongly adsorbed to their respective substrate in the cuticle under degradation, so that the exuvial fluid, just before ecdysis, contains very little enzyme. This explains how the re-adsorption of the hydrolysis products from the exuvial fluid into the epidermis takes place through the new cuticle being synthesized, with no damage from hydrolases (Jeuniaux, 1975).

The chitinolytic enzyme activities during larvae developmnet of the silk-worm *Bombyx mori*, and the localization of these enzymes in the subcellular fraction of the integument (Table 4.4) have been studied by Kimura (1973). Chitinase and chitobiase activities, during the 4-th instar of the silk-worm, begin increasing 36 hr after ecdysis. The values at 96 hr, before the new ecdysis, are twelvefold those observed at 36 hr.

Table 4.4. Subcellular localization of chitinase and chitobiase in the
integument of *Bombyx mori* (percentage of the enzyme activity). Kimura, 1973.

Fraction	Chitobiase		Chitinase	
	72 hr	84 hr	72 hr	84 hr
Cell debris (600g, 10 min)	4.7	8.3	0.0	17.2
Mitochondrial (10,500g, 20 min)	5.0	4.3	19.5	2.3
Microsomal (105,000g, 60 min)	7.6	3.5	23.0	7.1
Supernatant (105,000g, 60 min)	82.5	83.7	57.4	73.3

Bade (1974) has noticed that the activities reported by Kimura (1973) are low
compared to the necessarily high in vivo chitinase activity during the molt.
For instance, the chitin content of *Hyalophora cecropia* larvae undergoing the
pupal molt diminishes abruptly from 4.8 % to 0.8 % of dry weight, equivalent
to approximately 120 mg of chitin broken down in 24 hr. Moreover, an
inconsistency was noted in the optimum pH for chitinase, around 5.2 and the
alkalinity of the molting fluid.

Chitinase activity is held tenaciously in the old cuticle of *Manduca sexta* and
it is not completely extracted; cuticle chitinase has high endogenous activity
in contrast to soluble chitinase. The results support the hypothesis that the
cuticle chitinase normally has endogenous substrate accessible to it and that
the enzyme has readier access to endogenous than to exogenous substrate.

In bacteria, the chitinase system is widespread: it was demonstrated in
organisms from soil and fresh water, in several organisms of medical and
phytopathological importance and in marine muds. There has been no systematic
examination of the ability to degrade chitin among the Actinomycetes, although
some of the best characterized chitinolytic enzymes are derived from this
group. Many filamentous fungi are known to produce chitinase; some of them
contain chitin in their cell walls and in the supernatants of their cultures;
some of the fruiting types are known to be rich sources of chitinases at
certain stages of their development.

Fermentation of certain microorganisms such as the bacterium *Streptomyces anti-
bioticus*, the fungi *Beauveria* sp., *Cordiceps* sp. and others, is the process
used to date by manufacturers to obtain large quantities of chitinases.

Seeds of plants belonging to one family in the monocotyledons and to ten in
the woody and herbaceous dicotyledons, possess chitinase and chitobiase
activity. The highest activity of chitinase was found by Powning & Irzykiewicz
(1965) in bean, wheat and cabbage and the highest activity of chitobiase was
found in bean, cabbage, almond and waratah.

Plant tissues produce a variety of fungistatic and fungicidal materials, and
it is conceivable that chitinases function in this capacity in seeds. Since
fungal cell walls contain chitin, it is possible that, during germination,

any fungal hyphae penetrating the outer seed coat would be arrested by the chitinase. Mitchell & Alexander (1962) showed that the treatment of soil with chitin stimulates chitinase-producing actinomycetes and depresses the phyto-pathogenic *Fusarium*.

CHAPTER 5

CHROMATOGRAPHIC APPLICATIONS

The information presented in the previous Chapters mentions characteristics of chitin and chitosan of great value for chromatographic supports: they are basic polymers and therefore they may behave as anion-exchangers; they possess amino and hydroxyl groups that can act as electron donors toward transition metal ions; they are made of glucosamine and N-acetylglucosamine units and therefore can interact with organic substances like proteins because of the chemical nature of the repeating units and the physico-chemical nature of the polymers.

Examples of general chromatographic applications of chitin have been reported by Iwata & Nakabayashi (1974) who removed colored substances from tea, coffee, apple juice, dry mushroom infusions, caramel, sugar and other solutions by column chromatography.

Thus, several fields of chromatography are open to applications of chitin, chitosan and their derivatives: ion-exchange chromatography, chelation chromatography, ligand-exchange chromatography, affinity chromatography, high--pressure liquid chromatography, gel chromatography and thin-layer chromatography.

While some of these branches of chromatography have been covered with examples of applications of chitin and chitosan, some others are still to be explored, principally high-pressure liquid chromatography and gel chromatography. Other chromatography-related techniques have found applications in the industrial field.

This Chapter focuses attention on uses of chitin and chitosan which have some importance in analytical chemistry; other chromatographic applications will be described in Chapter 6.

ION-EXCHANGE CHROMATOGRAPHY.

Acids.

Chitosan undergoes slow dissolution in acidic media where anions like acetate, formate or chloride are present, especially at pH values below 2. While it is possible to perform chromatographic separations on chitosan at low pH values, particularly when sulfuric acid or oxalic acid are present, a simple way to obviate the partial collapse of granules, which alters the flow-rate of the columns and makes them unsuitable for prolonged operations, is to use epichlorohydrin chitosan, whose pK_a is identical to that of chitosan. Epichlorohydrin chitosan is insoluble in acidic media as well as in strong alkali (see page 124).

Epichlorohydrin chitosan is a white and stable powder that does not shrink or swell and can be regenerated.

Chitosan and epichlorohydrin chitosan are strong bases toward hydrochloric acid as the primary amino groups of these polymers, whose pK_a is 6.3, easily form quaternary nitrogen salts at low pH values; thus, in acidic solutions they have high anion capacity. At higher pH values, however, they are weak bases because the primary amino groups are not protonated and therefore do not interact with anions and do not dissociate neutral salts: for instance, they do not retain chloride from sodium chloride neutral solutions. This is a peculiar feature of chitosan and epichlorohydrin chitosan, in so far as they would be classified as strongly basic anion-exchangers with no dissociation capacity for neutral salts.

The chloride capacity of epichlorohydrin chitosan, according to Nuguchi, Arato & Komai (1969) is 4.93, 4.06 and 3.93 $mmol \times g^{-1}$ for hydrochloric acid concentrations 0.5, 0.2 and 0.1 N respectively. The breakthrough curves are reported in Fig. 5.1.

When DL-mandelic acid is passed through a column of epichlorohydrin chitosan, the effluent containing DL-mandelic acid is enriched with L-mandelic acid, while some D-mandelic acid is retained on the column; in fact, when elution is

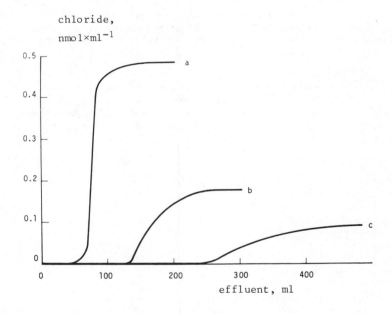

Fig. 5.1. Breakthrough curves for hydrochloric acid at the concentrations of: a = 0.5 M; b = 0.2 M and c = 0.1 M, for 45 × 1.65 cm columns of epichlorohydrin chitosan 45 ÷ 50 mesh, at the flow-rate of 1.5 $ml \times min^{-1}$. Noguchi, Arato & Komai, 1969.

performed with ammonia, the DL-mandelic acid which saturates the column is
enriched with D-mandelic acid, as shown in Fig. 5.2. The capacity for mandelic
acid is 3.7 $mmol \times g^{-1}$.

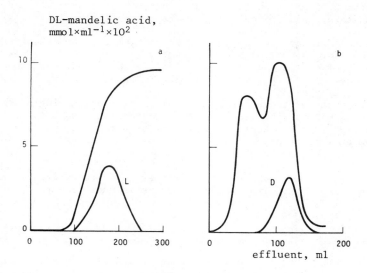

Fig. 5.2. Breakthrough (a) and elution (b) curves for 0.1
M DL-mandelic acid on 4.5 g epichlorohydrin chitosan
column (1.53 cm ϕ), flow-rate 0.5 $ml \times min^{-1}$, elution performed with ammonia 0.1 M. Symbols D and L refer to
enrichment performed, expressed in $mmol \times ml^{-1} \times 10^3$.
Noguchi, Tokura, Inomata & Asano, 1965.

Chitosan and epichlorohydrin chitosan have optically active conformation of D
type and therefore they can distinguish D and L forms of substances in solution. With acids, the interaction due to the ionic charges is prevalent
compared to the interactions due to the stereochemical configuration and
therefore the isomer separation is limited. The double peak in Fig. 5.2 is due
to the presence of primary and secondary amino groups, as the epichlorohydrin
chitosan used in this experiment was prepared without protecting the amino
group.

Nucleic acids.

Townsley (1961) anticipated that chitin can be used for nucleic acid chromatography. However, ten years had to pass before reading results concerning
nucleic acids on chitosan. In the meantime thin-layer chromatography of
nucleic acid constituents has been accomplished on diethylaminoethylcellulose,
ecteola cellulose and polyethyleneimine cellulose, on which clean and rapid
separations of minute amounts of these constituents were possible.

Powdered chitosan (2.4 g) was dissolved in 2.5 % formic acid (60 ml) and
diluted to 300 ml with distilled water and then filtered through a glass
filter. A suspension of cellulose powder (15 g) in the resulting 0.8 %
chitosan solution (60 ml) was homogenized in a glass homogenizer for 25 sec.
After deaeration with suction, the suspension was spread evenly on glass
plates to give 0.25 mm thick layers. The coated plates were allowed to dry
at room temperature and stored in a desiccator containing silica gel.

Layers of free chitosan can be prepared by soaking the chitosan formate layers
in 1 % ammonia for one hr, washing with water and drying in air. As far as the
chromatography of nucleic acids constituents is concerned, the two types of
layers were equivalents (Nagasawa, Watanabe & Ogano, 1970 and 1971).

For the development of these chitosan layers, solutions having pH values as
low as 2.5 were used as chitosan layers tolerate acidic solutions. The separation of nucleotides was achieved using pyridine formate system at pH 4.4 and
0.25 M ammonium formate at pH 4.0 and 6.5, and other solutions including 0.05
M formic acid at pH 2.5. Results are reported in Tab. 5.1.

The role of chitosan in lowering the R_f values is quite evident when a
comparison of microcrystalline cellulose alone is made with the same impreg-

Table 5.1. $R_f \times 100$ values for nucleosides and for nucleic bases on chitosan +
cellulose layers in: a = water; b = methanol + water, 1:1; c = ethanol + water
4:1; d = 0.05 M formic acid, pH 2.5; e = 0.5 M Pyridine, pH 9.2; f = 0.5 M
triethylammonium carbonate, pH 7.3. Nagasawa, Watanabe & Ogano, 1970.

Compound	Solvent					
	a	b	c	d	e	f
inosine	60	45	40	64	63	74
adenosine	42	39	51	43	51	44
guanosine	45	37	31	46	54	63
cytidine	63	56	47	71	70	69
uridine	71	57	60	75	72	77
deoxyadenosine	40	44	63	43	51	45
deoxyguanosine	46	43	50	48	56	64
deoxycytidine	63	62	67	74	69	69
deoxyuridine	70	61	61	77	70	77
tymidine	70	66	66	79	72	72
adenine	28	46	73	29	32	33
hypoxanthine	42	55	62	50	50	58
guanine	21	33	39	23	29	21
cytosine	49	64	67	70	64	65
uracil	58	66	80	70	63	67
thymine	58	71	92	70	65	65

Table 5.2. R$_f$×100 values for dinucleoside-3'→5' phosphates in pyridine + formic acid and in water + formic acid, on chitosan + cellulose layers. ¶ indicates tailing. Nagasawa, Watanabe e Ogano, 1971.

Compound		Pyridine + formic acid at pH values:			Aqueous formic acid 0.2 M
		2.6	3.0	4.4	
3'-adenylyl ester of	adenosine-5'	61	45	33	67
	guanosine-5'	37	30	41	39
	cytidine-5'	61	56	44	76
	uridine-5'	47	41	43	41
3'-guanylyl ester of	adenosine-5'	39	34	36	39
	guanosine-5'	16 ¶	32 ¶	32 ¶	10 ¶
	cytidine-5'	52	48	41	50
	uridine-5'	27	26	38	10
3'-cytidylyl ester of	adenosine-5'	63	63	50	73
	guanosine-5'	53	59	47	52
	cytidine-5'	71	71	57	76
	uridine-5'	66	58	53	56
3'-uridylyl ester of	adenosine-5'	54	52	49	49
	guanosine-5'	24	28	44	11
	cytidine-5'	66	62	58	58
	uridine-5'	38	43	57	13

nated with different amounts of chitosan.

For the separation of nucleotides, 0.25 M pyridine + formate at pH 2.6 was the best solvent system, as from data in Tab. 5.2.

Takeda & Tomida (1972) used chitin powder as a chromatographic support in thin layer chromatography, for the separation of nucleic acid derivatives. The resolution of chitin layers was almost equal to that of cellulose powder for nucleic acid bases (adenine, cytosine, guanine, thymine, uracil and hypoxanthine), nucleosides (adenosine, inosine, guanosine, cytidine and uridine) and nucleotides (monophosphates of adenosine, cytidine and uridine). Higher R$_f$ were reported for chitin thin layers than for cellulose layers, as shown in Tab. 5.3 and 5.4.

Chitin layers permitted a reduction of the developing time. The method was applied in the identification of nucleic acid bases in ribonucleic acid hydrolysate from yeast.

Amino acids.

Chitin was also used in thin-layer chromatography for the separation of amino acids by Takeda & Tomida (1969). Chitin thin layers have about the same

Table 5.3. $R_f \times 100$ values for nucleic acid bases and for nucleosides on chitin and cellulose thin layers, in: a = saturated ammonium sulfate aqueous solution + 7 M sodium acetate + 2-propanol, 40 + 9 +1; b = iso-butyric acid + aqueous ammonia + water, 33 + 1 + 16. Takeda & Tomida, 1972.

Compound	Solvent a		Solvent b	
	Chitin	Cellulose	Chitin	Cellulose
adenine	33	14	99	96
hypoxanthine	65	37	76	52
guanine	53	20	75	53
cytosine	77	59	88	88
uracil	80	58	76	56
thymine	68	47	86	73
adenosine	39	18	92	82
guanosine	59	38	65	41
inosine	68	50	68	42
cytidine	71	64	72	61
uridine	77	68	65	43

Table 5.4. $R_f \times 100$ values for nucleotides on chitin and cellulose thin layers in isobutyric acid + aqueous ammonia + water, 33 + 1 + 16. Takeda & Tomida, 1972.

Compound	Chitin	Cellulose
2'(3')-adenosine monophosphate	80	63
2'(3')-cytidine monophosphate	64	50
2'(3')-uridine monophosphate	53	34
5'-adenosine monophosphate	72	50
5'-inosine monophosphate	47	30
5'-cytidine monophosphate	61	44
5'-uridine monophosphate	49	32

resolution as silica gel layers; the basic amino acids, however, travel much faster on chitin and the R_f values of the acidic ones are decreased, as shown in Tab. 5.5.

Phenols.

Chitin is superior to silica gel and polyamide for thin-layer chromatography separation of some phenols: the results reported by Takeda & Tomida (1969) are in Fig. 5.3.

Table 5.5. R$_f$×100 values of some aminoacids on silica gel and on chitosan thin layers in n-propanol + water, 64 + 36 (w/w). Takeda & Tomida, 1969.

Aminoacid	Observed		Referred to leucine	
	Silica gel	Chitin	Silica gel	Chitin
glycine	32	54	58	68
alanine	37	61	67	76
valine	45	75	82	94
proline	26	61	47	76
methionine	51	76	93	95
cystine	32	29	58	36
aspartic acid	33	17	60	21
glutamic acid	35	26	64	33
arginine	2	48	4	60
lysine	2	51	4	64
histidine	20	58	36	73
phenylalanine	58	80	105	100
tyrosine	57	70	104	88
serine	35	57	64	71
leucine	55	80	100	100

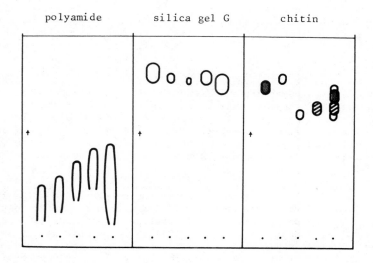

Fig. 5.3. Chromatograms on polyamide, silica gel G and chitin with water + acetone (4:1). From the left: resorcinol, hydroquinone, pyrogallol, phloroglucinol and mixture. Takeda & Tomida, 1969.

Inorganic oxyanions.

Muzzarelli & Rocchetti (1974) have experimentally verified that the sulfate anion is normally involved in the collection of transition metal ions when present in their solutions and helps in obtaining narrower chromatographic bands of the transition metal ions. When sulfate + sulfuric acid mixtures are used to perform elutions, a metal ion can be eluted from a chitosan column; the column can then be recycled.

Columns (1 × 24 cm) were prepared with chitosan (6 g) or sulfate-conditioned chitosan, then 0.05 M copper chloride or sulfate solutions (20 ml) were passed through and water (150 ml) was added for washing. It was found that the shortest bands (6 cm) of copper appeared in the sulfate-treated chitosan columns, while the washing with water produced a broadening of the band down to the bottom of the column in the case of absence of sulfate. As Tab. 5.6 shows, the best retention (no band broadening, based on the blue color developed by copper) is when sulfate accompanies copper, while the sulfate present in the conditioned columns ensures minimum band length. Moreover, it can be seen that 90 % of the sulfate added with the copper is retained in the column, against 76 % of the chloride. When sulfate and chloride are together, the chloride is completely washed out. This is the reverse of the current rules for anion-exchangers and stresses the particular physico-chemical requirements of chitosan.

Sulfuric acid in admixture with ammonium sulfate is quite a useful agent for enhancing the collection of transition metal oxyanions on chitosan, as shown in Tab. 5.7.

Elution of molybdate from chitosan can be carried out partially with carbonate with a yield of 32 %, while from p-aminobenzylcellulose and from diethylamino-ethylcellulose the elution yields are 95 and 89 %, respectively.

Muzzarelli & Rocchetti (1973) have proposed a new method for the determination of molybdenum is sea water, based on the combined collection of molybdate on chitosan microcolumns from as little as 50 ml of sea water and the atomic absorption determination of molybdenum on a homogenized column fraction. In view of the exceptionally strong interaction of molybdate with chitosan, which represents more than anion-exchange, but also polyoxysalt formation, conditioning of the chitosan column can be omitted. A comparison of methods is in Table 5.8.

The method based on the use of chitosan for the determination of molybdenum in sea water compares favorably with other methods. When ten determinations were done on aliquots of one sea water sample, the mean value was found to be 10.42 μg Mo×ml^{-1} with a standard deviation of ±0.13 μg. The columns are small and do not require conditioning, the sea water volume is very small and simplifies collection and transportation of a large number of samples, no chemicals need be be added except for the acid of the pH adjustment, elution is optional and much simpler. This method can be easily applied to routine determinations on large numbers of samples with limited glassware and personnel.

Chitosan has also been used by Muzzarelli and Rocchetti (1974) for the determination of vanadium in sea water at the concentration of 0.52 μg V×l^{-1},

Table 5.6. Anion release from a 6 g, 1×24 cm chitosan column (1 mm grain size) treated with 20 ml of 0.05 M copper solution and washed with 150 ml of water. Muzzarelli & Rocchetti, 1974.

solutions passed	$CuCl_2$	$CuCl_2$ + $(NH_4)_2SO_4$	$CuCl_2$	$CuSO_4$
column conditioning	none	none	$(NH_4)_2SO_4$	none
copper band height, cm				
before washing	12	12	16	12
after washing	24	12	18	12
eluate pH	8	8	8	8
ratio $\dfrac{\text{anion in column}}{\text{anion in eluate}}$	3.297	0.099	0.087	9.000
		chloride		sulfate

Table 5.7. Metal oxyanion collection on 200 mg of chitosan, expressed as fraction (%) extracted from 50 ml of 0.5 mM solutions at various pH values. Muzzarelli & Rocchetti, 1974.

pH	Time, hr	Ammonium sulfate absent			Ammonium sulfate 0.1 M		
		VO_4^3	CrO_4^2	MoO_4^3	VO_4^3	CrO_4^2	MoO_4^3
		Chitosan free base			*Chitosan free base*		
1.0	1	5	0	0	12	18	0
	12	12	0	0	23	18	0
3.0	1	89	88	67	78	34	6
	12	93	90	67	84	46	17
5.0	1	50	36	18	78	18	0
	12	70	38	27	82	25	10
		Chitosan conditioned in 0.1 M H_2SO_4			*Chitosan conditioned in 0.1 M H_2SO_4 + 0.1 M $(NH_4)_2SO_4$*		
1.0	1	0	12	10	0	17	15
	12	0	12	12	0	17	0
3.0	1	95	91	100	75	17	85
	12	100	95	100	60	20	85
5.0	1	100	92	100	30	28	30
	12	100	93	100	20	35	0

in form of columns made with 500 mg of chitosan. Vanadium, which occurs in sea water as vanadate, can be easily separated from the large amounts of salts present in sea water, and can be determined by atomic absorption spectrometry on the chitosan itself. Atomic absorption spectrometry and x-ray fluorescence spectrometry permit the direct determination of metals on chromatographic supports, provided that they are homogenized before analysis.

Table 5.8. A comparison of methods for the determination of molybdenum in sea water. Muzzarelli & Rocchetti, 1973

	Chitosan	p-aminobenzyl-cellulose	Dowex-A-1	Dowex 1- 8	Dowex 1- 8
Amount of polymer, g	0.5	0.5	-	10.0	10.0
Column conditioning	none	none	nitrate	chloride	thiocyanate
Column dimensions, cm	1×3	1×3	1×6	2.5×3.5	2.5×3.5
Sea water volume, ml	10 to 50	50	3000	1000	1000
pH adjustment	2.5	2.5	5.0	0.1 M HCl $+ 6$ % H_2O_2	0.1 M HCl $+$ SCN$^-$
Washing, ml	15 water	15 water	200 water	50, 0.5 M NaCl	100, 1 M H_2SO_4
Eluate, ml	none	10, 1 M $(NH_4)_2CO_3$	24, 2 N NH_4OH	60, 0.5 M NaOH $+$ NaCl	60, 0.5 M NaOH$+$NaCl
Further treatments	none	none	required	required	required

CHELATION CHROMATOGRAPHY

Chitosan is a most suitable polymer, compared to Dowex A-1, diethylaminoethyl-cellulose and other polymers, for the collection of copper from solutions, brines and natural waters (see page 140). Chelation chromatography can be easily demonstrated for technical and teaching purposes: for instance a system of two aquaria connected through a chitosan filter unit has been on display at the Sep-Pollution Exhibition in Padua, Italy (1972). The upper aquarium contained copper at a concentration toxic to several fishes, while in the lower aquarium a fish very sensitive to copper, *Alburnus albidus*, could survive because the copper, coming from the upper aquarium, was retained in a visible blue band on chitosan (Muzzarelli & Marinelli, 1972). The separation of copper from sodium on chitosan columns has been elaborated as a teaching experiment which has been carried out by hundreds of students (Muzzarelli, unpublished).

Eluents for copper and other transition metals should fulfil several require-ments: they should be effective enough to take copper out of chitosan in a narrow chromatographic band; they should not attack or degrade the polymer; they should be sufficiently free of the metal under study to yield negligible background signals on atomic absorption spectrometry or other analytical technique adopted or they should be low-priced enough for industrial purposes.

The data for several complexing agents are reported in Tab. 5.9: they are relevant to copper and nickel, two of the most efficiently retained cations. 1,10-Phenanthroline as a 1 % solution was tested and found quite suitable for elution when applied at 50°. The disodium salt of EDTA can elute copper even more easily as can ethylenediamine, but they are not suitable because 0.1 M EDTA contains so much copper that it prevents the determination, while ethylenediamine slightly attacks the polymer. Dimethylglyoxime is also a good eluting agent for copper but not for nickel. 0.1 M Sulfuric acid combines good

Table 5.9. Determination of copper and nickel in eluates from chitosan columns. Muzzarelli & Rocchetti, 1974.

Eluent	Copper elution yield, %	Nickel elution yield, %
ethanol	0	0
methanol	0	0
8-quinolinol, 0.1 %	0	0
bipyridine, 0.1 %	0	0
dimethylglyoxime, 1 %	100	partial
1,10-phenanthroline, 0.1 % at 50°	partial	partial
1,10-phenanthroline, 1 % at 20°	100	80
sulfuric acid, 0.1 to 1 M	100	100

elution ability with purity and low price ; it is especially suitable when
used in admixture with ammonium sulfate. Details on column operation are to be
found on page 216.

Determination of copper in sea water.

Chitosan has been found suitable for metal ion collection by chelation in the
prevention and monitoring of inland sea-water pollution by Muzzarelli (1972)
and by Muzzarelli & Rocchetti (1974). Among transition metal ions, copper can
be easily collected from sea water by chitosan. Copper is mainly present in
sea water as Cu^{2+}, $CuCl^+$ and $|Cu(HCO_3).2OH|^-$. However, little information is
available on organic complexes in sea water, and on colloidal forms of copper;
most of the anaytical methods for the determination of copper in sea water
include the addition of acids in order to destroy any organic complexes.

Studies on the isotope exchanges of ^{54}Mn, ^{60}Co and ^{65}Zn showed that spikes do
not go into equilibrium with the natural forms of these elements in sea water
for many hours. Research on model systems seems to indicate that amino acids
can play a role in the complexation of metal ions.

For the sea water analyses, small columns containing 100 mg of chitosan were
used. When one-l fractions of a single sea water sample were analyzed by
collection of copper on these chitosan columns and elution with 1,10-phenan-
throline solution, the reproducibility was good. For ten separate determinations, the average value was found to be 3.39 µg $Cu×l^{-1}$, with a standard
deviation of ±0.25 µg $Cu×l^{-1}$. Of course these results refer to certain
chemical species of copper only.

As copper in sea water occurs also in the above mentioned complex forms which
may escape fixation because of their stability, some measurements were made on
the sea water after treatment with persulfate; this oxidative treatment
destroys the organic matter which complexes part of the copper present in sea
water and thus makes it available for subsequent interaction with the
chelating polymer. When this procedure was applied to the above sea water
sample, the average value of ten determinations was 6.06 µg $Cu×l^{-1}$, the
standard deviation being ±0.56 µg $Cu×l^{-1}$. Copper could also be eluted from a
chitosan column with 0.1 M sulfuric acid at room temperature, with an overall
yield of 100 %. A comparison of methods is presented in Tab. 5.10.

According to the works of Muzzarelli & Rocchetti (1974) and of Florence &
Batley (1975 and 1976) it is apparent that Dowex A-1, the only man-made
chelating resin so far marketed, cannot be used in the manner described by
Riley & Taylor (1968) for the quantitative concentration of zinc, cadmium,
lead and copper from sea water. Even at high pH values a powerful chelating
agent similar to EDTA would be bound by the relatively high concentrations of
magnesium, calcium and iron present in sea water.

It is unfortunate that the Dowex A-1 resin, otherwise called Chelex-100, has
found numerous occasions of use by several authors in the field of chemical
oceanography; they have assumed, without sufficient analytical background,
that the chelating power of the resin was great enough to overcome the natural
equilibria existing in sea water between metal ions and "weak" complexing
agents; on the contrary these equilibria are very difficult to shift in the
favor of the formation of a chelate with Dowex A-1: collection of only part of

Table 5.10. Comparison of methods for the determination of copper in sea water.

	Method of Riley & Taylor, 1968	Methods of Muzzarelli & Rocchetti, 1974	
Polymer	Dowex A-1	Chitosan	Chitosan
Column size, mm	60×12	15×3	15×3
Sea water volume, l	1	1	1
Washing, ml	250	15	15
Eluent	30 ml, 2 M HNO_3	20 ml of 1 % 1,10-phenanthroline	20 ml of 1 M $\div$ 0.1 M H_2SO_4
Elution temperature	room	50°	room
Instrumental detn.	Atomic absorption	Hot graphite atomic absorption	

the metals takes weeks or months of contact.

The results obtained independently by Muzzarelli & Rocchetti (1974) and by Florey & Batley (1975 and 1976) have shown that all the metal ion concentrations in sea water obtained with the use of Dowex A-1 (Chelex) so far published, are not correct, or at least, they refer to a minor ionic portion of the transition metal ions under examination.

The fact that such a situation in the scientific field could take place and persist for several years, underlines the scarce attention that has so far been given to natural equilibria, in contrast with an even exaggerated endeavour to demonstrate that an artifical polymer shows chelating ability in artificial solutions.

AFFINITY CHROMATOGRAPHY.

Several proteins exhibit specific affinity toward glucosamine, chitobiose and other oligomers up to chitohexose. These oligomers obtained from chitin have been used to study the mechanism of action of lysozymes. A N-acetyl-β-D-hexosaminidase has been prepared by Rafestin, Obrenovitch, Oglin & Monsigny (1974) and β-galactosidase, β-N-acetylglucosaminidase and α-mannosidase have been purified by affinity chromatography on a modified Sepharose including p-aminobenzyl-1-thio-2-acetamido-2-deoxy-β-D-glucopyranoside. Several authors have reported on the derivatives of 1-thiochitobiose. Delmotte & Monsigny (1974) have reported the synthesis of p-nitrobenzyl-1-thio-β-chitobioside and 1-thio-β-chitotrioside, and p-nitrophenyl-β-chitotrioside.

Lysozymes.

Chicken egg lysozyme has had numerous clinical applications, as an anti-bacteriolytic agent against infectious diseases, as a therapeutic agent in the treatment of wounds and as a potentiator of many antibiotics. However,

applications of chicken lysozyme are limited because of immunological
responses in humans. Human lysozyme is not easily available because there are
difficulties in obtaining it in a pure enough form. Large amounts of human and
animal lysozymes will enable expanded research on use of lysozyme bound to
antibiotics which has a longer lasting effect and requires much less antibiotic concentration.

Affinity chromatography achieves enrichment of lysozyme to a concentration in
the final eluate of 1500 to 10,000 times that in tissue homogenates, not only
from human tissues but also from phages, bacteria, insects, plants, bird eggs
and animal tissues.

Chitin can be used as a support for affinity chromatography of homogenates
from human tissues: while the lysozyme is retained in the column, other
proteins flow through and traces of them are washed out with distilled water.
Chitin was used as a specific adsorbent for lysozyme by Kravchenko (1967) and
other authors. To improve the holding capacity and reproducibility of the
adsorbent, carboxymethylchitin and deaminated chitin were used by Cherkasov &
Kravchenko (1969). It was found that chitin binds lysozyme tightly at pH
between 7 and 9 under low ionic strength; the bound lysozyme was eluted
quantitatively with 0.2 M acetic acid. Similar results were obtained with
chitin columns.

Lysozyme containing solutions have been passed through deaminated squid chitin
columns at 20 ml$\times$cm$^{-2}\times$hr^{-1}; 120 column volumes of water were used for washing.
2 N Hydrochloric acid or 1 mM sulfuric acid were used for elution; 10 mM
acetic acid allows rapid elution due to the solubility of lysozyme in this
medium and offers the advantage of easy removal by evaporation or lyophiliza-
tion. Many experimental details are given on this technique by Katz, Fishman &
Levy (1975).

The pH dependence of the complex formation between lysozyme and partially
hydrolyzed glycolchitin with or without buffer solutions of constant ionic
strength, shows a peak centered at pH 6. Very similar results were obtained
with chitin by Hayashi, Hamasu, Doi & Funatsu (1973).

The pH dependence of adsorption of lysozyme on chitin columns is shown in Fig.
5.4: the maximum adsorption occurs at pH 9, but in the presence of 0.5 M
sodium chloride, the adsorbed amount was not pH-dependent in the rage of pH
2 $\div$ 8. The pH dependence of adsorbed lysozyme on a column of insoluble substrate showed only a plateau, when a limited amount of lysozyme was introduced
into the column; but when an excess amount of lysozyme was applied, the pH vs
adsorption profile showed again a peak possibly associated with two types of
complexes with lysozyme.

Imoto & Yagishita (1973) prepared chitin-coated cellulose columns, by treating
alkali chitin with cellulose and neutralizing with acetic acid and water. Up
to 5 mg of lysozyme were specifically adsorbed per gram of this polymer and
desorbed quantitatively under mild conditions (0.1 M acetic acid). On the
other hand, ovalbumin was no retained and was washed out in two free column
volumes; washing with 6 $\div$ 8 free column volumes is sufficient for the total
removal of ovalbumin because it is not expected to react with the acetylated
amino groups, and the primary amino groups are few.

Imoto & Yagishita (1973) added to filtered sample solutions 6.5 ml of chitin-

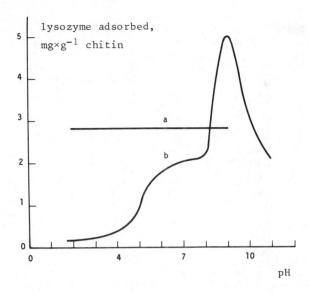

Fig. 5.4. pH-Dependence of adsorption of lysozyme on chitin columns; a = 0.1 M buffer solution containing 0.5 M sodium chloride; b = 0.1 M buffer solution. Hayashi, Hamasu, Doi & Funatsu, 1973.

coated cellulose equilibrated with 0.1 M phosphate buffer at pH 8 containing 1 M sodium chloride. After stirring for 20 min, the polymer was packed in a column 1 × 8 cm. The column was eluted first with 30 ml of the same buffer and then with 0.1 M acetic acid, at the flow-rates of 10 and 3 ml×hr^{-1}, respectively. The resulting elution curves for saliva, extracts of taro and yam are in Fig. 5.5. Enzyme activity was measured against glycolchitin and bacteria substrate. Chitin-coated cellulose was proposed as a good adsorbent for the investigation of the distribution of lysozyme-like enzymes in nature and for their isolation and preparation on a large scale.

Cherkasov & Kravchenko (1970) used a chromatographic column of finely powdered chitin to study the binding of lysozyme. They defined the affinity coefficient $k_a = V_r / V_0$ on the basis of the column retention volume V_r which is the difference between the peak elution volume and the column void volume. For evaluation of the binding constant K_b the concentration dependence of k_a was determined by passing various amounts of lysozyme (2 ÷ 30 mg) through the 0.9 × 19 cm column and by plotting $\log(k_a^\circ - k_a) / k_a$ vs concentration, as from Fig. 5.6 , where k_a° is the value corresponding to zero lysozyme concentration. The slope of this plot is unity, confirming the formation of a one-to-one complex. The values of K_b at pH 4.7 and 8.4 were 1.15×10^5 and 8.95×10^5 M^{-1}, respectively . Also the values of the free energy and entropy were calculated. Cherkasov & Kravchenko (1970) suggested that at pH 8.4 a pseudoproductive

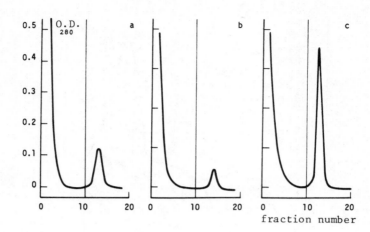

Fig. 5.5. Chromatography on chitin-coated cellulose
columns (1 × 8 cm) in phosphate buffer at pH 8 containing
1 M sodium chloride; a = saliva; b = extracts of taro;
c = yam. The elution was performed with 0.1 acetic acid.
Imoto & Yagishita, 1973.

complex takes place, with all six binding sites, without hydrolysis taking
place provided that the chain length is sufficient. This can also give
reasonable explanation to the discrepancies between the binding data for
chitin and those for chitotriose at pH 8 ÷ 9.

In fact, using the method of affinity chromatography of chitosaccharides on
chitin-immobilized lysozyme, Cherkasov & Kravchenko (1975) showed that this
complex includes 4-membered parts of chitin, since mono-, di- and trisaccharides do not bind to lysozyme at pH 8.4.

Chitin columns have also been used by Jensen & Kleppe (1972) both for the
concentration of the T4 lysozyme and for its purification. The columns, which
allowed a 150 fold purification, could be used repeatedly after conditioning
with buffers. As chitin is scarcely digested by lysozyme, a column can be used
daily for at least one month to bind lysozyme without any apparent change in
capacity. T4 Lysozyme binds at neutral pH values; 60 % can be eluted with
distilled water, and the remaining 40 % with 0.1 M acetic acid. The hen egg
white and the T4 lysozyme behave differently with respect to binding and
elution from chitin. The hen egg-white lysozyme binds at neutral pH only when
the ionic strength is 0.05 or higher, whilst T4 lysozyme binds only when the
ionic strength is low. Most T4 lysozyme can be desorbed from chitin at neutral
pH, by increasing the ionic strength above 0.1 and complete desorption can be
achieved by lowering the pH and increasing the ionic strength. Lysozyme can be
eluted from chitin not only by large amounts of buffers but also by mono-
ethylamine hydrochloride.

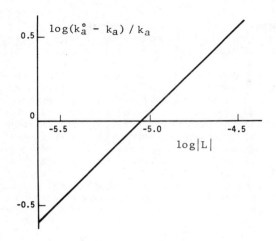

Fig.5.6. Determination of the binding constant of lysozyme
to chitin. Chitin column 0.9 × 19 cm; chitin particles
50 μm. Samples of lysozyme applied: 2, 4, 5, 6, 8, 10 and
15 mg. V_o = 18 ml; k_a^o = 4; pH 4.7; 25°. Cherkasov &
Kravchenko, 1970.

Immobilized enzymes.

Enzymes have been so far used mainly in solution as homogeneous catalysts. For
many applications this is not the most efficient way to use these catalysts.
The advantages of using immobilized enzymes are that enzymic reactions can be
stopped at any desired time by simply separating the solution under treatment
from the immobilized enzyme; a certain amount of enzyme can be used to treat
large amounts of substrate solutions without removing it from its support; the
enzyme is not lost in the product; there is no product inhibition to limit the
extent of the reaction; the enzyme does not undergo self-digestion; enzymes
from pathogenic organisms can be used and operations at altered pH optima are
made possible by modifying the charge characteristics of the support.

The choice of chitin and chitosan as solid supports for enzyme immobilization
is a novel approach to the use of enzymes: chitin has been recently used to
immobilize acid phosphatase, α-chymotrypsin, glucose isomerase, β-galactosidase
and D-glucose oxidase by Stanely, Watters, Chan & Mercer (1975), Stanley &
Watters (1975), Stanley, Watters, Kelly, Chan, Garibaldi & Schade (1976),
Kennedy & Doyle (1973), Masri, Randall & Stanely (1975) and Muzzarelli, Barontini & Rocchetti (1976). The amino groups present on these polymers can be of
help for anchoring bridge molecules like glutaraldehyde for the purpose of
establishing covalent bonds with the protein. Chitosan lends itself to the
formation of buffer systems and insolubilizing an enzyme produces a shift in

its optimum pH, generally to a lower value; these aspects are very important as they make certain operations easier since most products of the food industry are acidic.

Microbial preparations which contain enzymes, typically, cultures or cells of yeasts, molds and bacteria can be immobilized as well. Other materials like pancreas, liver, insect parts, papaya etcetera are susceptible of similar use.

The immobilization extent depends on the time of contact between support and enzyme, on the pH of the medium, on the presence of salts and transition metal ions, on the chemical derivatization of chitosan and on the technique used for the immobilization (shaking, percolation or contact). The best conditions for the immobilization of α-chymotrypsin are as follows: chitosan (1.5 g) is introduced into a jacketed column, washed with phosphate buffer at pH 8.5 until the effluent reaches the same pH value; 1.5 ml of solution containing 300 μg of α-chymotrypsin are introduced at the top of the column, together with 1.5 ml of buffer or with 1.5 ml of a 4 % glutaraldehyde solution, if desired. These 3 ml, corresponding to one free column volume, are allowed to penetrate slowly through the column and left standing for 1 hr. Washing is done with 50 ml of buffer.

The immobilization percentages relevant to 1.2 mg of acid phosphatase on a column of chitosan preconditioned at pH 6.5 with 0.1 N sulfuric acid are 100 %. The chitosan + chitosan sulfate + sulfuric acid system behaves as a buffer medium and therefore the use of a buffer solution can be avoided. The acid phosphatase solution is introduced at pH 6.5 and kept in contact with the support for 1 hr, as indicated above. Whilst the presence of copper brings about an immobilization yield decrease, the use of glutaraldehyde is superfluous (Muzzarelli, Barontini & Rocchetti, 1976).

The best conditions for the immobilization of acid phosphatase would be at pH 6.5; however, in view of the activity of the immobilized acid phosphatase, the following conditions should be adopted: chitosan (1.5 g) is suspended in water and 0.1 N sulfuric acid is added slowly until the pH is 5.0. This powder is then used to prepare a chromatographic column. The enzyme solution (1.5 ml), containing 1.2 mg of acid phosphatase and the sulfuric acid solution (1.5 ml) are introduced at the top of the column and allowed to penetrate it. After 1 hr contact at 30°, the column is rinsed with sulfuric acid at pH 5.0 (50 ml).

Fig. 5.7 shows the immobilization of acid phosphatase as a function of pH. The best pH interval for immobilization is between 6 and 7. The presence of phosphate buffer would sensibly lower the immobilization yield at these pH values. Acid phosphatase can be eluted from the column at pH below 4 and above 9. Fig. 5.7 further shows that the best pH interval for immobilization for α-chymotrypsin is between 8 and 9: beyond these values, a minor part of the enzyme brought into contact with the polymer can be immobilized. Below pH 5 and above pH 11 the complete elution of the enzyme becomes feasible.

The activities of the immobilized enzymes have been measured by Muzzarelli, Barontini & Rocchetti (1976) by stepwise percolating free column volumes of substrate solutions. The activities of the immobilized enzymes depend on the pH at which the enzymic reaction occurs. Fig. 5.8 shows the activity of a constant quantity of enzyme immobilized in various columns of chitosan at different pH values, at constant temperature. The highest activity of immobilized α-chymotrypsin is at pH 8.5, while the highest activity of acid

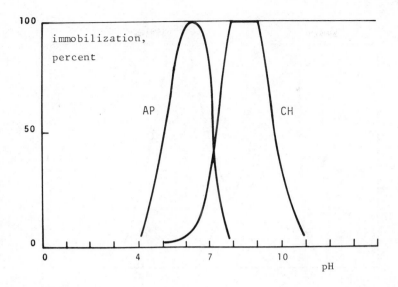

Fig. 5.7. Immobilization percent *vs* pH curves for acid
phosphatase (AP) and for α-chymotrypsin (CH), 1.2 mg and
300 μg, respectively, on 1.5 g chitosan columns.
Muzzarelli, Barontini & Rocchetti, 1976.

phosphatase is at pH 5.0. These pH values are practically the same as those
reported by Stanley, Watters, Chan & Mercer (1975) for these enzymes
immobilized on chitin. The shift towards more acidic pH values of the maximum
activity pH for immobilized acid phosphatase is quite sharp, compared to the
usual working pH for dissolved acid phosphatase. These pH values of maximum
activity allow working with enzyme-chitosan columns; at these pH values there
is no washing off of the enzyme.

Enzyme activity depends on the time of contact between immobilized enzyme and
substrate. This is quite important in planning the best column operation
conditions. It has been observed that the transformation of the substrate is
9 % when 1 mM ATEE solution is continuously percolated through the column at
the flow-rate of 0.6 ml×min^{-1}; on the contrary, if a free column volume is
rapidly introduced into the column and left standing there, the activity of
the immobilized enzyme reaches a plateau after 15 min of contact at 65 % of
the substrate transformation extent for acid phosphatase and 25 % for α-chymotrypsin. The values reported by Stanley & Watters (1975) for chitin + glutaraldehyde columns were expressed in terms of percentage of the original soluble
enzyme activity: 20 % for acid phosphatase and 14 % for α-chymotrypsin.

No loss of enzyme occurs during protracted column operation and no progressive
inactivation of the enzymes takes place after extended use, on the basis of

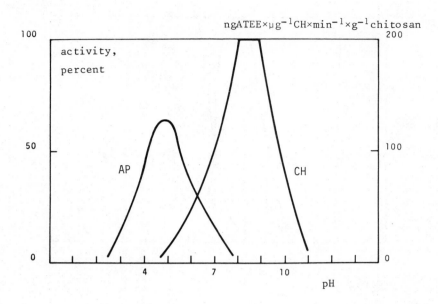

Fig. 5.8. Enzyme activity *vs* pH curves for acid phosphatase
(AP) and for α-chymotrypsin (CH), expressed respectively
as percentage of the soluble acid phosphatase activity
and as velocity of substrate transformation by the action
of immobilized α-chymotrypsin. Muzzarelli, Barontini &
Rocchetti, 1976.

the available data.

Chitosan is quite a suitable support for the immobilization of α-chymotrypsin
and acid phosphatase; these enzymes keep most of their activity and chitosan
enzyme columns can be submitted to extended use with no enzyme inactivation or
loss. Immobilization of these enzymes on chitosan can be either performed
through a covalent link with the help of a reagent like glutarladehyde, or
through such an interaction that permits eventually the enzyme recovery. The
regular distribution of primary amino groups in the polymer seems to be of
importance. The activities of the enzymes immobilized on chitosan are sharply
higher than those of the corresponding enzymes immobilized on chitin with
glutaraldehyde.

The activity of the enzymes immobilized on chitin with glutaraldehyde is also
retained over long periods of use. For instance, lactase immobilized on chitin
was used continuously for over two weeks and the lactase still retained more
than 80 % of its original activity (Stanley & Watters, 1975).

Other immobilization techniques can also be used in the case of chitosan: for instance one could take advantage of the insolubilization of dissolved chitosan for entrapping enzymes. This could be done by neutralization with sodium hydroxide of chitosan solutions in acetic acid where the enzyme has been added or by precipitation of chitosan sulfate upon addition of ammonium sulfate to a similar solution. Crosslinking and gel formation according to the reactions described in Chapter 3 are also possible, for enzyme entrapment.

The research on the immobilization of enzymes on chitin and chitosan started very recently and it is anticipated that important developments in this field will appear soon.

Wheat germ agglutinin.

Affinity chromatography on chitin has been used to purify wheat germ agglutinin, a plant lectin which has importance in the study of the role of carbohydrates in the structure and functions of cell membranes. It is highly specific for N-acetylglucosamine and its specific binding has been the basis of several procedures for its purification by affinity chromatography.

Ovomucoid, a glycoprotein rich in N-acetylglucosamine was used as the immobilizing ligand; affinity chromatography was also performed on Sepharose- -2-acetamido-N-(ε-aminocaproyl)-2-deoxy-β-D-glucopyranosylamine by Lotan, Gussin, Lis & Sharon (1973), on Sepharose-asparaginyl-N-acetylglucosamine by Wang, Sevier, David & Reisfeld (1975) and on Sepharose-6-amino-1-hexyl-N- -acetylglucosamine.

The influence of tri-N-acetylchitotriose binding on luminescence properties of wheat germ agglutinin was studied by Privat & Monsigny (1975) who analyzed the results in terms of steric tryptophan-ligand relations.

Chitin, on the other hand, when ground and packed into a column, serves both as matrix and ligand in affinity chromatography of wheat germ agglutinin. In order to follow such a purification scheme, Bloch & Burger (1974) determined how much wheat germ agglutinin could be bound per gram of chitin. This was investigated in two ways: by titrating a solution of agglutinin with increasing amounts of chitin; by saturating a small amount of chitin with a concentrated solution of agglutinin, washing the chitin free of unbound agglutinin, and then eluting the bound protein with 0.05 N hydrochloric acid. The results of these experiments indicate that 1 g of chitin binds approximately 10 mg of wheat germ agglutinin.

One kg of crude wheat germ was ground in a mill and extracted with 0.05 N hydrochloric acid (10 l) for one hr at room temperature. The mixture was centrifuged at 2000 g for 10 min and the extract was brought to 35 % saturation with ammonium sulfate. The precipitate was collected at 10,000 g for 15 min, suspended in 0.05 N hydrochloric acid (one l) and dialyzed extensively against water and then against 0.01 M Tris-HCl, pH 8.5. The soluble material was mixed with DEAE-cellulose (400 ml) contained in a sintered glass funnel. The polymer was washed with 0.01 M Tris-HCl, pH 8.5, until only negligible agglutinating activity could be eluted. All fractions containing activity were pooled and applied to a chitin column.

The chitin column was first developed with 0.01 M Tris-HCl, pH 8.5, containing 1 M sodium chloride. This solution elutes protein which either does not bind

to the column, or binds nonspecifically. The column is then washed with 0.01 M
Tris-HCl to remove sodium chloride. If this step is omitted, very little wheat
germ agglutinin is recovered in the following step, presumably because the
protein precipitates on the support. Finally, the column is eluted with 0.05 M
hydrochloric acid: wheat germ agglutinin is eluted as the pH decreases, as
shown in Fig. 5.9.

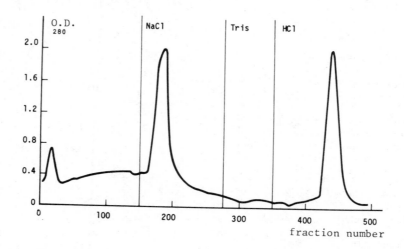

Fig. 5.9. Affinity chromatography of partially purified
wheat germ agglutinin on a chitin column. The effluent
(2.4 l) from DEAE-cellulose was applied to the chitin
column (250 g, 35 × 5.4 cm), buffered at pH 8.5. Flow-
rate one l×hr^{-1}, fractions of 15 ml. The agglutinating
activity strictly corresponds to the second peak. Bloch
& Burger, 1974.

The simplicity, speed and economy of chitin affinity chromatography suggests
that it should be useful for large scale purification of wheat germ agglutinin.
This procedure has several advantages; it does not rely on crystallization as
a final purification step; it entails neither the preparation of ligands nor
their coupling to a resin, and is considerably more economical. It is quite
rapid, and can easily be adapted for large scale purifications. Chitin affinity
chromatography is also used for the separation of inactive protein and modifying reagents from modified but active agglutinin. Iodinated and fluorescein
labeled wheat germ agglutinins have been prepared using this technique.

Tobacco mosaic virus.

Chitin columns (17 × 1.3 cm) equilibrated with hydrochloric acid solutions at
pH 6.8 were used in chromatography of the tobacco mosaic virus by Townsley
(1961). Tobacco mosaic virus solution (0.5 ml) containing 0.3 ÷ 2.0 mg of

virus in 10 mM sodium phosphate buffer at pH 6.8, was added to the column and
the virus washed into it with three 1-ml aliquots of hydrochloric acid. The
column was then connected for gradient elution, from water to 0.5 M K_2HPO_4.
The eluate was collected at 10-min intervals and monitored for optical density
at 254 nm.

The material eluted at pH 7.2 was four to ten times as infective as that
eluted at pH 7.7. Chitin was found to be very useful for this separation as
the treatment for adsorption and elution is very mild.

FUTURE DEVELOPMENTS.

Future uses of chitosan in chromatography can be easily imagined in the
various branches of this analytical and industrial technique.

Ligand-exchange chromatography on chitosan could follow the same lines of
development as with substituted celluloses, for instance in the insolubilization of active antibiotics (Kennedy, Barker & Zamir, 1974), for amino acid
separations, for steroid purifications (Pellizzari, Liu, Twine & Cook, 1973)
or for separations of organic compounds (Kunzru & Frei, 1974).

Triethylaminoethylcellulose has been used under a pressure of 210 Kg×cm^{-2} and
at temperatures as high as 80°, at flow-rates of 3 ml×min^{-1}, therefore chitosan, which is more rigid than celluloses should be suitable (Morozowich, 1974).

Various kinds of particles have been encapsulated with chitosan (Hallan &
Hinkes, 1975) including glass beads: thus chitosan is also available in a
pellicular form.

A variety of gels of chitosan are also easily prepared, having various degrees
of cross-linking depending on the nature and the proportions of the cross-
linking agent and on the molecular weight of chitosan. Chitosan powders
obtained by liophilization of precipitated chitosan are suitable for preparation of thin layers, as such or in admixture with other polymers.

Chelation chromatography, as it will be apparent from the next Chapter, will
certainly find applications in radioisotope separations as soon as more
information will be reached on transuranium elements and fission products; in
water monitoring and treatment; for the preparation of water and other
solvents completely free of metals and for the recovery or removal of traces
of toxic or precious metals.

CHAPTER 6

INDUSTRIAL PRODUCTION
AND APPLICATIONS

Chitin and chitosan are presently industrially produced in the United States of America and in Japan. Marine Commodities International, Inc., Star Route Box 140, Shrimp Harbor, Brownsville, Tx 78520, turns out nearly 12 tons per month of chitin and 9 tons per month of chitosan, for exclusive distribution by Hercules, Inc., Wilmington, De 19899. The raw material comes from the shrimp fishing activities in Texas and Mexico. A smaller, pilot-plant unit is also producing chitin and chitosan from crab shells: Food, Chemical & Research Laboratory, Inc., 4900 Ninth ave. NW, Seattle, Wa 98107, operates a 3-ton per month facility; the commercial distribution is done by the Kypro Co., P.O. Box 239, Haverford, Pa 19041.

Kyowa Oil & Fat Co., a joint venture of Nippon Suisan Kaisha, Ltd. and Nippon Soda Co., Tokyo, Japan, pioneered commercial production of chitin and chitosan in 1972. Kyokuyo Co., Tokyo, Japan, also established a plant in 1974 at Narita City in Chiba Prefecture. The total output of the two companies exceeds 55 tons per month of chitosan, all of which is used domestically.

Chitin and its derivatives are also available through dealers of chemical products. Among others, the following catalogs list them: Sigma Chemical Co., chitin flakes, chitin powder, chitosan and chitinase; Phanstiehl Laboratories, chitin and chitosan; J.T. Baker, chitin and chitosan; Schuchardt, chitin; Eastman Organic Chemicals, chitin; ICN Pharmaceuticals, Inc.: chitin; K.E.K.: chitin, deacetylated chitin, chitin nitrate and chitosan; P.L. Biochemicals Inc., chitin and γ-chitin; Polyscience Inc., chitin and chitosan; Calbiochem, chitin, γ-chitin, γ-chitin-red and chitosan.

INDUSTRIAL PRODUCTION OF CHITIN AND CHITOSAN.

Resources and availability.

The extension of the resources of raw material, the evaluation of the costs and the operations involved, and the final assessment of the product price, under reasonable assumptions related to model systems, are necessary information usually requested as soon as the industrial applications of chitin are mentioned.

Two fundamental statements should be made preliminary to further considerations; first, chitinous organisms, mainly crabs, shrimps, prawns and krills are very abundant all over the world and a limited part of these resources is being exploited by the marine food industry which produces canned meat, crab meal, proteins, and of course, shell wastes. Second, the production of chitin is possible only as a secondary activity related to the marine food industry. These points may be illustrated separately.

The regions of the world where crustacea are abundant are especially the
United States of America, India, Thailand, Malaysia, Philippines, South Africa
and Mexico. To give an idea of the resources available, we can mention that
the wet waste from shrimp in Thailand raised from 60,000 tons×year^{-1} in 1968
to 110,000 in 1976; prawn waste in India is presently (1976) 59,000 tons×
year^{-1}. As for the USA, data from a very recent report on the industrial
prospects for chitin from sea food wastes, by Hattis & Murray (1976) indicate
that the total waste from crab and shrimp is such as that the amount of chitin
that can be obtained is 4747.9 ÷ 7869.8 tons×year^{-1}, as shown in Tab. 6.1.

These data are certainly high enough to encourage industrial applications of
chitin as far as the resource availability is concerned. Imported material
(see for example Tab. 6.2) and other resources should also be considered.

Little is known about the Antarctic krill *Euphausia superba* abundance, which
is certainly impressive, as a consequence of an ecological disequilibrium due
to the disappearence of the whales (of which mankind is responsible). Russian
and Japanese fleets are starting to fish krills and they need by-products to
offset the high cost of operating in those waters.

It may be underlined here that the exploitation of chitin would never
represent an ecological threat because it is only a consequence of the fishing
activity. Any concern about the depletion of the existing abundant resources,
should be carefully taken into account in the context, however, of the protein
and canned meat industrial production. It is foreseen that a 30-fold increase
in the shrimp meat market will take place in the next 20 years. Mariculture is
quickly developing to assist the production of the resources, especially for
prawns and it is expected that the advances in mariculture will prevent
further ecological tragedies.

Chitin can also be obtained from sources other than marine wastes: large
quantities of fungi currently grown in fermentation systems producing organic
acids, antibiotics and enzymes, constitute a potential source of chitin. The
single chemical prepared in fermentation systems on the largest scale is
citric acid. The 1973 production of citric acid was 90,000 tons, and it is
realistic to estimate the *Aspergillus niger* waste thus produced at a few tens
of thousands tons. Three fungal species, *Mucor rouxii*, *Phycomyces blakesleeanus* and *Zygorhynchus moelleri* have been found to contain significant quantities of chitosan which may thus provide a convenient source of the deacetylated form of chitin.

As for the second point, Hattis & Murray (1976) make a number of assumptions
on whose basis several considerations are developed. It is estimated that a
chitin production from dockside to customer is unfeasible or in any case would
be a mistake; therefore, the primary producer of raw material is expected to
sell a shell waste residual to a chitin producer.

Let us examine the typical situation assumed as a model for the United States:
shellfish processors possess the waste and any practical schema to utilize
such material must have sufficient attractiveness to them to obtain their
active participation. There is a general pressure from community and environmental authorities for shellfish processors to find alternatives to disposal
of shellfish wastes (Environmental Protection Agency, 1974 and 1975). The
chitin business offers an opportunity to these firms to both help eliminate a
waste disposal problem and upgrade the value of their output. On the other

Table 6.1. Potential products in differents areas of the United States of
America, from shrimp and crab wastes, expressed as tons of potential products
of different types per year. Hattis & Murray, 1976.

Area	Adventitious protein	Total protein	Total chitin
Maine	168.2 ÷ 326.5	244.0 ÷ 467.1	124.2 ÷ 250.8
Massachusetts	53.5 ÷ 104.3	78.4 ÷ 147.4	39.4 ÷ 77.5
Maryland + Delaware	424.1 ÷ 825.5	707.6 ÷1029.6	306.1 ÷ 441.3
Virginia	498.9 ÷ 975.2	834.6 ÷1215.6	161.5 ÷ 521.6
North Carolina	379.6 ÷ 730.2	621.4 ÷ 920.7	269.8 ÷ 397.3
South Carolina	101.6 ÷ 196.4	161.9 ÷ 257.6	71.6 ÷ 114.3
Georgia	334.7 ÷ 625.9	544.3 ÷ 798.3	249.4 ÷ 361.9
Florida I	69.4 ÷ 133.3	114.7 ÷ 167.8	50.8 ÷ 73.5
Florida II	40.8 ÷ 67.5	63.9 ÷ 88.9	34.0 ÷ 46.2
Florida Keys	52.6 ÷ 85.7	76.6 ÷ 123.8	35.8 ÷ 57.6
Florida W I	98.4 ÷ 158.3	152.4 ÷ 209.1	82.0 ÷ 110.2
Florida W II	171.9 ÷ 342.4	269.4 ÷ 453.5	137.8 ÷ 240.4
Alabama + Mississippi	217.7 ÷ 480.8	343.3 ÷ 648.6	166.4 ÷ 328.4
Louisiana I	412.7 ÷ 970.6	625.9 ÷1347.2	312.5 ÷ 684.9
Louisiana II	172.3 ÷ 357.8	278.5 ÷ 467.1	126.0 ÷ 214.0
Texas I	155.1 ÷ 270.3	245.8 ÷ 352.8	119.7 ÷ 168.7
Texas II	106.1 ÷ 170.5	162.3 ÷ 229.5	86.1 ÷ 119.2
Texas III	68.1 ÷ 92.9	97.5 ÷ 133.8	59.4 ÷ 79.3
California I	100.6 ÷ 196.8	145.6 ÷ 282.5	74.8 ÷ 152.8
California II	78.9 ÷ 153.3	124.2 ÷ 203.6	57.6 ÷ 97.9
California III	33.5 ÷ 65.3	53.9 ÷ 85.7	24.5 ÷ 40.4
Oregon + Washington	132.9 ÷ 258.9	206.3 ÷ 348.8	97.0 ÷ 171.0
Washington	151.4 ÷ 297.0	239.4 ÷ 395.5	111.1 ÷ 190.5
Alaska	2694.3 ÷5238.9	4431.5 ÷6649.6	1950.4 ÷2930.1

Table 6.2. Three examples of U.S. imports in 1975: shrimp from Columbia and
Ecuador; shrimp, prawn and lobster from Thailand, expressed as tons. NOAA
NMFS, 1976.

Month, 1975	Colombia	Ecuador	Thailand
January	70	198	899
February	70	143	1,027
March	138	183	984
April	283	169	1,058
May	180	332	977
June	292	543	1,007
July		511	
August		424	
September		363	
October		267	

hand, companies in this category are not in the chemical business, as they are food processors, and it would require them to acquire new skills and modify their overall corporate policy.

It should be recalled here what the shell waste looks like: shrimp shell waste weights only 22 $kg \times m^{-3}$ and it is $75 \div 80$ % water; crab waste is 60 % water; this corresponds to 6 % chitin in wet waste. Protein is present in two forms: adventitious protein, in remnants of flesh and connective tissue attached to the shell and may be recovered through separators, and protein complexes with chitin and calcium carbonate as integral part of the shell.

From the above description it becomes apparent that a shellfish processor faces dewatering as a major problem. From the raw shell waste, he can obtain high quality protein, which is a valuable product, and use it to compensate the cost for drying the shells, to produce shell waste residual.

Economical aspects.

The cost of the shell waste residual, on the assumption that one third of the total waste in each area is available to the shellfish processor, and that dried protein will sell for $51 \div 66$ $\cancel{c} \times kg^{-1}$ ($23 \div 30$ $\cancel{c} \times lb^{-1}$) is given in Tab. 6.3 , along with the cost of chitin in shell waste residual.

From this operating revenue realized by the sale of protein, one must deduct the operating costs of acid, labor and fuel for protein precipitation and drying. While the cost of the acid is not high but regionally variable, drying a moist (80 % water content) high-protein to a salable product (<10 % water) is not without difficulties. The process must be designed to minimize or tolerate the tendencies of the material to foam and stick, and at the same time the protein must not be heated to temperatures where amino acids are degraded and nutrient value lost.

The first column in Tab. 6.3 gives the summation of the various components (labor, transport, credit for sale, shell waste residual drying, depreciation and return on capital investment) and represents the estimated ranges of the long-run market price which would be needed to induce a rational shellfish processor in each area to institute production. Many of the numbers of the left-hand, optimistic side of this column are in parentheses, indicating negative numbers. For those areas, the apparent result is therefore that, at least under the most optimistic set of assumptions used, a shellfish processor holding at least one third of the waste available there, may be able to earn a sufficient rate of return on the protein alone, to justify the required investment without any need for revenue from the sale of shell waste residual. The results of the right-hand, pessimistic side of the same column indicate that although there may be no areas where the protein production can pay on its own, there are a number of areas where the minimum required return on investment would be earned with shell waste residual prices of less than 33 $\cancel{c} \times kg^{-1}$ (15 $\cancel{c} \times lb^{-1}$).

The second column of Tab. 6.3 recomputes the previous results on the basis of the percentage of chitin content expected for the shell waste residual in each area.

The reason of the above discussion about processing the raw shell waste to

Table 6.3. Costs of shell waste residual and of chitin in shell waste residual in U.S. ¢×kg⁻¹. Values on the left for an optimistic set of assumptions, values on the right for a pessimistic set of assumptions. Numbers in parentheses are credits. Hattis & Murray, 1976.

Area	Residual		Chitin in residual	
Maine	(1.5)	÷ 9.7	(4.4)	÷ 29.9
Massachusetts	3.3	÷ 29.4	9.9	÷ 107.5
Maryland + Delaware	(2.8)	÷ 2.9	(13.6)	÷ 16.3
Virginia	(2.9)	÷ 2.5	(14.0)	÷ 14.5
North Carolina	(2.6)	÷ 3.2	(12.2)	÷ 20.8
South Carolina	(0.3)	÷ 11.7	(1.1)	÷ 60.7
Georgia	(2.1)	÷ 3.7	(9.5)	÷ 19.5
Florida I	1.3	÷ 24.9	5.7	÷ 138.3
Florida II	6.1	÷ 30.4	22.7	÷ 128.8
Florida Keys	5.6	÷ 32.6	17.7	÷ 114.3
Florida W I	0.6	÷ 12.2	20.8	÷ 52.2
Florida W II	(1.6)	÷ 6.4	(5.7)	÷ 28.6
Alabama + Mississippi	(2.2)	÷ 5.4	(7.5)	÷ 24.9
Louisiana I	(2.5)	÷ 4.8	(8.1)	÷ 20.4
Louisiana II	(1.7)	÷ 6.5	(7.2)	÷ 34.1
Texas I	(1.1)	÷ 7.5	(4.4)	÷ 35.8
Texas II	0.4	÷ 12.3	1.3	÷ 49.9
Texas III	4.0	÷ 24.0	11.3	÷ 68.9
California I	0.2	÷ 16.8	0.5	÷ 54.0
California II	9.5	÷ 15.0	3.5	÷ 75.7
California III	6.7	÷ 35.4	26.3	÷ 177.8
Oregon + Washington	(1.0)	÷ 13.1	(3.2)	÷ 59.8
Washington	(1.2)	÷ 8.2	(4.4)	÷ 39.4

produce protein, is that the chitin production should rely on the meat and protein production, to be economically feasible. It is therefore clear that to establish a shell waste residual production plant on existing enterprises which have long established systems for picking up and handling the waste (so far turned into a very low value crustacea meal) is a logical approach to the chitin production problem. The contrary, operating a fleet of trucks to pick up shell waste from the processing houses does not seem to be convenient because the material has to be taken at times and in quantities that suit the suppliers and that usually introduce spoilage of the wet material.

In order to extract chitin, the fact that the by-products can generate sizeable amounts of revenue should be accepted and kept present. For example, blue crab wastes (on a dry basis) is approximately 50 % calcium of one form or another, 35 % protein and 15 % chitin. Each ton of dry material will generate a potential 350 kg protein, a large quantity of calcium salts and a certain amount of valuable carotenoids. If these are extracted wisely and sold at maximum value, they can actually exceed the value of the chitin portion, yet just a few people recognize this, even in a protein-hungry world.

Chitin and chitosan production process.

Stoichiometry requires two moles of hydrochloric acid per mole of calcium
carbonate dissolved, when calcium carbonate removal from shells is planned.
Since commercial concentrated hydrochloric acid is about 12 M (with density
1.18), the minimum requirement is 153 ml per mole of calcium carbonate, thus
it is calculated that the dissolution of one kg of calcium carbonate requires
2 kg of said acid. A production flow-sheet is in Fig. 6.1.

It is not clear how much more than this minimum might be used to perform the
demineralization in actual practice. Until chitin and chitosan standards are
set, the amount of hydrochloric acid is usually taken empirically. However, if
it is assumed that the demineralization step requires twice as much hydrochloric acid as would be expected from stoichiometry, then a crucial parameter
to be taken into consideration is the ratio of calcium carbonate to chitin;
usually crab shells contain more carbonate than shrimp shells. Acid demineralization cost for shell waste residual from Maryland and Delaware, an all-crab
area, contribute $73 \div 88$ ¢$\times$kg^{-1} $(33 \div 40$ ¢$\times$lb$^{-1})$ to the cost of chitosan,
whereas the shell waste residual from Texas, all-shrimp area, adds a demineral-
ization cost of only $29 \div 30$ ¢$\times$kg^{-1} $(13.1 \div 13.7$ ¢$\times$lb$^{-1})$ of chitosan made.

As for the deacetylation, as the price of caustic soda is currently (1976)
about U.S. \$ $140 \div 180$ per ton, the deacetylation cost is $77 \div 99$ ¢$\times$kg^{-1} $(35 \div
45$ ¢$\times$lb$^{-1})$ of chitosan. Of course these costs may be lower, if caustic soda
recovery is planned and the whole chemical process is optimized not only in
terms of temperature, time, concentrations etcetera but also in terms of
capital investments, general managing and effective capacity of output.

Fig. 6.2 shows the chitosan production costs under two sets of assumptions.
Curves relevant to optimistic and pessimistic estimates show not only the
range of the chitosan cost, but also the dimensions of the single production
facilities. An optimum size interval and a limit dimension of the single
chitin and chitosan production plants appear from Fig. 6.2. Hattis & Murray
(1976) anticipate that plants for chitosan production of between 500 and 2000
tons per year definitely make feasible a $20 \div 40$ % return on capital investment at chitosan selling prices between \$ 1.98 and 4.84 per kg $(0.90 \div 2.20$
\$$\times$lb$^{-1})$.

While no detailed engineering design has been made to evaluate in detail the
capital costs and labor requirements of the various types of process equipment
necessary, the flow sheet of the Marine Commodities International, Inc. plant
is in Fig. 6.1 and it is reported here to illustrate the chitin and chitosan
production process: this plant has a capacity of 250 tons per year.

While the assessment of the possibility that the market would absorb the
chitin + chitosan production is beyond the scope of this book, it can
certainly be said that the price and the amounts available are very attractive
for predictable applications of chitin and its derivatives in selected fields
where they show unique properties or are competitive with other polymers.

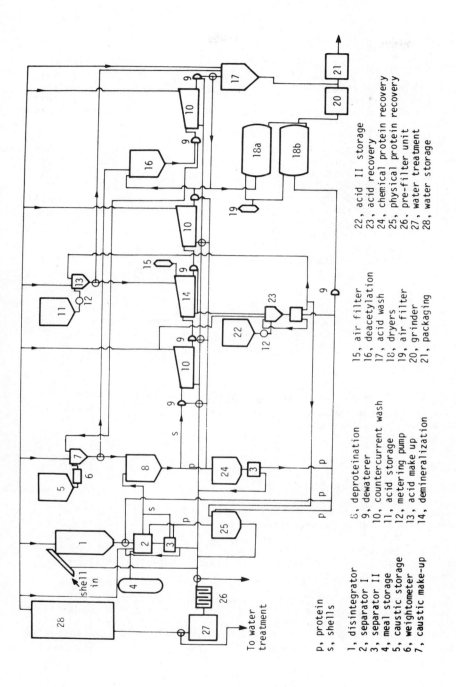

p, protein
s, shells

1, disintegrator
2, separator I
3, separator II
4, meal storage
5, caustic storage
6, weightometer
7, caustic make-up

8, deproteination
9, dewaterer
10, countercurrent wash
11, acid storage
12, metering pump
13, acid make up
14, demineralization

15, air filter
16, deacetylation
17, acid wash
18, dryers
19, air filter
20, grinder
21, packaging

22, acid II storage
23, acid recovery
24, chemical protein recovery
25, physical protein recovery
26, pre-filter unit
27, water treatment
28, water storage

To water
treatment

Fig. 6.1. Chitosan production flow-sheet. Courtesy of Marine Commodities International, Inc.

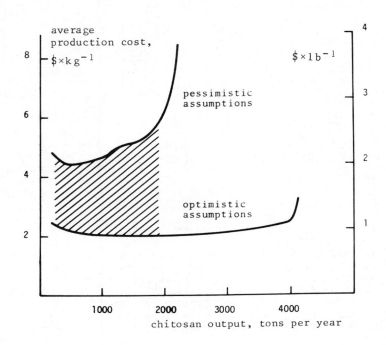

Fig. 6.2. Definition of the size of single plants producing chitosan, under two sets of assumptions. The shaded
area shows the best plant dimensions for the most convenient, long-run average production cost. Hattis &
Murray, 1976.

CHELATING POLYMERS FOR HARMFUL METALS.

In recent years, considerations of public health and environmental and
economic pressure have made it desirable to reconsider the methods employed
for treatment and disposal of toxic wastes. This is true, especially for waste
waters containing toxic or radioactive metals, since the metals cannot be
destroyed; certainly their fate is to abnormally increase the metal content of
the environment where they are deposited.

Certain metals, like mercury and cadmium, present special hazards because they
are accumulated biologically, but metals like copper, nickel, chromium and
zinc, widely used in metal finishing, are also of concern because their
discharge in waste waters may contaminate sewage sludge to the extent that it
becomes unsuitable for farm land, reduce the efficiency of sewage-treatment
processes and harm aquatic life.

As an alternative to traditional treatment processes, chitosan may be an answer for the prevention of water pollution by mercury, copper and many other elements, as stable or radioactive isotopes (Muzzarelli, 1972).

Nuclear fission products.

As many fission products are transition metals, it would seem that chitosan is well suited for the removal of certain radioisotopes from waters. The resistance of chitin and chitosan to gamma irradiation when applied in chromatography of radioactive solutes has been reported by Muzzarelli & Tubertini (1972).

Titanium, zirconium and hafnium have been studied in this connection by Muzzarelli, Rocchetti & Marangio (1972). These metal ions are very efficiently collected, particularly zirconium. Zirconium is not only collected from chloride solutions, but also from thenoyltrifluoroacetone solutions; the latter is a very strong complexing agent for zirconium. The physical appearance of chitin and chitosan after collection of zirconium is different and more work should be carried out to elucidate the nature of the reaction product.

Niobium and ruthenium are also collected by chitin and chitosan; ruthenium is known to form nitrosyl complexes which are very hard to collect. Samples of various waters or solutions contaminated with nitrosyl ruthenium have been percolated through chitin and chitosan columns with rather satisfactory results, as shown in Tab. 6.4 (Muzzarelli, 1970), especially when compared to results obtained with other methods. The nitrosyl complexes may be disrupted by oxidation, thus leaving ionic ruthenium which is very easy to collect on chitosan with which it forms a black complex.

Table 6.4. Nitrosyl ruthenium collection percentages on chitin and chitosan columns (12 × 1 cm) from one l of water containing 1.32 μg Ru. Flow-rate, 2 ml×min^{-1}; temperature 20°. Muzzarelli, 1970.

Water	Chitin	Chitosan
nitric acid solution, pH 3.1	100	100
tap water with HNO_3 at pH 3.1	100	96
sodium nitrate, 4 g×l^{-1}	100	90
disodium carbonate, 4 g×l^{-1}	86	80
aluminum nitrate, 250 mg×l^{-1}	45	80
sea water	75	25
waste water from Marcoule Nuclear Center	42	60

The results mentioned above have been extended to low-level activity waste waters from the Marcoule Nuclear Center, France. The results for these waters, after chromatography on chitin and chitosan columns are reported in Tab. 6.5. From these waters, cesium cannot be removed, as it is an alkali metal.

Table 6.5. Chromatographic decontamination of waste water from the Marcoule Nuclear Center, France, on chitin and chitosan, without treatment and after oxidation, pH 3. Values are expressed in $mCi \times m^{-3}$. Muzzarelli, Rocchetti & Marangio, 1972.

Radioisotope	Before chromatography	After chromatography on chitin	After chromatography on chitosan		
			Direct	$KMnO_4$	$KMnO_4 + Na_2S_2O_8$
$^{95}Zr + Nb$	2.4	1.1	0.33	~ 0.1	~ 0.1
^{103}Ru	3	2	1.8	~ 0.8	~ 0.2
^{106}Ru	23	18	13	6	1.4
^{137}Cs	3	~ 2.6	3	3	3
^{134}Cs	0.5	0.5	0.5	0.5	0.5
^{144}Ce	~ 3.5	~ 3.5	< 1	< 1	< 1
^{141}Ce	0.5	0.2	0.2	0	0
^{125}Sb	1	1	1	~ 0.5	~ 0.3

For analytical and radiochemical purposes, the separation of cesium from accompanying radioisotopes of other elements, is often sought. The separation of ^{137}Cs from ^{106}Ru, ^{95}Zr, ^{95}Nb and ^{144}Ce has been performed after oxidative treatment on a column of chitosan by Muzzarelli, Rocchetti & Marangio (1972).

Little is known about lanthanons and other fission products. The lanthanides studied generally show a slower collection rate and both chitin and chitosan have limited capacity for them, at least under the conditions studied; data for cerium in two oxidation states are presented in Fig. 6.3. Data for terbium, europium and thulium are listed in Tab. 6.6.

As far as other elements of nuclear interest are concerned, for uranium only little information is available. Manskaya & Drozdova (1968) published results on the collection of uranium on chitin from fungi and animals and reported that chitins from different sources are not equivalent (Tab. 6.7). ^{233}Uranium has been used by Muzzarelli, Raith & Tubertini (1970) to study the behavior of uranium towards chitosan in sea water and other solutions. On the basis of liquid scintillation measurements reported in Tab. 6.8, the adsorption of uranium on chitosan can be considered total and rapid. Uranium can be eluted from chitosan columns with sodium carbonate solutions.

Mercury and copper.

Muzzarelli & Isolati (1971) showed that methyl mercury acetate can be removed from industrial waters containing small amounts of acetic acid and acetaldehyde, with the aid of chitosan. Organometalic compounds are not collected so effectively as the simple ions and elution can be performed with hot water. Chitosan offers a good volume reduction of waste water, which is an essential requirement for water reclamation and pollution prevention.

Ionic mercury removal from tap water (37°F hardness) has been studied by

Table 6.6. Collection percentages of three lanthanides from 1.76 mM solutions at pH 6.0; 50 ml with 200 mg of chitin or chitosan powder, 100-200 mesh. Muzzarelli, Rocchetti & Marangio, 1972.

Contact time, min	Chitin		Chitosan	
	25°	40°	25°	40°
$^{152+154}$*Europium*				
15	5	3	30	26
60	7	5	33	30
120	8	5	37	34
240	9	6	45	38
^{170}*Thulium*				
15	3	0	31	40
60	4	0	35	42
120	4	0	37	42
240	5	0	38	43
^{160}*Terbium*				
15	9		32	
60	6		36	
120	4		39	
240	4		43	

Table 6.7. Collection of uranium from uranyl sulfate solutions, on chitin from fungi and from crabs. Manskaya & Drozdova, 1968.

Chitin, g	pH	Time, hr	Uranium collected, %
Chitin from fungi			
0.5	4.0	4	60
0.5	4.0	90	65
0.5	4.0	90	65
Chitin from crabs			
0.5	4.0	90	20
0.5	6÷7	90	25
1.0	6÷7	90	40
0.5	8.0	90	60

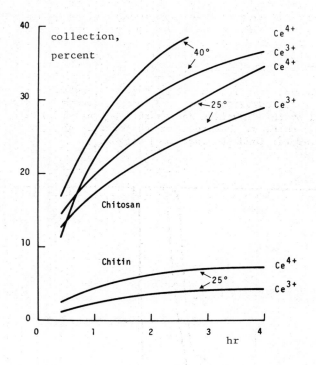

Fig. 6.3. Rates of collection of Ce^{3+} and Ce^{4+} ions from
1.76 mM solutions (50 ml) on chitin or chitosan (200 mg),
at pH 6.0 and at 25° and 40°. Muzzarelli, Rocchetti &
Marangio, 1972.

Tab. 6.8. Collection of ^{233}uranium from aqueous solution (50 mL, containing
4.2 µg of uranium) on chitosan (200 mg) at pH 5.5. Muzzarelli, Raith &
Tubertini, 1970.

Reference solutions, cps×g^{-1}	Treated solution, cps×g^{-1}	Shaking time, hr	Collected, %
27.2 ± 3.0	1.7 ± 0.5	1	94.3
31.7 ± 1.3	1.2 ± 0.5	2	96.0
30.2 ± 0.8	0.9 ± 0.5	18	97.0

Muzzarelli & Rocchetti (1974) at three pH values, 4, 7 and 8. A capacity of chitosan for mercury of 114 mg of mercury per gram of chitosan was calculated from the breakthrough curves. pH Values higher or lower than 7 anticipate the breakthrough point. No effect due to the chloride and acetate anions was observed. Zinc and copper ions that are fixed on the column under the same conditions do not disturb the fixation of mercury as verified at concentrations of 4.4 and 0.1 ppm.

When tap water containing 1.0 ppm of mercury was used, the curves in Fig. 6.4 were obtained: these curves are for chitosan columns of different sizes. They are not altered by the flow-rate and exhibit a typical shoulder. For the longer, narrower columns, the breakthrough point is higher and, in fact, the reduction in diameter leads to increased capacity.

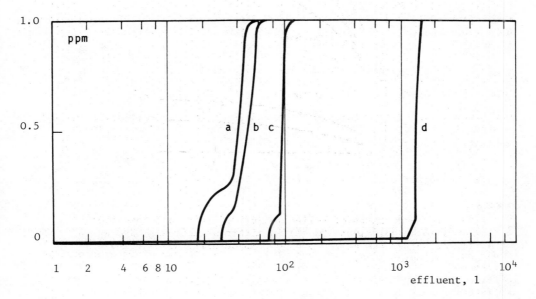

Fig. 6.4. Breakthrough curves for mercury at the 1 ppm concentration, from chitosan columns; a = 1 g, 6 × 1 cm; b = 1 g, 12 × 0.4 cm; c = 2 g, 12 × 1 cm; d = 15 g, 20 × 2 cm. Mercury concentration in the effluent vs liters of effluent. Muzzarelli & Rocchetti, 1974.

For 12 × 1 cm columns, containing 2 g of chitosan, the slope of the curve is steeper, as can be seen in Fig. 6.4, curve c, because of the more favorable length to diameter ratio, and mercury can be detected in the effluent after 80 l have passed. The reduction of mercury concentration is from 1.0 to 0.02 ppm.

For 20 × 2 cm columns containing 15 g of chitosan, the breakthrough point is reached after 1100 l have passed. This column contains 7.5 times more chitosan than the 12 × 1 cm column containing 2 g, but it purifies thirteen times more water. These results demonstrate that not only the amount of chitosan but also the dimensions of the columns must be taken into account for optimum results.

When treating water containing 0.1 ppm of mercury, the effluent contains a constant concentration of 0.02 ppm of mercury; for 6 × 1 cm columns containing 1 g of chitosan, the breakthrough point is at 120 l and saturation is reached after 140 l have passed. Waters containing 1.0 ppm of mercury in the form of ethylmercuric chloride are not appreciably purified, as mercury is present in the fourth liter of effluent.

The complete elution of mercury can be performed with 10 mM potassium iodide solution; 15 ml of this solution are enough for the complete washing of a 6 × 1 cm column containing 1 g of chitosan. The recovery of mercury after elution with 0.01 M solutions of potassium cyanide is also as high as 98 %.

Sulfuric acid is also a good eluent for removing several metals from chitosan, but mercury can only be partially eluted. The first regeneration of chitosan is complete, but the mean recovery for the subsequent cycles is about 43 %.

Organic complexing agents were found to be of no use for the elution of mercury: 1 % solutions of succinimide, 2-aminopyridine and diphenylcarbazide gave recoveries as low as 21, 15 and 5 %, respectively.

Wheatland, Gledhill & O'Gorman (1975) applied several polyelectrolytes for the treatment of metal-bearing effluents including chitin; this work is being continued by Bourne (1976).

Muzzarelli (1974) reported the separation of manganese and ferrous ions from nickel and copper on chitosan. A solution containing excess manganese (12.9166 g) and minor amounts of the other metals (0.2936 g Ni and 0.1761 g Cu) all in form of sulfates, was diluted to 100 ml with 0.1 M ammonium sulfate at pH 4.5, and passed through a chitosan sulfate column 8 × 1.5 cm. At the column outlet the manganese concentration was unvaried while nickel and copper were absent. Upon elution with 0.1 M ammonium sulfate + sulfuric acid, at pH 1.0, copper could be recovered in the fraction 21 ÷ 40 ml while nickel is present in the fractions 31 ÷ 50 ml. Elution and recovery yields are close to 100 % even after many column cycles. Similar results have been obtained with ferrous sulfate in the presence of hydrogen peroxide, whose behavior under these conditions is similar to that of manganese. The selected conditions simulate solutions obtainable from manganese nodules. In view of the interest in recovering copper, nickel and cobalt from marine nodules, this technique seems to provide an answer to a chemical problem which so far has not been solved in a satisfactory way by other approaches.

PAPER AND TEXTILE ADDITIVES.

Applications of chitosan in papermaking.

Chitosan has attracted the attention of the paper industry for a long time: in 1936, Rigby reported that a coating of a 3 % aqueous solution of medium viscosity chitosan acetate can be applied to one surface each of two sheets of paper, and the sheets then joined by their coated surfaces. Chitosan can be insolubilized by exposure to ammonia vapors and drying. The material thus made is resistant to water.

Coating for sizing was also reported by Sadov & Markova (1954) who used 2 % solutions of chitosan acetate; they observed that the wear resistance of chitosan sized articles is increased because this size is not washed off.

Chitosan was used as a thickener for printing and for finishing various fabrics by Viktorov & Maiofis (1940); they observed that the chitosan was not washed out from the fabric by alkaline solutions. Sadov & Vil'dt (1958) further reported on chitosan acetate as a thickener and a fixing agent in the preparation of printing pastes.

A coating for lithographic paper plates, which accepts and holds both water and ink, can be based on chitosan, colloidal silica or clay fillers, a cross-linking agent such as glyceraldehyde or N,N'-ethylenedimethylolurea and a catalyst, such as zinc acetate. This composition has been used by the Oxford Paper Co. (1963) for paper coating to prepare printing plates; one plate could produce one thousand copies. Cardboard was prepared by Fel'dman, Makarov-Zelyanskii, Reutov & Goldovskii (1959) by sizing pulp, cotton linters and chrome leather shavings with chitosan acetate.

The applications of chitosan in the paper industry have been also mentioned in a report by Brine & Austin (1974) and by Foster (1972) who has considered the papermaking system, the drainage in the presence of flocculants including chitosan and problems of side-effects when using drainage and retention aids. Chitosan formate impregnated papers have been used for ion-exchange chromatography (see Chapter 5).

Polymers can interact with fibers at many energy levels. The resultant effects in order of increasing enegetics are usually classified as van der Waals forces, hydrogen-bonding, ionic interactions and covalent bond formation. One convenient way of changing the energy of bonding in paper is to supplement low-energy hydrogen bonds ($\Delta H = 4.5$ kcal$\times$mol^{-1}) by higher energy ionic ($\Delta H = 10 \div 13$ kcal$\times$mol^{-1}) or covalent ($\Delta H = 70 \div 84$ kcal$\times$mol^{-1}) interfiber linkages.

In the ionic bonding studies, negative sites can be introduced onto the fibers using fiber reactive dye technology. Up to 10 % by weight of a modifier can be covalently attached to the fiber so that its surface becomes rich in anionic groups; the latter, however, are not available for interfiber bonding because they are mostly located within the microporous structure of the cell wall. Bridges can be constructed in any case with polyethylenimine with improved tensile strength with increasing molecular size. An alternative is to deposit an acidic polymer like alginic acid within the cellulose fiber, to provide anionic carboxylate bonding sites. Subsequent treatment with a large polyamine gives improved properties. Several authors have coated papers with a swelled but insoluble layer of alginate and with a polyethylene layer to produce packaging material. In certain instances, the intrinsic anionic nature of the fibers, like groundwood fibers, makes feasible the polyamine treatment, with improved characteristics of the product. Treatment of softwood groundwood with polycations, by virtue of the abundance of anionic sites due to the lignin content, also gives improvement in the physical properties of the resulting sheet.

Artificial polyamines, however, do not represent the ideal reagent for the
creation of interfiber bonds. Cationic sites within the macromolecules are
unable to reach appropriately the anionic moieties on the fiber surface.
During drying, there may be formed a pancake of polymer whose cohesive
strength is low because of the mutually repulsive groups acting within it
together with its hydrophilic nature. Moreover, polyethylenimines are not
film-formers, and thus they reduce the strength of a sheet according to the
percentage of artificial polymer added.

Theoretically, the preferred polymer for enhancing paper strength should be
linear, to allow complete accessibility to its functional groups; film-forming
and of high molecular weight, for good cohesive strength and for the ability
to span interfiber distances; polycationic for ionic bond formation, and in
addition, it should be capable of forming hydrogen bonds with nonionic areas
of the fiber surface.

In principle chitin ought to be readily bondable to fibers within the
cellulosic fiber network. However, even pulp fibers do not bond well reciprocally, unless the area of contact has been increased by the mechanical process
of beating which makes the tubular structures flexible and fibrillates the
exterior surface of the fibers. So, a modification of the chitin surface is
required to perform this task. Since beating is inefficient for chitin, the
modification of the chitin surface can be easily achieved by chemical means.
Deacetylation has the effect of exposing free amino groups readily available
in fiber-chitosan bonding. As shown in Fig. 6.5, the higher the degree of

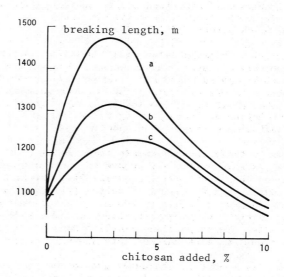

Fig. 6.5. Effect of chitosan addition (< 44 μm particles)
on tensile strength of handsheets. Deacetylation: a = 60
min; b = 30 min; c = 15 min. Allan, Friedhoff, Korpela,
Laine & Powell, 1975.

deacetylation, the better the improvement of the tensile properties.

Chitosan therefore meets not only the requirements for a suitable surface from both the chemical and the physical points of view, but meets equally well the criteria established above for an ideal polymer applicable in the paper industry.

Sheet formation procedures.

The effectiveness of any polymer in increasing the paper strength depends on the method of incorporation into the cellulose, thus several methods have been worked out, by Allan, Crosby, Lee, Miller & Reif (1972), by Allan, Friedhoff, Korpela, Laine & Powell (1975) and by Allan, Crosby & Sarkanen (1976).

Equilibrium adsorption: a vigorously stirred suspension of fibers (5 g) in water (400 ml) is adjusted to pH 5 with acetic acid and treated with an aqueous solution of chitosan acetate (2 % solution, pH 5). The consistency is adjusted to 1 % and stirring is continued for 1 hr at 25°. The suspension is then diluted to 0.25 %, divided into four equal parts, and formed into hand-sheets in accord with TAPPI Standard T-205 m-58.

Precipitation: the procedure for precipitatig chitosan onto the fiber surface is identical to that described above, except that after stirring for 1 hr, the pH of the suspension is adjusted to 10 with 0.5 M sodium hydroxide. Stirring is continued for 10 min at this elevated pH prior to sheet formation.

Spray application: sheets of untreated fibers are formed in the usual manner, couched from the sheetmold and, while still wet and on the blotter, are sprayed with a chitosan solution (1 %, pH 5).

Chitosan composites and films: the chitosan + sulfite fiber composites are prepared using a series of sulfite handsheets ranging in weight from 0.6 g to 1.2 g and a series of solutions (60 ml) varying in chitosan content from 0 to 1.2 g. Composites are formed by pouring the chitosan acetate solution (30 ml) onto a polished metal plate held in a TAPPI drying ring, placing the appropriate sheet on the plate and then pouring the remainder (30 ml) of the chitosan acetate solution on top of the sheet.

Some data relevant to these techniques are illustrated in Fig. 6.6. The chitosan used for these investigations had an average molecular weight of 93,000 daltons and had been obtained from crab shells. The chitosan acetate solutions were prepared by stirring suspensions of chitosan in water and adding acetic acid: for instance 50 g of chitosan in 200 ml of water were stirred with 25 ml of acetic acid and finally diluted with water to 2.5 l to obtain 2 % solutions. The improvement of tensile properties of well beaten and lightly beaten kraft handsheets after treatment with chitosan are shown in Fig. 6.7.

Commercial pulp has some anionic character due to the presence of carboxyl groups originating from the hemicellulosic uronic acid residues or from oxidation associated with the bleaching process. Ionized chitosan is probably retained by the formation of ionic bonds between the amino groups in chitosan and anionic functions present on the fiber surface. Chitosan is very efficient in neutralizing all the anionic sites on the fiber.

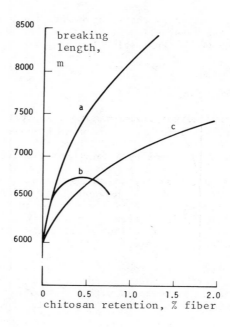

Fig. 6.6. Tensile strength of sulfite handsheets treated
with chitosan; a = by the spray method; b = by the
equilibration method; c = by the precipitation method.
Allan, Crosby & Sarkanen, 1976.

In the light of the above description, the equilibrium adsorption of chitosan
gave low retention levels on a weight basis.

Precipitation of chitosan onto the fibers prior to sheet formation prevents
fiber flocculation and allows the production of celluloses with higher chitosan content. In any case, the best procedure for the application of chitosan,
as suggested by Allan, Crosby & Sarkanen (1976) is the spray technique.
Chitosan is sprayed onto preformed sheets thus avoiding sheet formation
problems, and is deposited in the ionized and extended form, the most suitable
for forming ionic bonds and for covering large surface areas.

Single ionic bonds are more resistant to the disruption by water than hydrogen
bonds, the expected differences between the spray and the deposition methods
in their respective ability to impart wet strength characteristics to paper,
are confirmed by experimental data, as shown in Figs. 6.5, 6.6 and 6.7. The
bonding of fibers is also visible in Fig. 6.8.

At the 2 % level, sprayed chitosan gives a wet strength value of 23 % based on
the dry strength of the same sheet. This value is increased to 33 % when the
calculation is based on the dry strength of the untreated paper. Similar

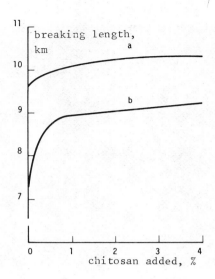

Fig. 6.7. Tensile properties of kraft handsheets treated
with chitosan: a = well beaten; b = light beaten. Allan,
Crosby & Sarkanen, 1972.

results were observed with unbleached sulfite pulp.

Chitosan sized paper has several times the wet and dry bursting strength of
unsized sheets. It has excellent resistance to water penetration and a good
surface for writing or printing upon with ink, while the unsized paper is
instantly penetrated by water and its surface is unsatisfactory for these
purposes (Du Pont de Nemours, 1936).

The appearance of the composition is not appreciably different from that of
the untreated paper but the new one has many advantages, for example higher
tensile strength and tear resistance as well as high resistance to water. The
chitosan sized paper possesses properties superior to those of paper sized
with rosin. For example, the rosin sized paper has lower wet and dry bursting
strengths, equal water resistance and poorer tear resistance on both aged
and unaged samples. Rosin sizing in paper requires precipitation in a slurry
which has a pH considerably below 6 and the acid remaining in the finished
sheets has degrading effects on the paper. Chitosan, on the other hand, can be
precipitated at a pH slightly above 7, hence no acid is left in the sheets to
weaken the cellulosic material.

Papers made from pulp sized with chitosan acetate, may be lacquered by any
suitable means with organic solutions of cellulose derivatives or resins.
Water dispersible materials may also be used in combinations with chitosan as
paper sizes; for example, rosin, starches, casein, glue, gelatin, water

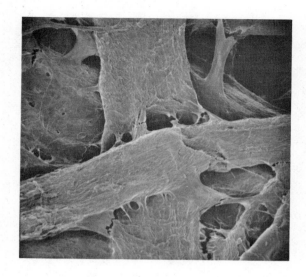

Fig. 6.8. Fiber bonding in kraft handsheet containing 25 %
chitosan. At lower addition levels, chitosan cannot be
distinguished because the bridging film merges with the
fiber surface. Courtesy of G.G. Allan.

soluble or emulsifiable polyhydric alcohol-polybasic acid resins. Chitosan
acetate was also used to size straw hat bodies and to obtain desirable
stiffness, gloss and water resistance. Unfinished, tanned soft neck leather,
cotton linters, cotton rags, pressed wood sawdust, woodflour, bagasse and
asbestos are also suitable for chitosan compositions.

Modified chitosan additives.

Further treatments of chitosan impregnated celluloses, pulp and other fibrous
materials can be carried out. It is known that chitosan can be acetylated
again to obtain regenerated chitin, or, generally speaking, acylated to yield
acyl derivatives having new physico-chemical properties. This was mentioned by
Merrill (1936) who used acetic anhydride to convert the chitosan impregnated
paper into a chitin impregnated paper, with very high resistance to oil,
grease, water and other substances, and improved insolubility of the coating.

Formaldehyde was also mentioned by Merrill (1936) as a precipitant for chitosan in formulations including cellulose. There is now no difficulty in
preparing special papers where chitosan, after impregnation and neutralization
is reacted with suitable carbonyl compounds like aldehydes. By cross-linking
chitosan, it may be possible to obtain unusual properties of special papers.

For many applications it is desirable to have paper products with good dry strength. In the paper industry there is at present a tendency toward reducing the basic weight of paper, especially that of publication-grade paper; this reduction would lead to lower mailing costs. Dry strength aids are therefore needed for lighter weight paper because as the basic weight is lowered, the dry strength of the paper decreases; by using dry strength additives, the production costs are reduced since less pulp and power are needed to make an equivalent sheet.

In the past, natural polymers like guar and locust bean gums and the native and modified starches have been the most commonly used dry strength additives. The performance of these natural polymers is difficult to control and hence somewhat inconsistent, and the use of starches involves lengthy preparations and further additives are required for adhesion to the fibers.

Several synthetic dry strength resins have recently appeared on the market. These compounds are polyacrylamides or modified cationic starch derivatives. These compounds do not function well in alkaline media and this may become important if the trend toward shifting from the present acid systems for paper making to neutral or alkaline pH systems continues, due to the detrimental effects of acids on machines and on the product which yellows with age. In alkaline systems, pigments like calcium carbonate can be used instead of the more expensive titanium oxide and aluminum oxide.

Slagel & DiMarco Sinkovitz (1973) have published information about a chitosan graft copolymer for making paper products of improved dry strength, performing well in low basis weight paper and in both acidic and alkaline media. Certain acrylic monomers, particularly 2-acrylamino-2-methylpropane sulfonic acid, can be grafted onto a chitosan substrate.

The method reported is based on the ceric salt redox system: it is known that ceric salts form a redox system when coupled with certain reducing agents like alcohols, aldehyde or amines. The reaction proceeds by a single electron transfer step, resulting in cerous ions and a partially oxidized reducing agent in free radical form, the free radical being formed on the chitosan macromolecule. If a monomer is present, polymerization occurs, and since the free radical is on the chitosan substrate, only graft polymers are formed and no other polymer. Other well known grafting techniques can also be applied.

The graft copolymers comprise from 5 to 99.5 % by weight of the chitosan substrate, the remaining percentage being derived from one or more of the above mentioned monomers. At least 2.5 % by weight of the monomers which are grafted onto the chitosan substrate is 2-acrylamido-2-methylpropane sulfonic acid.

For the preparation, chitosan (20 g), acrylamide (30 g) and 15 % acetic acid solution (450 ml) are heated at 30° and stirred for 1 hr under nitrogen. The ceric catalyst is then added; the catalyst is a 0.1 N ceric ammonium nitrate in 1 N nitric acid solution. The ceric solution (5 ml) is added and the reaction mixture stirred at 30° for 3 hr. The polymer is then diluted with water and precipitated with acetone. The graft polymer is then dried under vacuum for 24 hr. The compound was evaluated by preparing a series of hand-sheets which were then tested for burst and tensile strength under standard conditions.

The dry strength additives may be added to the cellulosic pulp suspension in amounts ranging from 0.1 to 5.0 % by weight based on the dry weight of the cellulosic fibers. Below 0.1 % no appreciable effect on the paper is noticeable and the use of concentrations in the neigborhood of 5 % is generally an overtreatment. The preferred range is from 0.2 to 1.0 %.

The exact pH and concentration at which the dry strength additives will be utilized in the papermaking process will vary from instance to instance. It will depend largely on the type of cellulosic fiber being employed, the other common paper making additives being used and the properties desired of the final product. Accordingly, in each instance, the optimum conditions can easily be found by simple laboratory trials. However, the dry strength additives reported by Slagel & DiMarco Sinkovitz (1973) are effective within the range of pH 3.5 ÷ 9.0 and in the concentration range mentioned above.

TEXTILE FINISHES.

Cellulose + chitin fibers can be spun by mixing chitin viscose and cellulose viscose prepared by the same low temperature method as for the chitin viscose, and under the spinning conditions for chitin fibers, but better results can be expected by mixing the chitin viscose with cellulose viscose prepared by the ordinary method and by spinning under the conditions of the viscose rayon method. Further improvement in the fiber quality can be expected by the addition of the hot roller treatment (110°) in the process. Different from the case of chitin fibers alone, the fusion between fibers does not occur for the mixed fibers even spun with the conditions of viscose rayon method, and rather a better quality of fibers can be obtained than spinning at the low temperature.

Chitin viscose and cellulose viscose prepared by the ordinary method were mixed by Noguchi, Wada, Seo, Tokura & Nishi (1973) in the desired proportion at 0 ÷ 5°. Urea was added, the viscose filtered, and then returned to room temperature for spinning after removal of bubbles. The following chemicals were used for the coagulation bath: 10 % sulfuric acid, 30 % sodium sulfate and 1 % zinc sulfate (at 45 ÷ 50°), and for the stretching bath, a saturated solution of sodium sulfate (at 80 ÷ 90°).

Immediately after stretching, the sample was treated with a hot roller at 110° and reeled onto a bobbin. After washing with water overnight, the sample was bleached in an aqueous solution of 1 % Silbright at 50 ÷ 55° for 30 min, then washed well with water and dried. Far stronger fibers can be obtained with these conditions than by the low temperature method. The spinning pressure was 1.2 $kg \times cm^{-1}$, with a 2 mm nozzle.

Chitin fibers were obtained with a high Young's modulus and enough strength against dryness for use, but they were inferior in strength against wetness and in knot strength. The cellulose + chitin fibers prepared by the low temperature method also have the same tendencies, but the ones prepared by the improved method using the conditions of the viscose rayon method show considerable improvement in those deficiencies. The dyeability of chitin fibers by acid and substantive dyes is far higher than that of viscose rayon, and the dyeability of cellulose + chitin fibers improves with the increase in chitin content. The fibers containing about 3 % chitin has a ramie-like feeling. Because the chitin viscose readily blends with cellulose viscose, it seems to be more promising to use it for improvement of cellulose fibers, such as giving the viscose rayon the ramie-like properties, than to use it as chitin fibers alone.

Aqueous solutions of blends of polyvinylalcohol and chitosan acetate were spun to introduce the amino groups of chitosan into "Vinylon" fibres by Itaya (1976). Chitosan acetate solutions are mixable at any ratio with aqueous solutions of polyvinylalcohol without separation. The aqueous solutions of the polymer blends were spun by almost the same process as "Vinylon". The manufacturing conditions of chitosan + polyvinylalcohol fibers are summarized in Tab. 6.9.

Table 6.9. Manufacturing conditions of chitosan + polyvinylalcohol fibers. Courtesy of M. Itaya.

Spinning solution concentration	15 % (chitosan 1.4 % and PVA 13.6 %)
Spinning solution temperature	80°
Gear pump output	2.36 ml$\times$min^{-1}
Nozzle	0.09 mm$\phi\times$50 hole
Coagulating bath	Na_2SO_4 400 g$\times$l^{-1}, $ZnSO_4$ 30 g$\times$l^{-1}; 45°; 2 m
Spinning speed, first goddet	9.3 m$\times$min^{-1}
Spinning speed, second goddet	30.4 m$\times$min^{-1}
Heat treatment	10 min in air at 100°, 1 min oil at 230°
Formalization	H_2SO_4 20 %, Na_2SO_4 25 %, HCHO 7 %; 65°; 90 min
Washing	petroleum, ligroin and methylene chloride
Bleaching	$NaClO_2$ solution 1 %; 65°; 25 min

The fiber obtained from the aqueous solution containing 1.4 % chitosan and 13.6 % polyvinylalcohol showed interesting properties as reported in Tab. 6.10 especially from the point of view of its electrical properties, because its electrostatic charge is much smaller than that of "Vinylon". The chitosan + polyvinylalcohol fiber can be dyed with direct or acid colors. It burns without melting. The touch is rather like linen.

Fabrics sized with chitosan salts have certain highly desirable properties, like a greater stiffness and fullness. The treatment also improves the appearance of fabrics by adding lustre and brightening their colors. The finish is reportedly superior in this respect to the ordinary sizing agents based on starch and gums. A very important quality is the resistance of chitosan to laundering, and in this respect it differs markedly from the ordinary sizing agents; in fact chitosan is precipitated on the fiber by alkali.

Fabrics can be laminated to each other or to paper and the like by means of chitosan, and they become impervious to the passage of liquids and gases. Non-woven fabrics of resin bonded regenerated cellulose have been described by Dabrowski (1967). Non-woven fabrics are composed of fibers which are

Table 6.10. Textile properties of polyvinylalcohol fibers and of chitosan + polyvinylalcohol fibers. Courtesy of M. Itaya.

	Polyvinylalcohol	Polyvinylalcohol + chitosan 10 %
Moisture content, %	4.4	5.7
Denier, D	2.54	3.00
Dry strength, $g \times d^{-1}$	2.76	2.27
Dry elongation, %	32.7	34.0
Wet strength, $g \times d^{-1}$	2.17	1.71
Wet elongation, %	34.0	35.4
Knot tenacity, $g \times d^{-1}$	1.61	1.80
Knot elongation, %	23.8	25.4
Young's modulus, $g \times d^{-1}$	28.5	29.4
3 % Elastic recovery, %	81.5	80.9
5 % Elastic recovery, %	65.0	64.2
Hot water shrinkage ratio at 80°, %	0	0
Hot water shrinkage ratio at 100°, %	0.7	1.1
Statical electrification, Volt	1380	373

mechanically and/or chemically bonded together; they are not made by knitting or weaving but are commonly produced by sheeting an aqueous slurry of the fibers in paper-making equipment, although they can also be produced as air-laid or carded non-woven webs.

Non-woved fabrics were made to contain 0.5 ÷ 4.0 % by weight of a water soluble salt of chitosan, and approximately the same amount of a thermosetting resin, preferably an aldehyde precondensate, with or without pigments or dye-stuffs; water repellent agents like wax emulsions, aluminum stearate and other salts of fatty acids, and fire retardant agents of the phosphate type, depending on the desired properties of the final products. The saturated bed was subjected to temperatures in the range 135 ÷ 175° to effect curing of the resin and reaction between the resin and the chitosan salt, to produce a cross-linked polymer which is uniformly distributed throughout the web. All the ingredients can be introduced in the same bath as that containing the chitosan salt and the aldehyde precondensate. The resulting products have higher tensile strength in the dry state and reasonably good tensile strength in the wet state, both appreaciably higher than the corresponding tensile strength of fabrics produced employing much greater amounts of other saturants like acrylic resins. In addition, under the temperature and pressure conditions employed in the embossing, the saturant does not stick to the surface of the fabric and the saturated fabric does not adhere to the embossing rollers.

An acid catalyst is employed in the bath; the preferred catalyst being 2-amino-2-methyl-1-propanol hydrochloride although other catalysts like ammonium chloride or tartaric acid can be used to catalyze the curing.

Impregnation of the unsaturated non-woven fabrics was effected by passing the

fabric through the bath, and then through a pair of press rolls which remove
the excess saturant so that the fabric leaves the press rolls with a wet
pickup around 100 %. Calendering can be effected immediately after impregnation and drying of the fabric and before the passage of the impregnated
fabric through the heating and curing zone where the reaction between chitosan
and the aldehyde precondensate takes place to produce a light matrix
surrounding the fibers.

For example, a non-woven fabric made from viscose staple of about 50 %
extensibility was saturated by passage through a bath containing 1 % chitosan,
1 % acetic acid, 3 % ethylene urea formaldehyde and catalyst. The residence
time in the bath was 2 sec and the web upon leaving the bath passed through a
pair of press rolls; the saturated web leaving the press rolls had a wet
pickup of 100 %. The saturated web was cured by passage through an oven at
170° for 2 min. The fabrics were then passed through embossing rollers under
pressure at the same temperature.

Natural fibers are receptive to dyes, thus the problem of dyeing natural
fibres is reducible to a rather specific chemical reaction and good products
involving high degrees of color stability can be devised when the proper
combination of dye and fiber is used.

Synthetic fibers, although they provide greater problems in dyeing than do the
natural ones, can be dyed with accurately controllable chemical manipulations.
However, with the advent of glass fibers and well developed asbestos fibers,
the problem of dyeing takes on an entirely new dimension, because the inorganic character of the fibers gives a physical structure different from natural
fibers and also one of an entirely foreign chemical identity. That is, no
naturally reactive dye centers are present.

Glass fabrics.

Attempts to dye glass fabrics using conventional dyes and conventional
techniques have uniformly led to complete failure, because the dyes are
merely superficially deposited physically and wash out simply by wetting with
water. When this condition is coupled with virtual incapacity of mineral
fibers to withstand laundering processes, it is apparent that some chemical
alteration of the medium is necessary. For this purpose, chitosan has been
found particularly advantageous.

Hurd & Haynes (1960) discovered that chitosan, when applied to glass fibers or
fabrics forms a permanent coating with many available sites for the adsorption
of dyes of a wide variety thereby creating a product with physical
characteristics inherent to glass fibers and textiles, enhanced with improved
chemical capacity of receiving dyes. In addition, the coating of chitosan,
when applied to fabrics, imparts a high degree of fiber to fiber bonding and
increases distortion stability, abrasion resistance and improves the general
appearance of the fabric.

The preferred chitosan salts are acetate, formate, diglycolate, tartrate and
adipate. Various softeners such as low molecular weight polyethylene, can be
incorporated into the chitosan salt solution before padding, as well as
plasticizers. The incorporation of an aldehyde or aldehyde donor in the
treating solution or as an after-treatment, improves the durability of the

the finish and the resistance to water. Such compounds include formaldehyde and formaldehyde donors such as paraformaldehyde, hexamethylenetetramine and monomethyloldimethylhydantoin. The padded fabric is then dried and cured at 135° for 5 min. After this step, the glass fabric can be dyed with a number of dyes, for instance Congo Red, Direct Fast Red 8 BL and other dyes.

Hamiter & Collins (1963) followed a different approach to the problem of dyeing glass fabrics. They mixed cationic flocculating agents with a dispersion of a pigment to produce a dispersion containing flocculated pigment particles and treating a fabric having diverse surface characteristics with this dispersion: the treatment results in the flocculated pigment particles dyeing the loosely constructed, floating, raised or bulky areas of the fabric.

When anionic pigment dispersions are preferred, any pigment which can be charged anionically can be used, like oxides, sulfites, sulfates of cobalt or other transition metals, phthalocyanine dyes, anthraquinone thioindigoid and indanthren vat dyes, azo coupling dyes and others.

Chitosan is the preferred cationic flocculant employed in the dyeing bath as pyruvate, acetate or lactate. Other cationic flocculating agents can be used, including for example polyalkylpolyamines, polyalkylpolyamine fatty acid reaction products and many more. The flocculating agent is included in the bath at concentrations of 0.005 ÷ 5 % based on the weight of the bath. The latter should be alkaline, because the agglomerate dispersions are more stable in basic pH conditions.

Resinous film-forming binders such as butadiene-acrylonitrile copolymer may be added to the bath for bonding the pigments to the fibers to form fast colors. Softeners, delustering agents, abrasive resistant agents and finishes may also be added.

For example, a bath may contain 1 % yellow iron oxide in 10 % water, 1 % chitosan solution, 0.5 % melamine formaldehyde resin and about 87.5 % methanol. The bath is produced by placing a dispersion of the dye in water in a container, adding the methanol and the chitosan, mixing until agglomerates result, and then adding the melamine resin. Upon padding an all-glass fiber fabric having bulked yarn through this dye bath and curing the fabric, a two-tone fabric results having the bulked yarns colored a bright yellow and the background areas an off-white shade.

Chitosan is preferably employed as itaconate because the shades produced with the chrome complex and chitosan itaconate have better brightness in general than those resulting from other salts of chitosan. This is due to the fact that on curing the film, itaconic acid reacts with the primary amino groups converting them into N-substituted γ-lactam, which causes less loss of color when curing. Moreover, the viscosity of the solution applied to the material is less when the same contains itaconic acid rather than acetic acid. The solution may be applied to the material to be colored in any desired manner, by padding, dipping, spraying etcetera.

The material pretreated in this way will not hydrolyze readily even when treated with hot water or dilute alkali. Acids, such as sulfuric or hydrochloric, effect stripping of the anchor layer. This feature enables the production of desirable effects by acidic treatment in selected areas. The dye types which can be used include dyestuffs of the monochlorotriazine group, dichlorotriazine group, vinylsulfone group and many others.

Plastic fabrics.

Other fibers, films, fabrics and yarns such as those made from olefins, particularly polyethylene and polypropylene containing (I) at least 50 % by weight of acrylic and related acrylic copolymers (such as Orlon, Creslan, Acrilan, Varel, Zefran, Darvan and Dynel), (II) vinylidene + vinylchloride copolymer fiber (such as Saran), (III) polyesters (ethyleneglycol and terephtalic acid condensation products, e.g., Dacron) and (IV) polyamide polymers (such as Nylon), are also difficult to dye employing commercially available dyes.

Fibers, fabrics and yarns of cellulose acetate (Fortisan), silk, wool and blends of two or more of the mentioned fibers are also difficult to dye with many available dyes; only limited selected dyes can be used, requiring special dyeing techniques employing either carriers or high temperature dyeing in enclosed systems.

According to Dabrowski (1962), the material to be colored is treated before dyeing or printing with an aqueous solution of chitosan and a Werner-type reactive chromium complex such as p-aminobenzoate chromic chloride, β-resorcylato, glycinato, tannato, glycolato, thioglycolato or sorbato chromic chlorides, and the thus treated material is dried and heated to produce a film of the reaction product bonded to the base material. This film has a surprising affinity for dyestuffs greater than the sum of the affinities for such dyestuffs of the individual constituents, indicating the influence of synergistic components in the mechanism.

The chemical bonds are probably due to electrostatic attractions and are in part covalent bonds; probably the hydroxyl, amino and amido groups present in the polymer form strong covalent bonds with the polymerized Werner-type reactive chromium complex. The resultant film or layer firmly bonded to the base surface shows remarkable affinity for commercial dyestuffs due, it is believed, to synergistic action between the two ingredients resulting in the formation of the durable, bonded crosslinked complex polymeric film exceptionally receptive to dyes.

Mody (1975) has proposed the use of chitosan in 1 % acetic acid solutions for the improved and simplified dyeing of polyester + cotton fibers. The amount of chitosan applied to the polyester + cotton fibres is below 5 % by weight. The polyester + cotton textile is first printed with the chitosan solution and it is dried with warm air, then it is colored. The following dyes have been used, among others: Procion Red M5B, Procion Orange M2R, mixtures of Procion Blue MR and Procion Yellow MHR and Procion Yellow M4G. This simple technique produces two nuances of color whose intensities are sharply different, the chitosan printed parts being more deeply colored than other parts. Chitosan solutions may contain pigments thus enabling the production of textiles with two colors.

Polymeric dyes.

Highly polymerized dyes, based on chitosan and other polymers, have been prepared according to a patent granted to Ciba Société Anonime (1962). Polymers carrying amino groups or imino groups, like chitosan, polyaminostyrene, polyvinylamine, polyethylenimine, casein, aminoethylcellulose, modified starch and dextrin can be used to prepare highly polymerized dyes, by reacting them with organic dyes or intermediates having chromogenic properties.

The latter react with the amino groups of the mentioned polymers to form a covalent bond through their epoxy, ethyleneimine, isothiocyanate, arylcarbamic ester, propiolamido, acrylamino and vinylsulfone groups. These dyes are in the groups of the oxazine, triphenylmethane, xanthene, nitro, acridonic, azoic, anthraquinone and phthalocyanine dyes. Among the intermediate compounds that are able to react with the said polymers, are the halopyrimidines, the halophthalazines, the haloquinazolines and the halotriazines.

As a preliminary step, the polymers are swollen in the reaction medium; the reaction can be carried out in aqueous suspensions of the polymers mentioned above, in the presence of mineral substances suitable for the neutralization of the acidic functions of the dyes, such as the alkali metal bicarbonates, carbonates, phosphates and hydroxides. The reaction temperature is normally between 20 and 100°.

When using chromogenic substances or other intermediates, the reaction products are further submitted to condensation, coupling or diazotization in order to transform them into dyes. The dyes thus produced are high molecular weight substances, where the dye residues are linked to amino or imino groups of a polymer chain, and therefore can be called polymeric dyes.

These polymeric dyes, which are water insoluble, can be milled to very fine powders, thus permitting their use for addition to viscose, synthetic resins or cuprammonia rayon, and very brilliant and resistant colors can be obtained. For example, a neutral suspension of chitosan freshly reprecipitated from acetic acid solutions can be mixed with a dye like the one represented here:

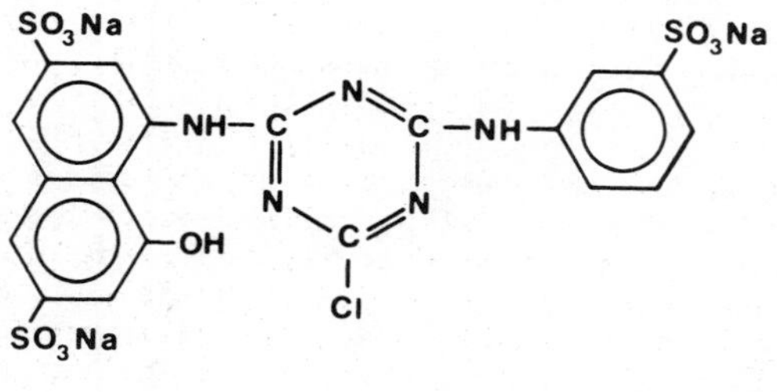

After stirring for 5 hr, heating at 60 ÷ 80° and neutralization, the solid reaction product is carefully washed and introduced into a cold bicarbonate solution of the diazoic derivative obtained from the 1-aminobenzene-2-sulfonic acid. After 5 hr, the reaction mixture is neutralized with hydrochloric acid and the red polymeric dye can be filtered.

A suspension of chitosan gel at -5° can be added to cyanuryl chloride in acetone. The pH value is maintained at a value between 4 and 5 with a dropwise addition of 2 N sodium hydroxide, and the temperature is kept below

5°. A neutral solution of 1,8-aminonaphthol-3,6-disulfonic acid is added, the temperature rises to 35° and the pH is adjusted to between 5 and 6. Aniline hydrochloride is then added, the temperature is increased to 70° and the pH is adjusted to 7 ÷ 8. After filtration and washing, the resulting cake is dispersed in water, and sodium bicarbonate is added. After cooling, diazotized 1-aminobenzene-2-sulfonic acid is added. The insoluble polymeric dye can be milled and used for coloring viscose and synthetic polymers. By chosing the appropriate coupling agent, a desired color can be obtained.

Batik dyeing.

According to the conventional batik dyeing method, a mixture of Japan wax, white wax, pine resin or similar is melted and painted onto a previously drawn design with brushes. When fine patterns are designed, these parts are coated with waxes all over, and after the wax has solidified, the coat of wax is scratched according to the patterns. Then the dyeing is carried out by a suitable method to produce the interesting unique effects of a batik dyed pattern.

The main purpose of batik dyeing is to produce resistant printed patterns of an aesthetic quality unique to this material by a handicraft method. However, this method presents certain problems because of the use of dye repellent material with waxes as the main ingredient. The wax has a low melting temperature and poor adhesion. As a result, it tends to come off the cloth. Difficulties in producing scratches in the desired pattern and in dewaxing have been experienced.

Adachi & Fujita (1970) paid attention to the fact that when chitosan salt solutions are in a viscous and pastelike state, they react with almost all types of dyes except cationic ones, producing water-insoluble precipitates, and when they are applied to a cloth and dried, a film with a strong resistance to peeling is formed. Since the film shows suitable plasticity, it can be cut and scratched as desired without causing its separation from the material.

The effectiveness of chitosan salts as print-resisting materials does not diminish when they are mixed with other pastes which are compatible with them. When chitosan salts are mixed with starches of rice, potatoes etcetera, or treated starches such as dextrin, natural gums such as gum arabic and tragacanth, or synthetic starches like methylcellulose and carboxymethyl-cellulose (all compatible with chitosan), the resultant print resistant starches are applied to paper, cloth or leather. The properties of the films obtained after drying, although they may present some differencies depending on the composition, the thickenss of the coat and the moisture content, invariably show a good print resisting effect and suitability for pattern imprinting. Especially when natural gums are combined with them, fine cracks suited for the batik dyeing are obtained.

Chitosan salts alone or mixed with compatible pastes are coated on knitted or woven fabrics of cotton, rayon, silk, wool, various synthetic fibers or blends of these, or leather, paper and other materials, with brushes or spatulas, and then the dried coat is scratched as desired. The paste may be applied using patterned sheets and spatulas, or by using screens. The use of screens makes it possible to obtain fine, delicate patterns.

In the dyeing process, direct, acidic, vat, reactive, disperse or other dyes are selected depending on the characteristics of the material to be used. Dye paste is prepared from the selected dye with the addition of some suitable auxiliary agents. It is then painted or sprayed on the print-resisting paste, or applied by overprinting.

The print resisting paste can be easily washed away with water and if this is not as efficient as desired, diluted acid may be used in addition to water. Using the above described operations, the dyeing of patterned batik becomes quite easy compared with the conventional batik dyeing method.

For example, a print resisting paste was prepared by mixing 10 parts of chitosan formate, 40 parts of gum arabic and 50 parts of water. With the use of a screen, this paste was applied to an acetate cloth. When the material was dried at 80°, fine scratches were produced on the filmy coat. A paste for acetate dyeing was sprayed or painted on, and after steam-heating and fixing the dye in this way, the cloth was immersed in a 0.1 % acetic acid solution, and washed with water. When dried, a cloth dyed in a pattern with characteristic fine lines was obtained.

Soil repellents.

The problems stemming from the surface soiling of manufactured articles are well recognized. In fabrics used for clothing and other textile products, such as carpeting and upholstery, surface soiling is not only unattractive but usually shortens the life of the fabric and often presents difficulties for proper cleaning.

To combat soiling problems, numerous approaches have been suggested. In general, attempts to render substrates soil repellent or soil releasable have involved an impregnation or coating of the substrate with a compound having itself some degree of one or both of these properties. Additives and coating agents which have been suggested in the main are polymeric compounds which range from strongly hydrophobic materials such as fluorinated polymers to more hydrophilic polymers containing carboxylic, phosphoric, and/or sulfonic acid functions. Certain of these compounds, while successful in providing protection to various substrates against certain types of soiling, generally have not been totally satisfactory in imparting soil resistance even to specific substrates against widely varying types of contaminants. In other instances, where impregnating or coating agents provide resistance against soiling by varying types of soiling agents, the additives may still be of limited use due to an inability to resist removal from the substrate when this is subjected to the wear it normally incurs. The permanence of an antisoilant on a substrate can be improved, in some instances, by chemically bonding the antisoiling agent directly or through crosslinking agents to the substrate. This approach, however, has its drawbacks mainly due to the chemical bonding interfering with or destroying other desired characteristics.

Among the agents capable of imparting improved antistatic, soil repellent and or soil release properties to textiles, the chemically modified chitosans deserve mentioning. It has already been mentioned in Chapter 3 (see page 128) that, in the presence of an alkaline catalyst, chitosan may react with sulfonium salts and hydrophobic polymeric materials (Owen & Sagar, 1967). Such materials include polyolefines, polyamides, polyesters, polymers and copolymers of acrylonitrile and cellulose esters.

It is well known that articles made from hydrophobic polymeric materials besides having a very limited capacity for absorbing moisture become electrified readily when subjected to friction. These properties are disadvantageous in that the electrified articles readily attract dirt and dust. Moreover, film composed of hydrophobic polymeric material which has become electrified is difficult to handle. Textiles composed of hydrophobic polymers which are used for clothing, besides having the disadvantage of soiling readily, tend to be unconfortable in wear owing to the limited capacity for absorbing moisture.

It is important to decrease the hydrophobic properties and reduce the tendency of the articles to become electrified: this can be done by treating them with chitosan. For instance, a nylon continuous filament fabric was impregnated with aqueous solution containing two parts of chitosan acetate, one part of 20 % aqueous solution of disodium tris(β-sulfatoethyl)sulfonium inner salt and 97 parts of water and squeezed so as to retain 50 % of its dry weight liquor. After drying at 60° the fabric was then impregnated with a solution containing 4 parts of sodium hydroxide and 96 parts of water and squeezed to remove excess liquor, rinsed with water and dried. The dry weight of the fabric had increased by 1.1 %. Compared to an untreated control fabric, the treated fabric had a reduced tendency to acquire an electrostatic charge and had increased affinity for the direct cotton dyestuffs.

The sulfonated polymeric alcohol derivatives described on page 127, are effective for use in providing antistatic, soil repellent and/or soil release coatings on a wide range of soil receptive substrate materials. They find effective use, for example as coatings for textiles, films or other articles made of glass, metals, ceramics, cellulosic materials, proteinaceous materials, polyesters, polyamides, polyurethanes and polymers of unsaturated compounds. These properties can be improved by providing in the ultimate coating one of the more highly insolubilized forms of the S-xanthogenate ester derivatives, which can be produced by heating them in the range 110 ÷ 180° and reacting them with cross-linking agents and converting the ester derivatives to heavy metal salts, during drying or curing treatments. The insolubilizing reagents can be incorporated into the bath or applied by spraying.

Cross-linking agents suitable for use in insolubilizing coating of S-xanthogenate esters are compounds with two or more functional groups reactive with hydroxyl groups and include diisocyanates, such as 1,4-toluene diisocyanate, aldehydes, polymethylol derivatives and epoxides, such as butadiene diepoxide. Heavy metal compounds suitable for this purpose are salts of aluminum, zinc, iron, chromium and zirconium.

Shrinkproofing of wool.

Chitosan exhibits very desirable characteristics for use in the shrinkproofing of wool. In the treatment of wool, it is suggested to use certain chitosan salts: the lower fatty acids serve very well for this purpose. The amount of chitosan salt which is deposited on the wool must be controlled so that it is sufficient to give shrinkproofing without interfering with the normal hand of the wool. A deposit equivalent to 2 ÷ 5 % of chitosan acetate is generally sufficient. The wool may be impregnated by padding and after drying the impregnated wool may be baked at 130° for 5 ÷ 20 min, in order to secure the

shrinkproofing effect. During the baking operations, the chitosan salts undergo loss of the volatile acids with the formation of a water insoluble layer of
chitosan. Enhanced softness and flexibility may be secured by incorporation of
1 ÷ 4 % of a polyhydric alcohol, such as glycerin, sorbitol, mannitol, or
carbowax. The carbowax additive have been found to partially plasticize the
baked chitin coating, and prevent it from flaking off. Polyhydroxyethers of
these polyhydric alcohols, such as allylsucrose, pentaerythritol give improved
results. Chitosan may be modified by reacting it with an aliphatic or an
aromatic halide to form modified chitin which tends to deposit in a more
flexible and water resistant film on the wool.

PHOTOGRAPHIC PRODUCTS AND PROCESSES.

In the field of photography, chitosan has found important applications since
processes for rapid development of pictures have been worked out, mainly by
the Technicolor Corp. and the Polaroid Corp. The choice of chitosan is mostly
due to its resistance to abrasion, optical characteristics, to its film
forming ability and to its behavior with silver complexes which are not
appreciably retained by chitosan and therefore chitosan can be easily penetrated by solutions carrying silver complexes from one to another layer of a film,
as described below. Furthermore, due to its regularly distributed amino groups
chitosan is suitable for forming mixtures with gelatin and for preventing
lateral diffusion of acidic dyes. While products containing chitosan seem to
be not yet marketed, it is clear that the photographic field is potentially
very important for chitosan applications.

Diffusion-transfer reversal process.

In photographic diffusion-transfer reversal processes a latent image in a
silver halide emulsion is developed and almost concurrently with such development a soluble silver complex is obtained by reaction of a silver halide
solvent with the undeveloped silver halide of the emulsion. This complex is
tranferred in solution from the emulsion to a suitable print-receiving layer
and the silver is there precipitated to form the desired positive print in
silver.

According to the process by Ryan & Yankowski (1961) a positive print is
obtained in a single step by suitable treatment of an exposed silver halide
emulsion containing a latent image with a processing composition. The processing liquid is in a viscous state and is spread in a liquid film between the
photosensitive element comprising the silver halide emulsion and the print-
receiving element, including a suitable silver precipitating layer. The
proces_ing compositon effects development of the latent image in the emulsion
and simultaneously forms a soluble silver complex, for example, a thiosulfate
or thiocyanate, with undeveloped silver halide. This soluble silver complex is
transported, at least in part, in the direction of the print-receiving element
and the silver is largely precipitated in the silver precipitating layer of
said element to form a positive image in silver.

The diffusion-transfer reversal processes may be substantially improved by the
addition of chitosan to the photosensitive silver halide emulsion. The
addition of chitosan provides, when the emulsion is contacted with an alkaline

photographic processing composition, a water insoluble and self-sustaining matrix. The chitosan matrix serves to maintain the wet photosensitive gelatin emulsion firm and undisturbed throughout processing operations especially when the processing operations take place at temperatures higher than normal. The concentration of chitosan disposed in the photosensitive silver halide gelatin emulsion may be of sufficient magnitude as to provide an integral single-layer element exhibiting sufficient internal support.

Addition of a sufficient amount of chitosan facilitates the utilization of a photosensitive silver halide gelatin emulsion, of a diffusion-transfer reversal process, after exhaustive removal of unexposed silver halide, as a suitable negative for the production of prints by conventional photographic procedures.

The concentration of chitosan admixed with the photosensitive silver halide emulsion may be varied over a wide range according to the degree of rigidity desired, during and subsequent to processing, and the thickness and character of the photosensitive emulsion employed. It has been found that good results may be obtained by the use of about 10 ÷ 35 ml of a 5 % solution of chitosan in 3 % acetic acid per 50 ml of a silver halide emulsion containing about 6 % gelatin and 5 % silver halide.

As illustrated in Fig. 6.9, a transfer process comprises a spreader sheet 0, a layer 1 of relatively viscous proceing agent, a photosensitive silver halide gelatin emulsion layer 2 containing chitosan, an image-receiving layer 3, preferably containing silver precipitating nuclei, and a suitable support layer 4. Support layer 4 may comprise an opaque material when a reflection print is desired or a transparent material when a transparency is desired.

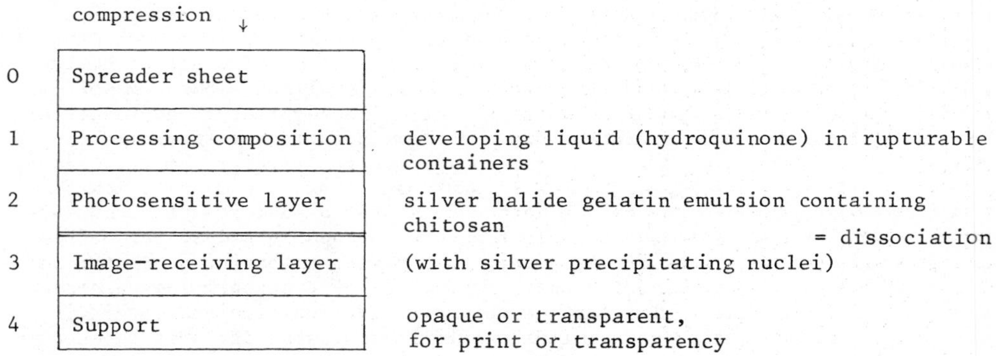

compression ↓

0	Spreader sheet	
1	Processing composition	developing liquid (hydroquinone) in rupturable containers
2	Photosensitive layer	silver halide gelatin emulsion containing chitosan
		= dissociation
3	Image-receiving layer	(with silver precipitating nuclei)
4	Support	opaque or transparent, for print or transparency

Fig. 6.9. Photographic unit based on chitosan. Ryan & Yankowski, 1961.

Liquid layer 1 may be obtained by spreading a photographic processing composition. The liquid processing composition may be disposed in a rupturable container so positioned in regard to the appropriate surface, of the photosensitive halide emulsion, that upon compression by a spreader sheet 0, a substantially uniform layer 1 of processing composition is distributed over the surface of the photosensitive emulsion 2, positioned distally in regard to image-receiving layer 3. The processing composition may comprise, for example, a developing agent such as hydroquinone, an alkali such as sodium hydroxide, a silver halide complexing agent and a high molecular weight film-forming thickening agent such as sodium carboxymethylcellulose.

In carrying out the transfer process, the photosensitive silver halide emulsion 2 is exposed to a predetermined subject to form a latent image. A substantially uniform layer of processing composition is distributed on the external surface of the emulsion, and permeates into the photosensitive emulsion contacting the chitosan disposed within it and simultaneously developing the latent image according to the point-to-point degree of exposure of this emulsion. As the same time as the development of the latent image, an imagewise distribution of soluble silver complex is formed from the unexposed silver halide within the emulsion. At least part of the silver complex, solubilized, is transferred by imbibition toward the print-receiving stratum 3 wherein it is reduced to elemental silver to provide a positive, reversed silver image of the latent image. Subsequent to precipitative deposition of the positive silver image in image-receiving layer 3, dissociation of this layer 3 from the emulsion layer may be effected.

The utilization of a photosensitive silver halide gelatin emulsion containing chitosan provides an emulsion exhibiting more adhesion capacity between the surface of the emulsion and a film-forming processing composition containing an acidic film-forming agent, as for example, sodium carboxymethylcellulose, than the adhesive bonding capacity exhibited between the image-receiving element and the processing composition. Thus, when print-receiving layer 3 is stripped from photosensitive emulsion 2 at the completion of processing, the film-forming processing composition remains substantially bonded to the photosensitive emulsion layer 2, thereby preventing the adherence of residual processing composition to print-receiving layer 3, which may result in possible stain formation or image degradation.

To obtain the low silver density required for films allowing non-removal of the negative image, silver precipitating centers like colloidal noble metals or copper sulfide may be introduced into the emulsion layer, according to Debruyn (1972). These centers suppress growth of the silver filaments beyond the size limits of the silver halide crystals and thus give rise to low silver covering power, whereas the silver of the positive image formed at the precipitating centers by physical development has high covering power. For this transfer process type, therefore, a very thin coating of chitosan containing such centers only is applied to the film. For instance, a cellulose acetate support was subbed with 1:1 gelatin + chitosan layer containing 4,6-diamino--o-cresol and coated with emulsion containing cadmium sulfide. The top coating consisted of chitosan (3 mg$\times$m^{-2}) and copper sulfide (1 mg$\times$m^{-2}). An additive tricolor screen can be inserted between film and emulsion layer.

Additive color reproduction.

Additive color photographic reproduction may be provided by exposing a photo-responsive material such as a photosensitive silver halide emulsion, through and additive color screen having a number of filter media or screen element sets each of an individual additive color, red, blue and green, and by viewing

the resultant photographic image, after development, through the same screen
element suitably registered. The photoresponsive material is permanently
positioned in direct continguous relationship with the color screen during
exposure, in order to maximize the acuity of the resultant image record.

The photographic film unit comprises a laminate, including a support carrying
on one surface, in order, a layer containing silver precipitating nuclei, an
inert non-nucleated protective layer, and a layer containing a photosensitive
silver halide emulsion. The purpose of the non-nucleated protective layer is
to provide a barrier over the layer carrying the transferred silver image,
after the emulsion layer has been removed subsequent to processing. The
material for the protective layer should be readily permeable to the processing composition and will not provide sites for the nucleation of the silver
forming the transferred image. The term "inert" refers to the inactivity of
the layer with respect to the photographic process. When the film unit is
employed as motion picture film, the protective layer contains a dispersed
friction reducing agent or lubrication compositon which, during processing,
will bloom to the outer surface of the protective layer imparting the
necessary slip properties subsequent to the removal of the emulsion layer. The
friction reducing agent, paraffin wax, polytetrafluoroethylene or similar, is
less dense than the protective layer and thus will appear at the outer surface.

The employment of chitosan as the protective layer is particularly
advantageous because chitosan provides non interference with the transfer and
formation of the silver image in the nucleated adjacent layer; chitosan forms
a water insoluble, abrasion-resistant matrix which can retain flexibility
after contact with alkali, i.e., the alkaline medium provided by the photographic processing composition. The use of chitosan as a protective layer
provides an initial material which offers no resistance to the passage of
image silver and processing compositions, but which will, subsequent to image
formation, provide a highly effective, abrasion-resistant layer for the
protection of the positive picture.

The stripping layer is employed adjacent to the photosensitive silver halide
emulsion layer and provides for the ready detachment of the emulsion layer
from the remainder of the film unit, and the material which is at least
softened by the processing composition which is employed in forming the photographic image. Materials which are attacked or softened by alkaline solutions
are suitable as stripping layer materials, for instance cellulose acetate
hydrogen phthalate and methylacrylate + acrylic acid copolymers.

The thickness of the non-nucleated protective layer is not critical, provided
the transfer of silver is not impeded; however the thickness is maintained at
a minimum, $0.1 \div 0.5$ μm in the case of chitosan.

Figure 6.10 shows an enlarged cross-sectional view of a film unit which
comprises a flexible transparent film base or support 0, carrying on one
surface, in order, a photoinsensitive layer 1, having associated silver
precipitating nuclei; an inert, processing composition permeable, non
nucleated, polymeric photoinsensitive layer 2, and a photosensitive silver
halide stratum 4 (Haberlin, 1972).

Figure 6.11 shows a particularly interesting film unit: member 0 carries on
one surface, in order, additive multicolor screen 8 positioned intermediate to
support 0 and photoinsensitive layer 1 which contains silver precipitating

0	Flexible film	transparent support
1	Photoinsensitive layer	with silver precipitating nuclei
2	Inert photoinsensitive layer	chitosan, permeable to processing comp.
4	Photosensitive stratum	with silver halide

Fig. 6.10. Photographic unit based on chitosan. Haberlin, 1972.

0	Flexible film	transparent support
8	Additive multicolor screen	
1	Photoinsensitive layer	with silver precipitating nuclei, in which the transfer image is formed
2	Inert photoinsensitive layer	chitosan
3	Stripping layer	
4	Photosensitive stratum	with silver halide

Fig. 6.11. Photographic unit based on chitosan and including an additive multicolor screen. Haberlin, 1972.

nuclei and in which the transfer image is formed. As in the previous Fig. 6.10 this film unit also shows the photoinsensitive protective layer 2, stripping layer 3 and photosensitive layer 4. Metallic sulfides and selenides are especially suitable as silver precipitating agents.

Chitosan is a matrix material apt for employment as layer 1 or 2; it is preferred to glues, gelatins, caseins, gums, starches, alginates, vinylalcohol, amides and acrylamides, cellulose ethers and esters and the like. For example, a gelatin subbed cellulose triacetate film base may be coated with a composition comprising chitosan and copper sulfide at a coverage of 0.4 $\mu g \times cm^{-2}$ chitosan and 0.24 $\mu g \times cm^{-2}$ copper sulfide. A second layer of chitosan, having 10 % of microcrystalline wax incorporated is coated thereon at a coverage of 2.5 $\mu g \times cm^{-2}$. On the external surface of the preceding layer a hardened gelatin silver iodobromide emulsion may be coated at a coverage of 20 $\mu g \times cm^{-2}$ gelatin, 10 $\mu g \times cm^{-2}$ silver and 0.4 $\mu g \times cm^{-2}$ algin. A film unit fabricated and processed as indicated above, exhibited a D_{max} of 3.8 and a D_{min} of 0.5.

In the making of color pictures by the dye-transfer process, it is necessary

to use a specially prepared dye receptive material which will carry the final
color image. This dye receptive material is a hydrophilic colloid, usually
gelatin, coated on a film or paper support and is often referred to as a
"blank". This hydrophilic colloid must be prepared or treated so that it will
not be physically damaged during the transfer process: this result is usually
obtained by incorporating a hardener at the time of the coating or by treating
the coated blank in a hardener solution.

In addition to this resistance to physical damage, the imbibition blank must
yield transferred images having good density, contrast, and also high definition. This latter condition is obtained by retarding the bleeding or
diffusion of the dyes in the colloid after transfer from the matrix by means
of a mordant distributed in the colloid. Mordants are introduced in the
colloid by mixing them into the coating dope or colloid prior to the coating
of the blank and prior to the dye transfer operation by impregnation of
the blank with a mordant solution.

Work by Delangre (1958) involved a premordanted dye imbibition blank comprising a support coated with a hydrophilic colloid containing an organic mordant
such as chitosan or chitin derivatives. Chitosan in a 3 % acetic acid solution
was mixed with a gelatin solution and the mixture was coated on a suitable
support, dried and hardened by processing for about 2 min in an alkaline
formaldehyde solution. The coated blank was rinsed in water for about 3 min
and dried for the dye transfer process. Transfer dye images made of this type
of blank are very sharp and of normal density and contrast. To obtain
transferred dye images of still higher definition or sharpness, the blank
prepared and hardened can be processed in the following solution for about 3
min: sodium sulfate (150 g), formalin (50 ml), hydrochloric acid (3 ml) and
water to one l.

The blanks prepared with chitosan are free from stain and opalescence. They
exhibit superior transfer behavior, give images of high definition, normal
shadow quality and normal density and are free from defects known as matrix
poisoning. The chitosan blanks are able to absorb dyes from the matrix much
faster and much more completely than other blanks, as they require 25 ÷ 40 %
less transfer time and their use results in the following advantages: the
definition of the transferred image is increased due to the fact that lateral
diffusion in the blank emulsion is greatly diminished in view of the shorter
time during which the blank is maintained in a wet condition; and the film
output of a given transfer machine can be increased by increasing the film
velocity, resulting in a lower contact time between the matrix and the blank.
Instead of decreasing the transfer time, the transfer temperature can be
lowered.

The effect of the addition of polymeric bases on fixing of acidic dyes in
gelatin layers was extensively studied by Pruglo, Bongard & Spasokukotskii
(1966 and 1972), Pruglo, Spasokukotskii & Bongard (1966) and by Bongard,
Pruglo & Spasokukotskii (1973). They observed that the essential reason of the
loss of acuity in the hydrotype process is the lateral diffusion of the dyes.
The path length of the dyes in a direction parallel to the layer surface
depends on many factors and is complicated by the interaction of the acidic
dyes with the basic groups of the gelatin. By introducing into the gelatin
layer certain compounds carrying basic groups, a better dye fixation is
favored. Many basic compounds have been tested, among which chitosan has been
pointed out as a very effective substance. For the fixation of the dyes in the

sensitive layer, rather large concentrations of basic groups are required, around 1×10^{-3} g.eq.$\times$g^{-1} gelatin, where for chitosan the equivalent is calculated on the basis of the monomer unit and on the basis of the deacetylation conditions. The polymer base fixative, at this concentration, compensates for the gelatin active groups.

Experiments carried out by the authors mentioned above, showed that in order to obtain enhanced print sharpness and optical density 2.5, the capacity of the 10 μm thick gelatin layer is used only at the extent of 10 %. An essential role of the fixative polymer chitosan is to offer a regular distribution at high concentration of basic groups to slow down diffusion. This is due mainly to the fact that the amino groups in the chitosan chain are regularly present on each ring. Chitosan was superior in this application to other polymers like polyethyleneimine, poly-(2-vinylpyridine), diethylaminoethylmethacrylate and other polymers. Monomeric materials such as derivatives of quinoline, pyridine and thiazole were not useful because of tanning and fogging.

Bongard & Pruglo (1973) prepared hydrotype color prints by using nine different azo dyes on films containing either chitosan or poly(vinylbenzyl-triethylammonium chloride) in the receptor layer as fixatives. One year storage of the obtained prints at ambient conditions and at high relative humidity and temperature indicated that migration of the dyes did not occur, on the basis of the intensity and spectra of the colors obtained.

Chitosan, like cadmium and gold ions, acts as a sensitizer, and influences the sensitivity of silver halides. Based on this concept, Steigman (1963) elaborated emulsions with properties varying between the gradation and speed of litho emulsions and those of negative emulsions.

CEMENTS.

Chitin and its derivatives have a role as adhesives not only because several authors have proposed them for cementing various materials together, but especially because they occur in nature in several systems quite well known for their properties of adhesion, even though not completely studied and understood.

For instance, Otness & Medcalf (1972) and Holland & Walker (1975) confirmed the presence of chitin in barnacle cement. Certain marine organisms like *Balanus nubilus* can secrete a cement to fix themselves to ships, rocks and other objects and it is an interesting fact that this cement becomes resistant and hard fairly rapidly, even in sea water. Cured or solidified barnacle cement is predominantly calcium carbonate in the crystalline calcite form, into a proteinaceous matrix. Thin sections of the solidified cement reveal a laminated array of matrices surrounded by calcite. The anionic groups on the matrix may serve as sites for nucleation during calcification. Barat & Scaria (1962) reported that eggs of hog lice are attached to bristles by an extremely hydrolysis-resistant substance that they showed to contain chitin and *p*-benzoquinone.

General materials.

The cementing action of chitosan salt solutions has been well known for many

years. The observations of this aspect of the physical properties of chitin
derivatives are a logical development of the efforts made over many years on
chitosan membrane casting and on the production of threads and shaped objects.

When used as a cement for electrically non-conducting surfaces, the chitosan
solutions have been found by Rigby (1936) to be general in application and to
develop decided strength and water resistance when thoroughly dried, heated or
properly chemically treated. When chitosan solutions are heated or allowed to
evaporate, they lose the acid if it is volatile enough; they can be treated
with ammonia or alkali to produce a definitely insoluble chitosan layer.
Chemical treatment includes washing with an acid solution which forms
insoluble chitosan salts, or treatment with a cross-linking agent, or with
carbonyl groups from small molecules like aldehydes, dialdehydes and anhydrides.

Athough chitosan films, even when not dissolved, are more or less swollen by
water, the amount of swelling is reduced by heating, formaldehyde treatment,
or even better, acetylation which produces chitin again.

When used as glue, chitosan and its derivatives possess good tensile strength
and superior water resistance, and particularly when insolubilized by a
further treatment.

It can be easily verified that chitosan cemented sheets of paper can be
separated only after soaking them for several hours in water. Common glues,
on the contrary, require only a few minutes soaking to separate. Paper sheets
cemented together with chitosan acetate and then treated with ammonia vapors,
require several days soaking to separate.

Heat tends to dehydrate the chitosan salt of certain organic acids by producing an amide; the chitosan amide is insoluble in water and in acids. The
probability that this kind of reactions occurs is shown by the fact that
heating organic acids of chitosan makes them insoluble in water and acids.
With some dibasic acids, imides may be formed; they are also insoluble in
water and acids. Chitosan dispersed in a sodium silicate solution forms a more
water-resistant joint than is obtained with the silicate alone.

Priming coats must often be applied in order to obtain perfect joints on a
number of materials. Satisfactory joints may be made of: paper joined to
regenerated cellulose, paper, wood, cloth, leather, glass etcetera; wood
joined to wood, cork, leather, glass, regenerated cellulose, rubber, cloth,
such as viscose rayon, canvas, glass etcetera; regenerated cellulose films
joined to cloth, and so on. This process patented by Du Pont de Nemours & Co.
(1936) was proposed for the manufacture of safety glasses, plywood, laminated
paper, wood veneer paper and furniture.

Further studies by Tsimehc (1938) showed that the advantages of chitin in the
field of textiles and adhesives are: intermiscibility with other dispersions
and emulsions; lack of foaming in preparation and use; stability to acidity
and to heat; film-forming properties; adhesiveness; water resistance after
thermal or chemical treatment; no mercerization with concentrated alkali
hydroxides and difficulty of esterification.

Kunike (1926) wrote that a thread can be spun wet or dry with differently
shaped cross-sections, having a tensile strength of 35 kg×mm^{-2}, whereas the

tensile strength of cellulose silk is only 25 kg×mm^{-2}. Films of chitin of large size, transparent and with high resistance to creasing were also obtained.

Du Pont de Nemours & Co. (1936) reported that asbestos treated with chitosan possesses better repellence to water and organic materials than do the untreated products. Chitosan is superior to water-proofing agents like waxes, resins and bitumen. Treatment with chitosan increases the dry and wet strength of asbestos sheets. The use of chitosan is particularly advantageous in the treatment of leather obtained from areas of the animal hide such as the neck or belly which are inferior in toughness to leather from other parts.

Patterson (1938) treated wood and plaster surfaces with a priming coating of chitosan and with a surface coating comprising a film-forming polyhydric alcohol ester of drying-oil acids. Du Pont de Nemours & Co. (1936) patented a process for stiffening of unsized straw-hat bodies; forming lacquers from which sheets may be cast; treating rosin-sized paper; they also mixed chitosan salt solutions with cellulose esters to prepare molding powders and other products.

Derivatives of chitosan also possess cementing ability. Chitin sulfate was proposed by Jones (1954) as a thickener in adhesives and drilling muds.

Wood and woody construction materials, due to their porosity, are customarily given protective or decorative coatings such as paint, enamel, lacquers or the like, after priming coats. These are required to give full protection and to avoid the formation of low gloss or non-uniform areas disfiguring the normal appearance of the properly applied coating. Patterson (1938) proposed that wood is coated with chitosan salt solutions because chitosan bonds strongly to wood even under adverse conditions. Chitosan is useful for the sealing and subsequent coating of rigid fibrous cellulosic surfaces in such a way that the adhesion to such surfaces of the finished coat is improved.

Chitosan is an effective sealer and primer not only for wood, asbestos-cement board and paper, but also for non-fibrous bases such as plasters, brick and tile. These construction materials are subject to deterioration in surface characteristics and appearance, due to the facility with which water, dirt, moisture, oils, grease, smoke and tar penetrate them. Once soiled, these surfaces are impossible to clean. Chitosan, when applied to these surfaces, decreases or prevents penetration of contaminants, thus prolonging the normal unsoiled appearance; it improves the strength of the surfaces and reduces the destructive effects of moisture.

Chitosan is an excellent binder for pelletizing fertilizers and feeds, if compared to conventional binders, like cellulose derivatives. Yoshida & Yamashita (1975) granulated a mixture containing ammonium sulfate, potassium chloride, urea and ammonium phosphate (520:2.75:100:74 parts) by spraying with a 0.5 % chitosan chloride solution. When chitosan is employed as a 100 mesh powder, on the contrary, it acts as an anticaking agent.

Reconstituted tobacco sheets.

A number of reconstituted tobacco compositions have found great commercial success as filler materials for cigarettes or cigars and as binders or wrappers for cigars. These compositions have generally utilized natural plant gums or their derivatives, such as galactomannans and cellulose derivatives. The art relating to these plant or vegetable adhesives is well advanced, however the cellulose derivatives and the galactomannans are not entirely satisfactory in creating reconstituted tobacco sheets which are completely water insoluble or which have superior tensile strength or have high ammonia stability. Numerous attempts have been made to overcome these drawbacks and chitosan has been found by Moshy & Germino (1969) quite suitable as an adhesive and a film-forming coating in the field of tobacco sheet production. Chitosan solutions, when mixed with tobacco and other optional ingredients, forms strong tobacco compositions, having good dry tensile properties, including wet and ammonia strength. When chitosan is used as a film-forming agent, it produces a sheet exhibiting hydrophobic and water repellency qualities. Using chitosan has also resulted in tobacco compositions having very good smoking characteristics such as burn, aroma, taste and ash.

For instance, a chitosan acetate solution is blended with some diatomaceous earth powder and polyglycerine. Tobacco is then added and blended. The final slurry contains 16 % solids. Hand sheets can be cast from this slurry on glass plates and dried in an electrically heated oven.

The glyoxal cross-linking of chitosan tobacco sheets was also performed: the amount of glyoxal used is very small. The glyoxal concentration used is related to the tensile strength of the product. Some tetraethylene glycol is added to a chitosan acetate solution, and blended with 2.5 % cellulose pulp and tobacco. After a few minutes, under stirring, glyoxal is added in the proportion of 0.2 % by volume, and the sheet is immediately cast.

Leather manufacture.

Babu, Rao & Joseph (1975) studied chitosan for use in tanning, paste-drying and finishing of leather. Chitosan added to the tanning drum with wattle bark extracts produces smooth, very tight and full leather and appears to have a good potential as a filling material for poor quality leather. When chitosan is used as an adhesive for paste-drying, the leather shows a full and tight grain, with no paste adhering to the leather. Chitosan can be also used as a primer for finishing formulations containing casein and other protein binders and as a filling coating instead of impregnation with resins.

COAGULANTS FOR SUSPENSIONS.

Chitosan may be used in treating turbid matter in aqueous solutions to help remove the turbid matter. Chitosan is used as a viscosity enhancer to settle solids suspended in liquids. Chitosan forms agglomerates comprising the impurities and chitosan itself which can be interpreted as a coagulant. The impurities are, for example, compounds of alkali-earth elements, vegetable matter and proteins.

Long molecules of dissolved chitosan wrap the solid particles of the turbid matter suspended in the liquid, and bring them together to form an agglomerate. The resulting agglomerate has to be discarded in all cases. In several cases, the action of chitosan can be compared to that of aluminum or ferric salts. For this process, there is interest in keeping the amount of dissolved chitosan used, as low as possible.

Chitosan has been shown to be effective for reduction of suspended solids in processing wastes from vegetable, poultry and egg-breaking operations. It is also effective on activated sludge, the biomass suspension which results from biological treatment of processing wastes. Chitosan is particularly effective in removing proteins from wastes, and the coagulated by-products could serve as sources of protein and animal feed.

Inorganic suspended solids.

Peniston & Johnson (1970) treated suspensions of montmorillonite with partially deacetylated chitin in conjunction with alum and lime: the deacetylated chitin was found to be as effective as Separan, a commercial flocculating agent. Some data are in Tabs. 6.11 and 6.12. Similar results were obtained with kaolinite clay suspensions.

Table 6.11. Settling of montmorillonite by chitosan and Separan; values are expressed in ppm. Peniston & Johnson, 1970.

	Montmorillonite samples					
Alum	100	100	100	100	100	100
Lime		2.5	12.5		2.5	12.8
Chitosan	1	1	1			1
Separan				1	1	
pH	4.3	6.4	4.7	4.3	6.6	4.0
Turbidity after:						
30 min	4	<1	<1	4	<1	<1
60 min	4	<1	<1	4	<1	<1

Table 6.12. Settling of montmorillonite by chitosans of various degrees of deamination and deacetylation; values are expressed in ppm. Peniston & Johnson, 1970.

Nitrogen, %	5.5	5.5	7.8	7.8	7.6	7.6
Deacetylation, %			53	53	39	39
Chitosan used,	1	0.5	1	0.5	1	0.5
Turbidity after:						
30 min	18	32	20	36	28	48
1 hr	18	26	11	27	25	27
2 hr	18	24	11	25	21	25
3 hr	12	22	8	22	15	22

A variety of chitosan obtained from partially deaminated chitin submitted to partial deacetylation was used to study the effect of nitrogen content and acetyl content on the coagulating ability of the polymer. Data in Tab. 6.12 show that high nitrogen content, low deacetylation samples, give the best results; they are used at very low concentrations: in fact there is interest in keeping as low as possible the amount of polymer.

Several applications were foreseen: treatment of waste water from mining operations and removal of tannin and polyphenolic materials from aqueous media and industrial wastes. The partially deacetylated chitin was then conceived as a clarification aid for rapid settling of suspended solids. Peniston & Johnson (1970) did not mention possible interactions with transition metal ions or compounds and the conditions reported do not allow the recovery of metal ions (Muzzarelli, 1972).

Vegetable canning wastes.

A few studies have dealt with coagulation of vegetable processing wastes with synthetic cationic and anionic polyeletrolytes, including chitosan. Bough (1975) has published data on the removal of suspended solids in vegetable canning waste effluents by coagulation with chitosan. As summarized in Tab. 6.13, chitosan can be accompanied by other coagulation aids, like carrageenan, alum, ferric chloride or calcium chloride.

Treatment with 10 mg×1^{-1} chitosan and 15 mg×1^{-1} carrageenan at pH 4.0 resulted in reduction in turbidity from 85 to 8.7; in suspended solids from 1624 to 6 mg×1^{-1}; and in COD from 2394 to 915 mg×1^{-1}. This indicates that treatment of soil-laden effluents from washing of vegetables and fruits by coagulation and separation of coagulated solids might allow recycling of such effluents. These wash waters could be treated and recycled in the washing operation and

Table 6.13. Conditions for coagulation of vegetable wastes with chitosan. Bough, 1975.

Vegetable	Effluent	Susp. solids in raw waste, mg×1^{-1}	Chitosan mg×1^{-1}	Salt or anionic polymer, mg×1^{-1}	Susp. solids in treated effluent, mg×1^{-1}
greens	washer	1624	10	carrageenan, 15	6
	filler	1747	5	carrageenan, 10 + alum, 40	125
	composite	143	10	none	15
spinach	composite	298	20	none	29
pimientos	peel	248	40	none	10
	core	32	10	none	5
	composite	75	30	none	8
gr. beans	blancher	116	5	CaCl$_2$, 80	6

significantly reduce the waste flow from the plant, provided that sanitary approval is granted.

The characterization of liquid waste from canning of leafy greens, indicated that 23 % of the COD load and 21 % of the wastewater flow originated with the dunker washers. Thus, segregation, treatment and recycling of this unit effluent would be an effective means for in-plant control of wastes.

Treatment of the composite effluent from spinach canning with 20 $mg \times l^{-1}$ chitosan reduced the suspended solids concentration from 298 $mg \times l^{-1}$ to 29 $mg \times l^{-1}$, as shown in Tab. 6.14.

Table 6.14. Coagulation of composite effluent from spinach canning, with chitosan and ferric sulfate at pH 5.0. Turbidity values are expressed in formazin turbidity units. Bough, 1975.

Chitosan, $mg \times l^{-1}$	Ferric sulfate, $mg \times l^{-1}$					
	0	20	40	80	160	320
0	60	41	39	36	36	40
5	39	41	40	37	39	44
10	14	28	34	37	39	44
20	3.4	3.3	3.9	10	15	20
30	1.4	1.1	1.1	1.2	2.5	4.5
40	1.1	1.0	1.0	0.9	1.2	3.5
50	2.3	1.0	0.6	0.6	0.7	2.1

The results for leafy greens effluent can be compared to those obtained with synthetic polymers. For treatment of the dunker washer effluent from leafy greens, a synthetic anionic polymer was effective at a lower concentration than chitosan. However, for the filler effluent, in contrast to only 5 $mg \times l^{-1}$ chitosan and 10 $mg \times l^{-1}$ carrageenan, 70 $mg \times l^{-1}$ Natron 86 and 10 $mg \times l^{-1}$ WT 3000 were required for effective treatment in an earlier study by Bough (1974). Chitosan was no more effective than the polyacrylamide (Natron 86) for treatment of the composite greens effluent.

For treatment of unit effluents from pimiento processing, chitosan was effective in lower concentrations than synthetic polymers previously tested. Comparable suspended solid reduction in the pimiento peel removal effluent was obtained with 40 $mg \times l^{-1}$ chitosan alone at pH 6 as with 50 $mg \times l^{-1}$ Natron 86 and 40 $mg \times l^{-1}$ alum at pH 4. In the core removal effluents, 60 $mg \times l^{-1}$ Natron 86 and 80 $mg \times l^{-1}$ alum were required for optimum treatment in contrast to 10 $mg \times l^{-1}$ of chitosan. Both treatments were at pH 6, the pH of the raw waste. In the pimiento composite effluent, chitosan at a level of 20 $mg \times l^{-1}$ was as effective as 30 $mg \times l^{-1}$ Natron 86 and 40 $mg \times l^{-1}$ ferric chloride (Bough, 1976; Bough, Landes, Miller, Young & McWhorter, 1976).

Protein recovery.

The solid phase separation of a microparticle suspension such as colloidal protein particles is an important process which is widely used in the food processing industries. When the final product is to be obtained as a concentrated solution, the concentrated solution should be as clear as possible, and it is important to improve the product yield by recovering the useful proteins from the treated solution. In this case it is necessary to separate the protein suspended in the form of microparticles.

Proteins are amphoteric and protein amino groups are progressively less protonated in the pH range above the protein's isoelectric point. In this range the proteins are anions. The isoelectric points of the proteins handled in the food industry are generally in the acidic range, and it is therefore preferable to handle the protein suspension in a neutral environment. In order to coagulate the proteins efficiently, cationic coagulating agents may be used to neutralize the negative ions effectively.

In order to simplify solid phase separation in its general applications to industries in addition to the food industries, e.g., waste water treatment, various coagulating agents have been developed; however these agents are not always suitable for the food industry. Cationic polymer coagulating agents are generally synthetic compounds which exhibit a high toxicity.

Since chitosan is not toxic, it can be safely used in the food industry. Chitosan is superior for use in food processing because it rarely spoils or molds. However, more data should be obtained in this field.

When a prescribed amount of chitosan acetate solution is added to the microparticulate suspension of protein, a rapid coagulation of the colloidal particles is induced. Solid phase separation occurs in a very effective manner.

Whilst, in general, a suspension containing colloidal protein particles cannot be separated by filtration because the particles adhere to the filter cloth, chitosan forms a large coagulate which is easily filtered. If the centrifugal pressure method is used, the separation will be remarkably improved. The recovery rate and clarity of the supernatant is also increased. The amount of chitosan to be added is dependent upon the concentration of the suspension, the concentration of coexisting ions and the environment of the suspension, such as the pH value. The added amount of chitosan should preferably be about 0.2 ÷ 10.0 % of the dry weight of the suspension.

Fujita (1972) subjected a concentrated fish solution to high frequency centrifugal separation to remove fish fragments: a large amount of protein microparticles remains suspended in the supernatant. A 0.5 % chitosan solution in 0.3 % acetic acid (35 parts) were added to this fish meat supernatant (1000 parts). Immediately after the addition of chitosan a coagulate formed, and it was separated by centrifugal precipitation. The supernatant was completely clear and the precipitate was added to the boiled fish meat with improved final yield.

Soybean milk was also treated with chitosan in a similar way. The coagulate was filtered with a filter cloth and a curdlike soybean protein coagulate was obtained. The clear filtrate contained at the end 0.95 % solids only.

A 1 % chitosan solution in 0.8 % lactic acid (100 parts) was added to skim
milk (500 parts) containing 3.12 % protein. After vigorous stirring a large
amount of protein coagulate was obtained. Casein (53 parts) with 73 % water
content was separated by centrifugal precipitation. The supernatant was
completely clear.

Chitosan tartrate was used for the treatment of filtrates from rice
fermentation in similar ways (Fujita, 1972).

In addition to poultry processing wastes and to egg breaking wastes, according
to Bough (1976), the waste water from packing plants which slaughter and
process meat can be effectively treated with chitosan and other polyelectrolytes. On the basis of turbidity data obtained by treatment with 1, 30 and 60
$mg \times l^{-1}$ of eight different polymers, chitosan was most effective at the pH of
the raw waste water (7.5). The results of jar tests employing different
concentrations of chitosan demonstrated that the optimum chitosan concentration was 30 ÷ 40 $mg \times l^{-1}$, as shown in Fig. 6.12.

The increase in turbidity observed at excessive polymer concentrations (above
40 $mg \times l^{-1}$) is a common effect of overdosing with polymers. The optimum
conditions developed were tested on a pilot-scale.

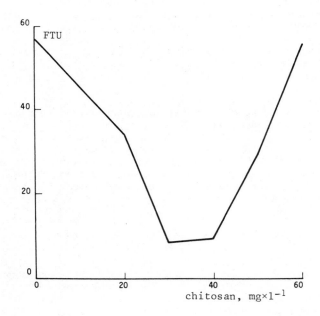

Fig. 6.12. Effect of chitosan concentration on reduction
of turbidity in meat packing waste water at pH 7.5.
Bough, 1976.

Coagulation of the waste water with 30 mg×l^{-1} chitosan and settling for 90 min reduced suspended solids by 89 %, from 485 to 49 mg×l^{-1}. The COD was reduced by 55 %. The coagulated solids recovered from the settling tank contained 41 % crude protein, 17 % fat and 11 % ash on a dry weight basis.

Bough (1976) has also compared chitosan to 21 commercially available polymers for removal of proteins in processing wastes from a shrimp canning plant and found it superior for abatement of turbidity and recovery of valuable protein.

Dewatering of municipal sludges.

As stricter regulations are applied to municipal and industrial waste water treatment, more efficient processes and equipment are being used to remove suspended and settleable solids, and other pollutants. The solids removed by a variety of solid-liquid separation methods in municipal and industrial waste water treatments include grit, screenings and sludge, and constitute the most important by-product of the waste water tretment process.

Sludges from settling tanks in the biological system are normally in the range of 0.5 ÷ 1.5 % by weight suspended solids with a specific gravity of 0.95 ÷ 1.04. Further gravity thickening of municipal activated sludge seldom produces concentrations greater than 3.5 %. The separation and concentration of suspended solids in waste water can be achieved by a variety of equipment including many types of centrifuges and filters.

The relatively recent addition of polymers or polyelectrolytes to chemical conditioning technology and their apparent success prompted the adaptation of centrifuges to the dewatering of municipal and industrial sludges and these have been used with varying degrees of success in the United States of America, Europe and Japan. Sludge dewatering by this technique is becoming increasingly popular due to the successful use of coagulants which permit one to handle sludges difficult to dewater. Asano, Suzuki & Hayakawa (1976) obtained results relevant to the addition of chitosan to various types of sludges; for raw primary sludges and activated sludges with added 0.5 ÷ 0.8 % chitosan they obtained a solids capture of about 98 %. Without the polymer addition, the solids capture was around 61 %. The relationship between the polymer dose and cake moisture was rather scattered in the range 60 ÷ 80 %.

Since organic sludge particles are most commonly negatively charged, a polymer like chitosan is expectedly effective for charge neutralization, agglomeration and releasing bond water in sludges.

CHAPTER 7

MEDICAL APPLICATIONS

Some of the data reported in the previous chapters have been originated from medical research. The interest of chitin and chitosan to the medical, ecological and pharmaceutical sciences is evident in many publications and has been clearly indicated in several instances. The information to be dealt with in this chapter is however scarcely homogeneous and the research projects from which it originates have been discontinuous and sometimes occasional.

Chitin and its oligomers have been of great help in the past for the progress of medical and biochemical knowledge: chitin oligomers were used to study the behavior and the chemical characteristics of lysozyme. Lysozyme is a very important enzyme present in the human body fluids; it dissolves certain bacteria by cleaving the polysaccharide component of the cell walls and makes them burst because of the osmotic pressure inside the cell.

The bacterial cell wall polysaccharide is very close to chitin from the chemical point of view, insofar as one out of two anhydroglucose units carries a lactyl group on carbon 3. It is currently described as a polysaccharide whose repeating unit is a disaccharide of N-acetylglucosamine and N-acetyl-muramic acid with β-1,4-glycosidic linkages: N-acetylmuramic acid and N-acetyl-glucosamine alternate in sequence. It seems also chemically correct to describe this polysaccharide as a chitin ether of lactic acid. A scheme of the peptidoglycan is shown in Fig. 7.1.

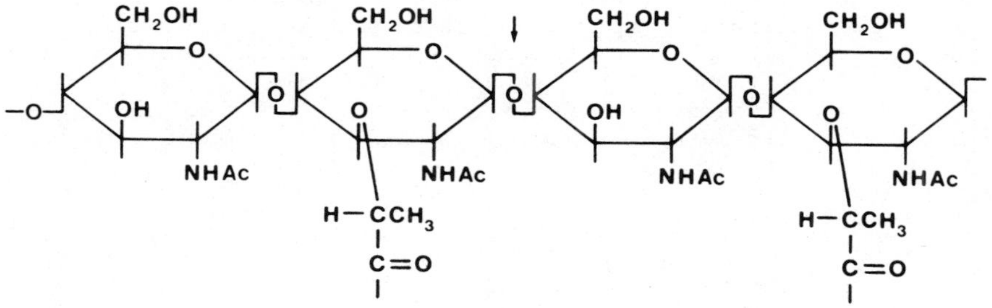

Fig. 7.1. Polysaccharide from bacterial cell wall, showing alternate N-acetylmuramic acid and N-acetylglucosamine units. This polysaccharide is an ether of lactic acid and chitin. The arrow points to the anhydroglucosidic bond hydrolyzed by lysozyme.

4-O-β-N-acetylmuramyl-N-acetylglucosamine and 4-O-β-N,6-diacetylmuramyl-N-
-acetylglucosamine have been isolated by Tipper & Strominger (1966) from the
cell walls of the *Staphilococcus aureus*. These disaccharides are isometric
to 4-O-β-N-acetylglucosaminyl-N,6-O-diacetylmuramic acid; all these
disaccharides contain beta linkages. The major part of the cell wall polysaccharide is a substituted chitin where every other unit of β-1→4-linked
N-acetylglucosamine carries a 3-O-lactyl substituent, and in which some
units are also substituted on the position 6 with an O-acetyl group.

Lysozyme hydrolyzes the glycosidic bond between C^1 of N-acetylmuramic acid
and C^4 of N-acetylglucosamine; the other glycosidic bond is not cleaved. It
is therefore not surprising that lysozyme cleaves also the chitin glycosidic
bonds; chitin, as well as the bacterial polysaccharide is hydrolyzed by
lysozyme.

In the same way as the bacterial polysaccharide (lactic acid ether of chitin)
has opened the route to the biochemical attack against microorganisms, chitin
itself offers at present the possibility of attacking insects, fungi and
other living organisms having a chitinous support. In chapter 1 we have
already described the insecticides which inhibit the chitin synthesis;
applications of the lysis of chitin to parasites in the human body are in
the following pages and it seems that in the near future much more work will
be carried out to understand the defense mechanisms against fungal infections
in various organs, especially kidneys. Corticosteroid therapy may render man
susceptible to systemic infections from opportunistic fungi such as *Candida
albicans* and *Aspergillus fumigatus*. The incidence of these infections is
increasing and part of the transplant recipients develop fatal aspergillosis.
Cortisone-sensitive resistance to mycelial growth is naturally present in the
mouse kidney and may be an important aspect of the resistance of mice to
fungal infection. At present, very sensitive analytical methods are available
for the quantitative determination of chitin in organs, and it seems possible
to develop new chemicals to attack the fungal chitin in vivo.

A new inhibitor of chitin synthesis, nikkomycin, a nucleoside-peptide
antibiotic consisting of uracyl, an amino hexuronic acid and a new aminoacid
containing a pyridine ring, has been recently described by Dahn, Hagenmaier,
Hohne, Konig, Wolf & Zahner (1976). It was isolated from the fermentation
broth of *Streptomyces tendae* ; its structure and identity were established by
optical and mass spectrometry, chromatography and radiochemical methods, as
shown in Fig. 7.2.

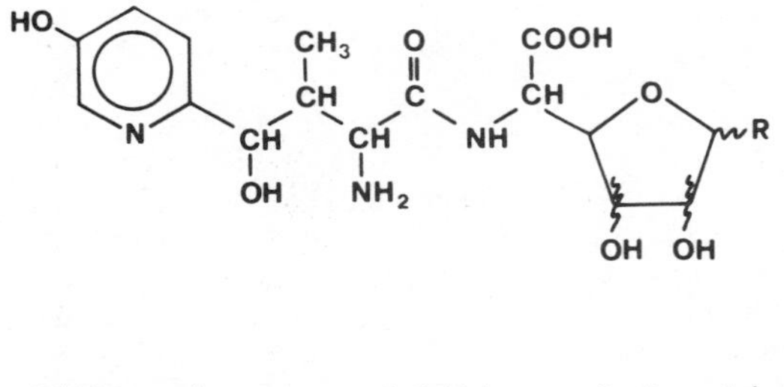

Fig. 7.2. Nikkomycin, a new inhibitor of the chitin synthesis. R = uracyl.

Nikkomycin was able to depress the synthesis of chitin not only in a buffer solution at pH 6.5 containing N-acetylglucosamine, ATP and UDP-$|^{14}C|$-N-acetyl--D-glucosamine together with chitin synthetase at 20°, but also in vivo, namely in a great number of fungi, especially Zygomycetes and Basidiomycetes: the effect of nikkomycin on the chitin synthesis of *Mucor hiemalis* is shown in Fig. 7.3.

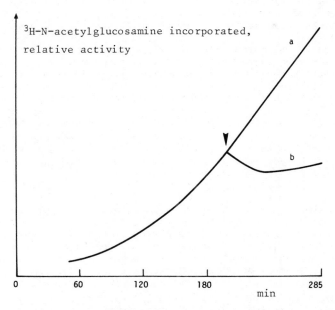

Fig. 7.3. The effect of nikkomycin on the chitin synthesis in *Mucor hiemalis*. Tritiated N-acetylglucosamine incorporated vs time. The antibiotic addition is indicated by the arrow. a = reference with no antibiotic; b = nikkomycin present at 0.5 µg×ml^{-1}. Dahn, Hagenmaier, Hohne, Konig, Wolf & Zahner, 1976.

ARTIFICIAL KIDNEY MEMBRANES.

Chitosan membranes have been proposed as artificial kidney membranes (see chap. 2) possessing high mechanical strength in addition to permeability to urea and creatine; they are impermeable to serum proteins and probably they might be unique in offering the advantage of preventing immission of toxic metals into the blood stream, as currently happens with other membranes. More research should be carried out in this interesting area.

As chitosan can be mixed with other high polymers, Mima, Yoshikawa & Miya (1975) have prepared chitosan + poly(vinylalcohol) membranes 30 µm thick. In water, at 25° the membranes absorbed 102 % water and exhibited favorable tensile strength, elongation and urea permeability coefficient values.

PREPARATIONS AGAINST PARASITES.

Many warm-blooded animals are subject to attack by parasites such as ticks,
mites and the like. Animal parasites as well as bacteria and viruses contain
antigenic substances. Porter (1971) has found that chitin from various sources
including dried insects is highly effective as an antigen when administered
to animals attacked by Arachnida or by certain bacteria and fungi.

For this purpose, finely ground chitin may be added to a suitable liquid
carrier such as physiological saline solutions; the mixture can be sterilized
according to pharmacopoeia procedures and can be administered subcutaneously.
Infested dogs received injections of chitin and after a few days they were
entirely tick-free in all areas. The hair coat of the dogs greatly improved
since the administered chitin exhibited antigenic activity against bacteria
and fungi. Observations on live fleas and ticks demonstrated that blood taken
from the injected dogs immediately affects the parasites. Female ticks aborted
within a few hours and then got comatose; all fleas were dead within one day.
The serum of warm-blooded animals, containing such antibodies developed by
chitin, is useful for immunization of other animals against parasitic attack
and the associated diseases.

Nematodes are worms that live in various human organs and produce diseases:
they possess a chitinous tegument. Beuzekom & Douma (1969) have found that
they can lose their tegument under the action of the molting hormones secreted
by the arthropods. Ecdysone, ecdysterone and synthetic analogs referred to the
formula in Fig. 7.4, for example 6-oxo-2,3-dihydroxy-cholest-7-ene, are
molting hormones. Several molting hormones have been isolated and characterized.

The hormones are administered to patients in the form of coated tablets when
the infested organs are the intestines; however injections are preferred when
muscular tissues are infested. The nematodes die after losing their integument
while no damage occurs to human tissues.

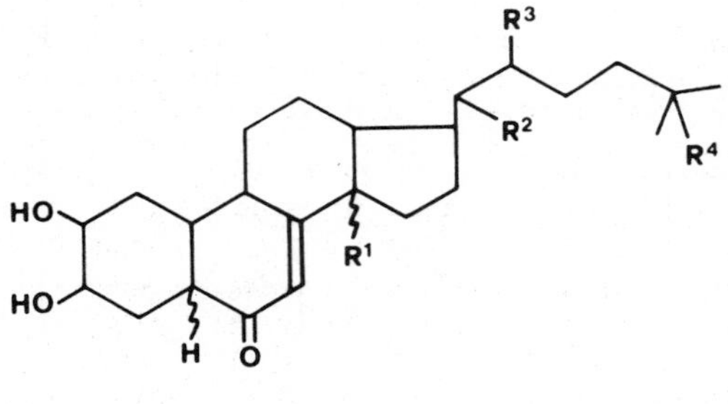

Fig. 7.4. Ecdysone hormones. R are H, OH or other groups.

BIODEGRADABLE PHARMACEUTICAL CARRIERS.

Chitin, 6-0-carboxymethylchitin, |6-0-(2-hydroxyethyl)chitin| and 6-0-ethyl-chitin in form of membranes have been used as enzymically decomposable pharmaceutical carriers.

The delayed release of drugs is a very important aspect of medical research. Chitin and its derivatives are appealing substances as carries because, as shown in previous parts of this book, they are degraded by lysozyme, an enzyme present in the human body, and the degradation products do not introduce any disturbance.

An interesting example of the applications of chitin membranes is that presented by Capozza (1975). Flexible and thin membranes of 6-0-carboxymethyl-chitin carrying pilocarpine nitrate were prepared by mixing solutions of the two compounds and, after evaporation on a glass plate, by soaking in 10 % alum for 5 hr (see Tab. 7.1.). Similar chitosan + pilocarpine membranes were soaked in 37 % formaldehyde. The 6-0-carboxymethylchitin membranes have been examined both in vitro and in vivo; after implantation in the eye the effect of pilocarpine was noticeable for 6 hr; longer lasting effect is anticipated for humans because pilocarpine in humans undergoes slower metabolic degradation than in rabbits. The membranes are slowly degraded and do not irritate the eye; they can be previously sterilized.

TABLE 7.1. Preparation of pilocarpine + 6-0-carboxymethylchitin membranes. The amounts used are expressed in mg. Capozza, 1975.

Pilocarpine	6-0-carboxymethylchitin	Content in 1.5 mg, 1×10 mm strips
9.1	90.0	0.10
19.4	80.6	0.24
33.3	66.6	0.50

The chitosan + formaldehyde membranes did not release initially any pilocarpine and even after more than 3 days 70 % of the pilocarpine was not released. Liquid scintillation spectrometry on tritiated pilocarpine confirmed that the chitosan + formaldehyde membrane is hardly degraded by lysozyme.

A similar approach is possible for the implantation under the skin of membranes carrying drugs to be released over an extended period of time. The chitin derivatives offer the advantage that they undergo enzymic hydrolysis and are not hydrolyzed by water or other substances. Antibiotics, like tetracycline, neomycin and bacitracin, antiviral products like idoxuridine, antiallergic products like chlorpheniramine, antiinflammatory products like hydrocortisone, prednisolone and its derivatives, decongestive substances like naphazoline and tetrahydrazoline, myotics and anticholinesterases like scopolamine, tropicamide and eucatropine, sympathomimetics like epinephrine and other drugs may all be incorporated into chitin membranes for delayed release.

BLOOD ANTICOAGULANTS.

Heparin, one of the most widely used blood anticoagulants, was isolated from the liver in 1918; during the following years it was submitted to a thorough purification and chemical analysis. One of the favorable characteristics of heparin is its non-toxicity on intravenous injections into animals as well as in man.

Heparin is an expensive product, presently in short supply. Attempts have been made to prepare a number of synthetic anticoagulants, but none of them has been found to be non-toxic as heparin. Piper (1945) remarked that synthetic sulfuric acid esters of cellulose, starch and chitin, all of which are anti-coagulants, can agglutinate the thrombocytes in vitro, so that these constituents are removed from the supernatant layer of plasma on the sedimentation of the red blood cells, the plasma thus becoming clear and translucent. The more or less pronounced toxicity of these compounds was attributed to their agglutinative effect on the platelets with the risk that the aggregates of thrombocytes may give rise to infarction. Astrup, Galsmar & Volkert (1944) noted that cellulose and starch sulfuric acid esters are toxic, while chitin disulfuric acid is slightly toxic.

Much progress has been made since then in the chemical characterization of these polysaccharides; among the many studies on this topic, are those by Moskowitz, Schwartz, Michel, Ratnoff & Astrup (1970) who studied the kinin--like activity induced by chitin disulfuric acid, heparin and other polymers at high concentrations; no kinin-like activity evolved when these agents were incubated with Hageman factor-deficient platelet-deficient plasma. The relatively high concentrations of heparin needed in order to induce kinin-like activity, indicated that this property of heparin is without biological significance, while the anticoagulant properties of chitin disulfuric acid precluded testing its action on Hageman factor.

Bernfeld, Nisselbaum, Berkeley & Hanson (1960) reported on the interaction of polyanions with human serum β-lipoproteins; the Upjohn Company (1960) published on the improved preparation of sodium O-sulfochitosan-N-sulfonate, chitosan-N-sulfonic acid and sodium O-sulfochitosan-N-sulfonate; Embery, Lloyd & Fowler (1971) studied the metabolic fate of the potassium salt of $(|^{35}S|$-sulfoamino)-chitosan and related substances and the conductometric determination of sulfate and carboxyl groups in heparin. Recently, Gainey, Malone & Rimpler (1975) reported that the protein moiety of chitin is responsible for inflammatory response when material containing chitin is injected into the tissues of higher animals, while injection of pure chitin does not give a detectable response.

Chitosan is a suitable starting material for the production of heparin-like blood anticoagulants. N- and O-sulfated chitosans have been tested and found to possess 15-45 % of the activity of heparin in vitro, by Doczi, Fischman & King (1953); Ricketts (1953), Wolfrom, Shen-Han & Summers (1953) and Warner & Coleman (1959). The latter authors selectively N-sulfated all the amino groups of chitosan but obtained a product possessing no anticoagulant activity. An increase in the sulfation to 18.3 % sulfur wherein a number of O-sulfate groups were present in the product, caused it to show an anticoagulant activity of 60 IU×mg^{-1}; this value is however still low, compared to the 110-150 IU×mg^{-1} exhibited by heparin, which has a sulfur content of only 12-13 %. Kochnev,

Molodkin, McHedlis & Plisko (1973) sulfated chitosan films and studied their anticoagulant activity in vitro.

Uronic carboxyl groups contribute significantly to the anticoagulant activity of heparin. Stivala & Liberti (1967) have stated that the binding character- istics of carboxyl groups should be considered in examining heparin activity. Whistler & Kosik (1971) introduced uronic acid carboxyl groups into chitosan sulfates. Chitosan, N-sulfated chitosan and N-sulfated-partly-O-sulfated chitosan were oxidized by dinitrogen tetraoxide or by oxygen over platinum catalyst. A range of products, differing in composition, were prepared and their anticoagulant activities were tested in vitro.

If the degree of sulfation is high enough to produce sulfamides of all the amino groups present, anticoagulant activity appears. This suggests that the anticoagulant activity is controlled by the presence of derivatized amino groups, or that free amino groups do not allow anticoagulant activity. As the level of sulfation further increases to produce O-sulfates, anticoagulant activity appears and rises rapidly in proportion to increasing sulfate content.

Introduction of uronic acid carboxyl groups increases the anticoagulant activity of sulfated chitosan. The effect of these groups is visible in Fig. 7.5. Insignificant activity is present until all amino groups are sulfated; for higher sulfation degrees the carboxylated polymers are more active. The anticoagulant potency of an unknown is determined by comparison of the minimum amounts of standard heparin and unknown which, when contained in a volume of 0.3 ml, will keep 1 ml of recalcified human plasma more than 50 % fluid for one hr at 37°.

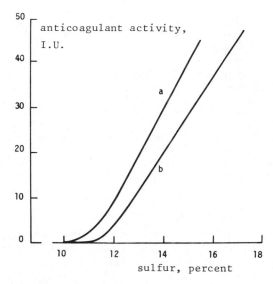

Fig. 7.5. Anticoagulant activity of chitosan derivatives; a = sulfated and oxidized; b = sulfated. Carboxyl contents were similar for all preparations (2.79 - 2.94 %). Whistler & Kosik, 1971.

The selective sulfonation of amino groups in chitosan has been fully
described by Warner & Coleman (1958); the colorimetric methods of estimating
the uronic acid content of polysaccharides have been described by Anderson &
Garbutt (1963).

The most recent development in this field appears to be the study by Horton &
Just (1974), already mentioned in Chapter 3. The chitosan derivative $(1 \rightarrow 4)$-2-
-deoxy-2-sulfoamino-β-D-glucopyranuronan sodium salt differs from heparin in
that the latter is an N-sulfated-partly-O-sulfated alternating polymer of
2-amino-2-deoxy-D-glucose and hexuronic acid residues. Nevertheless the
chitosan derivative was found to have an anticoagulant activity of 25.8
$IU \times mg^{-1}$. Its acute toxicity in the mouse (LD_{50}, intraperitoneal) was 237
$mg \times kg^{-1}$, that is about 3 times that of heparin. At the concentration of 100
$\mu g \times ml^{-1}$ it inhibited the growth of leukemia L-1210 cells by 50 %.

Maximal anticoagulant activity of heparin is known to depend to an important
extent on fractionation to a narrow range of molecular weight. Heparin
fractions of higher molecular weights have low or variable anticoagulant
activity, thus the low activity of the $(1 \rightarrow 4)$-2-deoxy-2-sulfoamino-β-D-gluco-
pyranuronan sodium salt may be ascribed to its high average molecular weight
and to the high degree of polydispersity. The complexes of sodium dextran
sulfate with chitosan reported by Kikuchi & Toyota (1976) are antithrombogenic
materials. Thrombous formation of the polyelectrolyte complex from high
molecular weight sodium dextran sulfate (640,000) was less than that of low
molecular weight (6,000). ($|^{35}S|$-sulfoamino)chitosan was selected as a poly-
meric model of the 2-deoxy-2-$|^{35}S|$-sulfoamino-D-glucose residues which occur
in a labelled heparin by Lloyd, Embery & Fowler (1971), for use in studies on
mammalian enzymes involved in the degradation of sulfamate groupings in
heparin.

AGGREGATION OF LEUKEMIA CELLS.

Malignant cells possess a net electronegative surface charge higher than that
of their normal cell counterparts; hyperanionicity of malignant cell surfaces
causes decreased intercellular adhesiveness, loss of contact, inhibition of
growth and movement, invasiveness and tissue disorganization.

The growth and invasive movement of cancer cells could be specifically
inhibited with positively charged polysaccharides. The adsorption of polycations to the surface of cells mediates profound biological changes: reversal
of electrophoretic cell potential, inhibited growth, cellular aggregation,
hemolysis, cytolysis and cytoplasmatic swelling of normal and malignant cells,
cell membrane permeability changes, inhibition of tumor cell transplantability
and induced invasion by virally transformed fibroplasts.

Normal blood cells, especially erythrocytes, also have a high anionic surface
charge: thus, the selctive binding of a polycation to malignat cell surface
over a wide range of concentrations is the real characteristic of a polycationic antitumor agent.

Specific adsorption of polyethylenimine to rat tumor cells was achieved;
similarly, a basic copolymer of ornithine and leucine (1:1) selectively
aggregated mammalian cells transformed by simian virus. N-formylchitosan poly-

sulfuric acid in a dose of 80 mg×kg^{-1} has been shown by Abbott, Monkhouse, Steiner & Laidlaw (1966) to produce marked reduction in the width of the zona glomerulosa of the rat adrenal gland without apparent effect on the other zones; this fact may provide an explanation for the prolonged inhibition of aldosterone secretion produced by N-formylchitosan polysulfuric acid in the human, in addition to influence on hypertension and electrolytes (Abbott, Gornall, Sutherland, Stiefel & Laidlaw, 1966).

Chitosan, as used by Sirica & Woodman (1971) at concentrations ranging from 6 to 50 neq×ml^{-1} selectively aggregates L-1210 leukemia cells in vitro, when compared to normal murine erythrocytes and bone marrow cells, as well as Sarcoma 37 and Ehrlich ascite cells, whilst poly-L-lysine and polyethylenimine at the same concentrations are not selective. Chitosan produces dense aggregations, cytoplasmatic swelling and/or cytolysis of the leukemia cells. Chitosan produces only a slight aggregation of the erythrocytes and does not significantly alter their morphology. Incubation of the leukemia cells with chitosan partially inhibits their transplantability; immunosuppression with cyclophosphamide does not alter this effect. Chronic chitosan treatment of mice, bearing intraperitoneal implants of the ascite tumor, retards the dissemination of the leukemic cells into the peripheral blood.

As the most consistently peculiar features of cancer cells are the physical and chemical properties of their surfaces, the aggregation endpoint pattern of a polycation for the normal versus tumor cells provides a means for investigating the characteristics of tumor cell surfaces. The studies so far available show that polycations can kill tumor cells in vitro and can alter the degree of tumor growth in vivo. Chitosan and its derivatives are suitable for the development of novel antitumor agents that act primarily against the cell surface.

WOUND HEALING ACCELERATORS.

Medicine has long been engaged in improving the healing of wounds. Patients suffering from diabetes or undergoing cortisone treatment show extremely slow rate of healing. Rapid healing of wounds is desirable for patients in tropical countries where the risk of infection is high. Rapid recovery after surgical operations resides on healing, and rapid healing is generally sought when immobilization is difficult or painful.

Cartilage from a wide variety of animal sources exhibits the striking ability of producing wound healing acceleration. Prudden, Migel, Hanson, Friedrich & Balassa (1970) observed that the differences between cartilage lots more or less active, consisted in the variable concentration of glucosamine in the active material. After various tests the authors turned to chitin as a wound healing promoting agent.

Certainly, there is considerable biochemical evidence linking N-acetylglucosamine with the metabolism of the hexosamines which are assumed to orientate and cross-link wound collagen. Uridine diphosphate N-acetylglucosamine is a key biochemical in the biosynthesis of chondroitin sulfate, hyaluronic acid and the glycoproteins.

Chitin is "physiologically" soluble because lysozyme, present in large amounts in recent and healing wounds, acts on it. Chitin derivatives suitable as healing accelerators are ethers formed with pharmaceutically acceptable groups and esters or salts with pharmaceutically acceptable acids. Examples are hydroxyethylchitin, carboxymethylchitin and its zinc salt, methylchitin and ethylchitin, chitin acetate, nitrate, citrate and phosphate, N-formyl-, N-acetyl-, N-propionyl- and other N-acyl derivatives. Partially depolymerized chitin is most suitable (Balassa, 1969, 1970, 1971, 1972, 1975).

As compared to the great variability in cartilage depending on the collection method, the animal and its age, chitin is a relatively uniform and easily obtainable material.

Chitin is topically applied as a finely divided powder, directly to the wound surface. Oral administration or implantation is also feasible. Colloidal chitin, or water-soluble derivatives of chitin in isotonic saline, can be administered intramuscularly, parenterally or intravenously. Chitin particle size is less than 150 μm, preferably less than 50 μm. Chitin of fungal origin is particularly suitable because grinding does not take much trouble. The fungal chitin obtained by fermentation of a fungus may be ground and used to promote healing, after suitable sterilization. Chitin can be applied topically as a dry aerosol powder, optionally together with other medicaments.

For this purpose, the entire fungal mat, after sterilization, is defatted by extracting with chloroform at room temperature. The defatted fungal material is treated with 1.0 N NaOH solution for 18 hr at room temperature and after acidification with hydrochloric acid, it is purified by dialysis.

Wound healing improvements, as measured by the method of Prudden, are as high as 46 % with *Penicillium* chitin. Surgical gauzes or pads to be applied to the wounds may also be impregnated with chitin. Chitin fibers obtained acccording to the chemical and textile technology mentioned in Chapter 6 may be used as sutures in surgery or included in bandages or other support bases for surgical dressing either in a woven or nonwoven fabric structure. Histologically, there is little reaction to chitin; the progressive elimination of the microparticle takes place over a period of a few months.

MICROBIOLOGICAL MEDIA.

Media containing chitin are highly selective for the isolation of actinomycetes in water and soil. The selectivity is based on the universal ability of actinomycets to hydrolyze chitin, since they produce chitinolytic enzymes, whereas few bacteria and fungi from soil can utilize it. Chitin induces the production of chitinase by actinomycetes while other sugars are not effective (Nakagami, Tonomura & Tanabe, 1969). Different preparations of colloidal chitin in agar for the isolation of actinomycetes have been tested by Hsu & Lockwood (1975) who compared chitin agar with other selective media currently in use. Chitin agar was prepared by mixing colloidal chitin with inorganic salts: after autoclaving, the melted agar medium was adjusted to pH 8 with 5 N sterile sodium hydroxide.

Colloidal chitin preparations support the development of more actinomycete colonies than media containing non-colloidal chitin. The zones of chitin utilization surrounding actinomycete colonies are clearer when purified chitin is used.

Chitin agar shows selectivity superior to that of several other media for
isolating actinomycetes from water and soil, by favoring these organisms and
suppressing development of most bacteria and fungi. In fact, Hiraga, Komatsu &
Takagi (1973) have reported the preparation of a bacteriolytic enzyme from
actinomycete cultures on chitin media. Whereas the lysis of microorganisms by
enzymes produced by actinomycetes has been the object of numerous studies, the
enzymes isolated from the chitin medium are much more powerful and exhibit a
wide antibacterial spectrum.

Two varieties of *Streptomyces albus* have been cultured at 30° for 30-72 hr,
under anaerobic conditions; at the end the bacteriolytic enzyme is in the
supernatant and it is precipitated with the aid of organic solvents. Optimum
activity is exhibited at 40-45°; incativation occurs after 30 min at 70°. The
enzyme was tested on *Tricophyton mentagrophytes* and *Micrococcus rhizoditicus*:
the lysis percentages, based on spectrophotometric readings, are reported in
Tab. 7.2.

Microbiological media from shrimp heads and abdomens have been prepared and

TABLE 7.2. Effect of the presence of chitin at various concentrations in the
culture media, expressed as lysis percentage. Hiraga, Komatsu & Takagi, 1973.

	Microbial tests			
Chitin %	*Micrococcus rhizoditicus*		*Tricophyton mentagrophytes*	
	sp 94	sp 107	sp 94	sp 107
0.0	18.0	4.0	0.0	0.0
0.5	47.2	20.0	14.2	20.3
1.0	91.9	40.2	19.0	26.2
2.0	90.0	41.5	40.2	28.5
3.0	50.6	30.4	19.0	21.2
Crab shell powder %				
0.5	40.0	10.0	5.0	10.4
1.0	61.2	26.7	10.5	18.5
2.0	72.4	33.8	18.2	20.5
3.0	40.0	12.1	8.6	16.2

evaluated by Stephens, Bough, Beuchat & Heaton (1976) for commercial distribution. Two peptones were extracted from raw shrimp waste after autolytic
digestion. Digests were evaluated for suitability in supporting growth of
microorganisms by measuring the total cell mass produced by five genera of
bacteria and five genera of fungi. Comparison to five commercially available
media showed that the shrimp digest was excellent in supporting growth of
fungi and bacteria. Sneh & Henis (1972) observed the inhibition of the saprophytic activity of *Rhizoctonia solani* in chitin amended soil; extracts of the
chitin amended soil inhibited the growth of many fungal species.

BIBLIOGRAPHY

1811
Braconnot H.: Sur la nature des champignons. Ann. Chi. Phys. 79, 265-304.

1823
Odier A.: Mémoire sur la composition chimique des parties cornées des insectes. Mém. Soc. Hist. Nat. Paris 1, 29-42.

1843
Lassaigne J.L.: Sur le tissu tégumentaire des insectes de differents ordres. Comp. Rend. 16, 1087-1089.
Payen A.: Proprietés distinctives entre les membranes végétales et les enveloppes des insectes et des crustacés. Comp. Rend. 17, 227-231.

1859
Rouget C.: Des substances amylacées dans le tissu des animaux, spécialement les Articulés (Chitine). Comp. Rend. 48, 792-795.

1878
Ledderhose G.: Ueber Chitin und seine Spaltungprodukte. Hoppe-Seiler Z. Physiol. Chem. 2, 213 -227.

1894
Hoppe-Seiler F.: Ueber Chitin und Zellulose. Ber. Deut. Chem. Gesell. 27, 3329-3331.
Gilson E.: Recherches chimiques sur la membrane cellulaire des champignons. Bull. Soc. Chim. Paris 3, 1099.

1898
Van Wisselingh G.: Mikrochemische Untersuchungen über die Zellwände der Fungi. Jahrb. Wiss. Bot. 31, 619-687.

1906
Von Furth O. & Russo M.: Ueber Kristallinische Chitosanverbindungen aus Sepienschulpen. Ein Beitrag zur Kenntnis des Chitins. Beitr. Chem. Physiol. Path. 8, 163-190.

1908
Offer T.R.: Concerning chitin. Biochem. Z. 7, 117-127.
Sollas I.B.J.: Identification of chitin by its physical constants. Roy. Soc. B-79, 474-481.

1909
Irvine J.C.: A polarimetric method of identifying chitin. J. Chem. Soc. 95, 564-570.
Lowy E.: Ueber Krystallinisches Chitosansulfat. Biochem. Zeitschr. 23, 47-60.
Scholl E.: The preparation of pure chitin from Boletus edulis. Monatsh. 20, 1023-1036.
Wester D.H.: Chitin. Arch. Pharm. 247, 282-307.

1910
Alsberg C.L. & Hedblom C.A.: Soluble chitin from Limolus polyphemus and its peculiar osmotic behavior. J. Biol. Chem. 6, 483-497.
Lowy E.: Crystalline chitosan sulfate. Biochem. Z. 23, 47-60.
Wester D.H.: Ueber die Verbreitung und Lokalisations des Chitins im Tierreik. Zool. Jahrb., Adt. System., Geogr. Biol. Tiere 28, 531-558.

1912
Brach H. & Von Furth O.: The chemical constitution of chitin. Biochem. Z. 38, 468-491.
Hamburger H.: Formation of levulic acid from glucosamine, chitin and chitose. Biochem. Z. 36, 1-4.
Von Lippmann E.O.: Occurrence of chitin. Ber. 44, 3716-3717.

1913
Alverdes F.: Chitinous bodies with concentric layers which occur in Branchipus grubbi. Zool. Anz. 40, 317-323.

1913 CNTD

Kotake Y. & Sera Y.: Ueber eine neue Glukosaminverbindung, zugleich ein Beitrag zur Konstitutionsfrage des Chitins. Z. Physiol. Chem. 88, 56-72.

1914

Kotake Y. & Sera Y.: A new glucosamine compound. Constitution of chitin: II. Formation of lycoperdine by the hydrolysis of bulffist and earthstar. Z. Physiol. Chem. 89, 482-484.

Van Wisselingh C.: Ueber die Anwendung der inder organische Chemie gebräuchlichen Reaktionen bei de phytomikromischen untersuchung. Folia Microbiol. 3, 165-198.

1915

Vouk V.: Microchemical chitin reactions. Ber. Deut. Botan. Ges. 33, 413-415.

1916

Hasz W.: Structure of the chitin of the arthropods. Arch. Physiol. 1916, 295-339.

Morgulis S.: Chemical constitution of chitin. Science 44, 866-867.

Van Wisselingh C.: The occurrence of chitin and cellulose in bacteria. Pharm. Weekblad 53, 1069-1078 and 53, 1102-1107.

1917

Morgulis S.: An hydrolytic study of chitin. Am. J. Physiol. 43, 328-343.

1918

Levene P.A.: Epimeric hexosaminic acids. J. Biol. Chem. 36, 73-87.

1919

Armbrecht W.: Chitose. Biochem. Z. 95, 108-123.

Levene P.A. & Matsuo I.: D-chondrosamino and D-chitosamino heptonic acids. J. Biol. Chem. 39, 105-108.

Suzuchi N.: Metabolism of the furan and hydrofuran derivatives in the animal organism. J. Biol. Chem. 38, 1-5.

1920

Schmiedeberg O.: Chitin and chitin compounds in animals and plants. Arch. Exp. Path. Pharm. 87, 76-85.

1921

Brunswick H.: The microchemistry of chitosan compounds. Biochem. Z. 113, 11-24.

Von Wettsteins F.: Das Vorkommen von Chitin und seine Vorwerting als systematisch- phylogenetisches Merkund im Pflanzenreich. Sitzber. Akad. Wiss. Wien 130, 3-20.

1922

Eidmann H.: The permeability of chitin in osmotic process. Biol. Zentr. 42, 429-435.

Karrer P. & Smirnoff A.P.: Polysaccharides. XVII. Chitin. Helv. Chim. Acta 5, 832-852.

1924

Herzog R.O.: The fine structure of fibrous materials. Naturwissens. 12, 955-960.

Karrer P., Schneider O. & Smirnoff A.P.: Polysaccharides. XXIX. Chitin. II. The configuration of glucosamine. Helv. Chim. Acta 7, 1039-1045.

Katz J.R. & Mark H.: Röntgenspectrometric study of the swelling of substances. Verslag Akad. Wetenschappen Amsterdam 33, 294-301.

1925

Kunike G.: Occurrence and detection of organic skeleton substances in animals. Z. Vergl. Physiol. 2, 233-253.

Levene P.A. & Lopez-Suarez J.: Hexosamines and Mucoproteins. Longmans, London.

Pantin C.F.A. & Rogers T.H.: An amphoteric substance in the radula of the whelk Buccinum undatum. Nature 115, 639-640.

1926

Gonell H.A.: Röntgenographic studies on chitin. Z. Physiol. Chem. 152, 18-30.

Knecht E. & Hibbert E.: Chitin. J. Soc. Dyers Color. 42, 343-345.

Mohring A.: Double refraction of natural cellulose and chitin fibers. Kolloid. Chem. Beiheft 23, 162-188.

Proskuriakow N.J.: The role of chitin forming the cell wall of fungi. Biochem. Z. 167, 68-76.

Von Veimarn P.P.: Filament production from plastic masses. Can. Chem. Met. 10, 227-228.

Von Veimarn P.P.: Universal method for converting fibroin, chitin, casein and similar substances into ropy-plastic state and into state of colloidal solution by means of concentrated aqueous solution of readily soluble salts, capable of strong hydration. J. Text. Inst. 17, T 642-644.

1927

Frankel S. & Jellinek C.: *Limulus polyphemus*. Biochem Z. 185, 384-388.
Von Veimarn P.P.: Solutions of colloids of large molecular compounds by a very easily soluble strongly hydrolized material. Kolloid Z. 42, 134-140.
Von Veimarn P.P.: Conversion of fibroin, chitin, casein and similar substances into the ropy-plastic state ancd colloid solution. Ind. Eng. Chem. 19, 109-110.

1928

Kuhnelt W.: Microchemical identification of chitin. Biol. Zentr. 48, 374-382.
Meyer K.H. & Mark H.: Constitution of chitin. Ber. 61-B, 1936-1939.
Schmid L.: Alkali compounds of polyhydric alcohols and carbohydrates. Monatsh. 49, 107-110.
Sumi M.: Chemical components of the pore of *Aspergillus oryzae*. Biochem. Z. 195, 161-174.
Von Veimarn P.P.: Dispersoidological investigations of latex. Bull. Chem. Japan 3, 157-168.

1929

Campbell F.L.: The detection and estimation of insect chitin and the irrelation of chitinization to hardness and pigmentation of the cuticula of the american cockroach *Periplaneta americana*. Ann. Entomol. Soc. Am. 22, 401-426.
Karrer P. & Francois G.V.: Polysaccharides. XL. Enzymic degradation of chitin. Helv. Chim. Acta 12, 986-988.
Karrer P. & Hoffmann A.: Polysaccharides.XXXIX. Enzymic degradation of chitin and chitosan. Helv. Chim. Acta 12, 616-637.
Merker E.: The transparency of chitin for ultraviolet light. Zool. Anzeiger 4, Suppl. 181-186.

1930

Anderson B.G. & Brown L.A.: Chitin secretion in *Daphna magna*. Physiol. Zool. 3, 485-493.
Karrer P.: The enzyme decomposition of native reprecipitated cellulose, artificial silk and chitin. Kolloid Z. 52, 304-319.
Karrer P. & White S.M.: Chitin. Helv. Chim. Acta 13, 1105-1113.
Von Franciis G.: Chitins, Lichenins and cellulose. Zurich.

1931

Bergmann M., Zervas L. & Silberkweit E.: The biose of chitin. Naturwissens. 19, 20.
Bergmann M., Zervas L. & Silberkweit E.: Chitin and chitobiose. Ber. 64-B, 2436-2440.
Rammelberg G.: Chitin from fungi and from crab shells. Bot. Arch. 32, 1-37.
Zechmeister L. & Toth G.: Hydrolysis of chitin with hydrochloric acid. Ber. 64-B, 2028-2032.

1932

Johnson D.E.: Chitin-destroying bacteria. J. Bact. 24, 335-340.
Khouvine M.Y.: X-ray of chitin from *Aspergillus niger, Psalliota campestris* and *Armillaria mellea*. Comp. Rend. 195, 396-397.
Yonge C.M.: The nature and permeability of chitin. I. The chitin lining of the foregut of decapod Crustacea and the function of the tegumental glands. Proc. Roy. Soc. B-111, 298-329.
Zechmeister L. & Grassmann W., Toth G. & Bender R.: The nature of the glucosamine residue linkage in chitin. Ber. 65-B, 1706-1708.
Zechmeister L. & Toth G.: Hydrolysis of chitin with hydrochloric acid. Ber. 65-B, 161-162.

1933

Abderhalden E. & Heyns K.: Demonstration of chitin in the wing remains of Coleoptera of the upper middle Eocene. Biochem. Z. 259, 320-321.
Elson L. & Morgan W.T.J.: A colorimetric method for the determination of glucosamine and chondrosamine. Biochem. 27, 1824-1828.

1934

May O.E. & Ward G.E.: Hydrolysis of the chitinous complex of lower fungi. J. Am. Chem. Soc. 56, 1597-1599.
Rigby G.W.: Substantially undegraded deacetylated chitin and process for producing the same. U.S. 2,040,879.
Rigby G.W.: Process for the preparation of films and filaments and products thereof. U.S. 2,040,880.
Shoruigin P.P. & Hait E.: Nitration of chitin. Ber. 67-B, 1712-1714.
Zechmeister L. & Toth G.: Comparison of plant and animal chitin. Z. Physiol. Chem. 223, 53-56.

1935

Benton A.G.: Chitinovorous bacteria; a preliminary survey. J. Bacteriol. 29, 20.
Benton A.G.: Chitinovorous bacteria; a preliminary survey. J. Bacteriol. 29, 449-464.
Bucherer H.: Microbiological action of chitin. Zentr. Bakt. Parasit. 93, 12-24.
Diehl J.M. & Van Iterson G.: The double refraction of chitin tendons. Kolloid Z. 73, 142-146.

1935 CNTD

Meyer K.H. & Pankow G.W.: The constitution and structure of chitin. Helv. Chim. Acta 18, 589-598.
Rigby G.W.: Emulsions. U.S. 2,047,225.
Rigby G.W.: Chemical process and chemical compounds derived therefrom. U.S. 2,047,226.
Shoruigin P.P. & Hait E.: Acetylation of chitin. Ber. 68-B, 971-973.
Shoruigin P.P. & Makarova-Semlyanskaya N.N.: Desamination of chitin and glucosamine. Ber. 68-B, 965-969.
Shoruigin P.P. & Makarova-Semlyanskaya N.N.: Methyl ethers of chitin. Ber. 68-B, 969-971.

1936

Aleksandrov V.Y.: The permeability of the chitin of some diptera larvae and a method for its investigation. Biol. Z. 3, 490-507.
Bergstrom S.: Polysaccharide sulphuric acids with heparin action. Z. Physiol. Chem. 238, 163-168.
Clark. G.L. & Smith A.F.: X-ray diffraction studies of chitin, chitosan and derivatives. J. Phys. Chem. 40, 863-879.
Castle E.S.: The double refraction of chitin. J. Gen. Physiol. 19, 797-805.
Diehl J.M.: Vegetable chitin. Chem. Weekblad 33, 36-38.
Du Pont de Nemours & Co.: Coating and impregnating compositions. Brit. 458,813.
Du Pont de Nemours & Co.: Emulsions, dispersions. Brit. 458,815.
Du Pont de Nemours & Co.: Process of cementing materials together. Brit. 458,818.
Du Pont de Nemours & Co.: Chitin compounds. Brit. 458,839.
Heyn A.N.J.: X-ray investigations on the molecular structure of chitin in cell walls. Proc. Acad. Sci. Amsterdam 39, 132-135.
Merril W.J.: Compositions containing partially deacetylated chitin and suitable for resisting laundering. U.S. 2,047,218.
Nikol'skii A.A.: Utilization of chitin for the manufacture of plastic masses. J. Appl. Chem. SSSR 9, 1308-1315.
Rigby G.W.: Chemical products and process of preparing the same. U.S. 2,072,771.
Rogovin Z.S.: Micelle structure of cellulose. Org. Chem. Ind. SSSR 2, 645-648.
Smith M.: The occurrence of chitin in microorganisms. Arch. Mikrobiol. 7, 241-260.
Stuart L.S.: A note on halophilic chitinovorous bacteria. J. Am. Leather Ass. 31, 119-120.
Van Iterson-Meyer K.H. & Lotmar W.: The fine structure of vegetable chitin. Rec. Trav. Chim. 55, 61-63, 130.
Yonge C.M.: Nature and permeability of chitin. II. The permeability of uncalcified chitin linings of the foregut of *Homarus*. Proc. Roy. Soc. B-120, 15-41.

1937

Brown A.W.A.: A note on the chitinous nature of the peritrophic membrane of *Melanopus bivittatus*. J. Exp. Biol. 14, 87-94.
Danielli J.F.: The activation energy of diffusion through natural and artificial membranes. Trans. Faraday Soc. 33, 1139-1147.
Gibson E.C.: Freeing edible crustacean flesh from chitin material. U.S. 2,080,263.
Meyer K.H. & Wehrli H.: Chemical comparison of chitin and cellulose. Helv. Chim. Acta 20, 353-362.
Rigby G.W.: Benzyl deacetylated chitin. U.S. 2,072,771.
Zobell C.E. & Rittenberg S.C.: Occurrence and characteristics of chitinoclastic bacteria in the sea. J. Bacter. 35, 275-287.

1938

Alshina V.I.: Destruction of chitin by sulfate-reducing bacteria, and changes of the oxidation-reduction conditions in the reduction of sulfate. Microbiol. SSSR 7, 850-859.
Patterson G.D.: Coated product and process whereby prepared. U.S. 2,128,961.
Patterson G.D.: Coated objects. U.S. 2,128,962.
Tsimehc A.: Chitin. Rayon Silk J. 14, 26-28.
Tsimehc A.: Chitin. Silk J. Rayon World 14, 27-28
Yonge C '.: Recent work on the digestion of cellulose and chitin by invertebrates. Sci. Progr. 32, 638-647.
Zechmeister L., Toth G. & Balant M.: Chromatographic separation of some of the enzymes of emulsin. Enzymologia 5, 302-306.

1939

Thor C.J.B.: Chemical compounds and products produced therefrom. U.S. 2,168,374.
Thor C.J.B.: Chitin xanthate. U.S. 2,168,375.
Zechmeister L.: Chitin and its cleavage products. Forshr. Chem. Org. Nat. 2, 212-247.
Zechmeister L & Toth G.: Chromatographic analysis of chitinase. Naturwissens. 27, 367.

1940

Hock C.W.: Decomposition of chitin by marine bacteria. Biol. Bull. 79, 199-206.
Picken L.E.R.: The fine structure of biological systems. Biol. Rev. Cambridge, Phil. Soc. 15, 133-167.

1940 CNTD

Tadokoro I. & Nisida M.: Strength of artificial fibers containing fish proteins. J. Soc. Chem. Ind. Japan 43 Suppl. b 296.

Thor C.J.B.: Process for producing articles of regenerated chitin and the resultant articles. U.S. 2,217,823.

Thor C.J.B. & Henderson W.F.: The preparation of alkali chitin. Am. Dyestuff Rep. 29, 461-464.

Thor C.J.B. & Henderson W.F.: Chitin xanthate and regenerated chitin. Am. Dyestuff Rep. 29, 489-491.

Toth G.: Preparative proof of chitin in mollusks. Z. Physiol. Chem. 263, 229-236.

Viktorov P.P.: Utilization of chitin in the textile industry. Khlopchatobumashnaya Prom. N° 11+12, 52-53.

Worden E.C.: Development in organic non-cellulosic fibrous materials. Rayon Text. Monthly 21, 527-528; 21, 609-610; 21, 680-682 and 21, 733-735.

1941

Dewitt Smith H.: Recent developments in synthetic fibers and fabrics. Rayon Text. Monthly, 22, 508-510 and 22, 590-591.

Hock C.A.: Marine chitin decomposing bacteria. J. Marine Res., Sears Found. 4, 99-106.

Izard E.F.: Base for making films and similar surfaces. Ger. 712,328.

Lafon M.:Composition of the shells of Crustacea. Comp. Rend. Soc. Biol. 135, 1003-1006.

Neuberger A.: Action of periodic acid on glucosamine derivatives. J. Chem. Soc. 1941, 47-50.

Sadov F.I.: A new material for a dressing (textile) which is fast to washing. Text. Prom. 2, 52-54.

Thor C.J.B.: Filaments formed of chitin and cellulose. U.S. 2,217,823.

Trim A.R.: Studies on the chemistry of the insect cuticle. Biochem. J. 35, 1088-1098.

Worden E.C.: Developments in organic non-cellulosic fibrous materials. Rayon Text. Monthly 22, 17-18; 22, 85-86; 22, 147-148; 22, 225-226; 22, 354-356; 22, 455-456 and 22, 517-518.

1943

Karrer P., Koenig H & Usteri E.: Blood-coagulation inhibiting polysulfuric acid and related compounds. Helv. Chim. Acta 26, 1296-1315.

Kishi Y.: Protein and chitin from silkworm pupae. Japan 157,624.

Lafon M.: Structure and chemical composition of the tegument of Limulus polyphemus. Bull. Inst. Oceanogr. N° 850.

Lafon M.: Recherches biochimiques et physiologiques sur le squelette tégumentaire des arthropodes. Ann. Sci. Nat. Ser. Bot. Zool. 11, 113-146.

Lassaigne J.L.: Sur le tissu tégumentaire des insectes de différents ordres. Comp. Rend. 16, 1087-1089.

Sumiki Y. & Yaita M.: The utilization of chitin. I. Preparation of tartaric acid from glucosamine. J. Agr. Chem. Soc. Japan 19, 723-726.

1944

Astrup T., Galsmar I & Volkert M.: Polysaccharide sulfuric acid esters as anticoagulants. Acta Physiol. Scand. 8, 215-226.

Billard L.: Histochemical study of the glycogen cycle in the course of the intermolt of decapod Crustacea. Bull. Inst. Oceanogr. N° 866.

1945

Piper J.: Influence of synthetic polysaccharide sulfuric acid esters on the thrombocytes in vivo and in vitro. Acta Physiol. Scand. 9, 28-38.

Rosedale J.L.: Composition of insect chitin. J. Entomol. Soc. S. Africa. 8, 21-23.

1946

Barsoe C. & Selso S.: Anticoagulant action of synthetic polysaccharide polysulfuric acids. Acta Pharmacol. Toxicol. 2, 367-370.

Dennel R.: Study of an insect cuticle: the larval cuticle of Sarcophaga faculata. Proc. Roy. Soc. B-133, 348-373.

Ogata A., Takagi K. & Mizutani A.: Components of charred insects. I. The red dragon fly. J. Pharm. Soc. Japan 66, 44.

Piper J.: The anticoagulant effect of heparin and synthetic polysaccharide-polysulfuric acid esters. Acta Pharmacol. Toxicol. 2, 138-148.

Piper J.: Toxicology of synthetic polysaccharide polysulfuric esters. Acta Pharmacol. Toxicol. 2, 317-328.

Purchase E.R. & Braun C.E.: D-glucosamine hydrochloride. Org. Synth. 26, 36-37.

Richards A.G. & Cutkomp L.: Correlation between the possession of chitinous cuticle and sensitivity to DDT. Biol. Bull. 1946, 90-107.

1947

Fraenkel G. & Rudall K.M.: The structure of insect cuticles. Proc. Roy. Soc. B-134, 111-143.

Picken E.R., Pryor M.G.M. & Swann N.M.: Orientation of fibrils in natural membranes. Nature 159, 434.

1947 CNTD

Richards A.G.: Studies on arthropod cuticle. I. The distribution of chitin in Lepidopteran
    scales and its bearing on the interpretation of arthropod cuticle. Ann. Entomol. Soc. Am.
    40, 227-240.
Richards A.G. & Korda F.H.: Electron micrographs of centipede setae microtrichia. Entomol.
    News 58, 141-145;
Stanier R.Y.: The non-fermenting myxobacteria. I. *Cytopaga johnsonae*, a chitin decomposing
    myxobacterium. J. Bact. 53, 297-315.
Tachibana T. & Nakagawa I.: Dispersion state of molecules in viscous solutions and structural
    viscosity. Kagaku no Ryoiki 1, 31-33.

1948

Abe Y.: Studies on the chemical composition of mold. I. The ether extracts and carbohydrates
    of *Penicillium notatum*. Extracting experiments and results. Proc. Fac. Eng. Keiogijuku
    Univ. 1, 42-47.
Lafon M.: New biochemical and physiological studies of the tegumentary skeleton of Crustacea.
    Bull. Inst. Oceanogr. N° 939.
Lord K.A.: Sorption of DDT and its analogs by chitin. Biochem. J. 43, 72-78.
Matsushima S.: Glucosamine from scales of crustaceans. Japan 175,981.

1949

Jones J.K.N.: The chemistry of cellulose, amylose and chitin. Fibre Sci. 1949, 20-25.
Picken L.E.R.: Shape and molecular orientation in lepidopteran scales. Trans. Roy. Soc. 234-B,
    1-28.
Richards A.G.: Arthropod cuticle. III. The chitin of *Limulus*. Science 109, 591-592.
Rust J.B.: Deacetylated chitin shrinkproofing of wool and product. U.S. 2,669,529.
Sumiki Y., Vatta M. & Matsushima S.: The utilization of chitin. II .. Preparation of tartaric
    acid from glucosamine. J. Agr. Chem. Soc. Japan 23, 6-7.
Vinogradova A.Z.: Chemical composition of invertebrates of the Black Sea. Dokl. Akad. Nauk.
    SSSR 65, 891-894.

1950

Astrup T. & Alkjaersig N.: Polysaccharide polysulfuric acids as antihyaluronidases. Nature
    166, 568-569.
Black M.M. & Schwartz H.M.: Estimation of chitin and chitin nitrogen in crayfish waste and
    derived products. Analyst 75, 185-189.
Darmon S.E. & Rudall K.M.: Infrared and x-ray studies of chitin. Disc. Faraday Soc. 9, 251-260.
Frey R.: Chitin and cellulose in the cell walls of fungi. Ber. Schweiz. Botan. Ges. 60, 199-
    220.
Hatta S., Kuwabara S., Miyamoto H., Aoyamak K., Utsunomiya N. & Tanji S.: Macramin, a new
    antibacterial substance derived from chitin. Japan Med. J. 3, 119-123.
Jeanloz R. & Forchielli E.: Hyaluronic acid and related compounds. 3. Determination of the
    structure of chitin by periodate oxidation. Helv. Chim. Acta 33, 1690-1697.
Jeuniaux Ch.: Evidence for a chitinolytic bacterial flora in the intestinal tract of the
    snail *Helix pomatia*. Arch. Intl. Physiol. 58, 350-351.
Kopp F.I. & Markianovich E.M.: Chitin destroying bacteria in the Black Sea. Dokl. Akad. Nauk.
    SSSR 75, 859-862.
Lotmar W. & Picken L.E.R.: A new crystallographic modification of chitin and its distribution.
    Experientia 6, 58-59.
Okimasu S. & Senju R.: Studies on chitin. 2. The molecular forms and the degree of polymeriza-
    tion of glycolchitin from the point of view of the viscosity of its aqueous solutions. J.
    Agr. Chem. Soc. Japan 23, 437-441.
Picken L.E.R. & Lotmar W.: Oriented proteins in chitinous structures. Nature 165, 599-600.
Richards A.G. & Korda F.H.: Arthropod cuticle. 4. An electron microscope survey of the intima
    of arthropod tracheae. Ann. Entomol. Soc. Am. 43, 49-71.
Senju R. & Okimasu S.: Chitin. 1. Glycolation of chitin and the chemical structure of glycol-
    chitin. J. Agr. Chem. Soc. Japan 23, 432-437.
Terayama H.: Water-soluble chitosan. Japan 563.
Terayama I., Miyamoto H., Koshinuma K. & Hosoda T.: Antibacterial mechanisms of the high
    molecular weight, positive ion, Macramine. Bull. Natl. Hyg. Lab. Japan 67, 189-196.

1951

Blumberg R., Southall C.L., Van Rensbrug N.J. & Volkman O.B.: South African fish products. 32.
    Rock lobster chitin production from processing wastes. J. Sci. Food Agr. 2, 571-576.
Campbell L.L. & Williams O.B.: A study of chitin-decomposing microorganisms of marine origin.
    J. Gen. Microbiol. 5, 894-950.
Hourvick A.L., Kreger D.R. & Roelofsen P.A.: Composition and structure of yeast cell walls.
    Nature 169, 693-694.
Jeuniaux Ch.: A method of determination of chitinase. Arch. Int. Physiol. 59, 242-244.
Matsushima Y.: Amino hexoses. 1. Oxidation of glucosamine and chondrosamine with sodium hypo-
    chlorite. Sci. Papers Osaka Univ. 31, 1-6.
Molteno C.C.: Byproducts of the fishing industry. S. African Ind. Chem. 5, 2-7.

1951 CNTD

Richards A.G.: The Integument of Arthropods. Univ. Minnesota Press. Minneapolis.
Roelofsen P.A. & Hoette H.: Chitin in the cell wall of yeasts. J. Microbiol. Serol. 17, 297-313.
Rozendaal H.M., Bellamy W.D. & Baldwin T.H.: Composition and structure of yeast cell walls. Nature 168, 693-694.
Tracey M.V.: Cellulase and chitinase of earthworms. Nature 167, 776-777.

1952

Myamoto H., Akama K. & Morita E.: Influence of positive and negative colloid ions on the hemolytic system and its mechanism. J. Med. Sci. 1, 35-44.
Senju R.: Kinetics of the hydrolysis of polysaccharides. 1. Theoretical considerations of the decomposition of high-molecular chain compounds. J. Agr. Soc. Japan 25, 227-230.
Smithies W.R.: Chemical composition of a sample of mycelium of *Penicillium griseofulvum*. Biochem. J. 51, 259-264.
Terayama H.: Methods of colloid titration: a new titration between polymer ions. J. Polymer Sci. 8, 243-253.
Tsao C.H. & Richards A.G.: Studies on arthropod cuticle. 9. Quantitative effects of diet, age, temperature and humidity on the cuticles of five representative species of insects. Ann. Entomol. Soc. Am. 45, 585-599.
Veldkamp H.: Aerobic decomposition of chitin by microorganisms. Nature 169, 500.
Zimmermann A.: Chitin xanthate. Ger. 856,146.
Zimmermann A.: Salts of high-molecular ether carboxylic acids. Swed. 136,717.

1953

Blank F.: The chemical composition of the cell walls of dermatophytes. Bioch. Biophys. Acta 10, 110-113.
Doczi J., Fishman A. & King J.A.: Direct evidence of the influence of sulfamic acid linkages on the activity of heparin-like anticoagulants. J. Am. Chem. Soc. 75, 1512-1513.
Foster A.B., Martelew E.F. & Stacey M.: Correlation of the rates of deamination of glucosamides with configuration at the glycosidic center. Chem. Ind. 1953, 825-826.
Hoogland P.L. & Hiltz R.S.C.: Glucosamine from lobster shells. Fish. Res. Board Can. Progr. Rep. Atlantic Coast Stas. 57, 6-8.
Houwink A.L. & Kreger D.R.: The cell wall of yeast: electron microscope and x-ray diffraction study. Antoine Leeuwenhoek J. Microbiol. Serol. 19, 1-24.
Karrer P., Koenig H. & Usteri E.: Zur Kenntnis blutgerinnungshemmendes Polysaccharid-poly-schwefelsäure-ester und ähnliches Verbindungen. Helv. Chim. Acta 26, 1295.
Leloir L.F. & Cardini C.E.: The biosynthesis of glucosamine. Bioch. Biophys. Acta 12, 15-22.
Lusena C.V. & Rose R.C.: Preparation and viscosity of chitosan. J. Fish. Res. Board Can. 10, 521-522.
Mangenot F.: The chitin of fungi. Bull. Soc. Sci. Nancy 12, 10-18.
Passonneau J.V. & Williams C.M.: Molting fluid of the *Cecropia* silkworm. J. Exp. Biol. 30, 545-560.
Ricketts C.R.: Anticoagulant activity of a chitosan sulfuric acid ester. Res. 6, 17s-18s.
Santoro A.: An attempt to adsorb optical antipodes selectively by means of deacetylated chitin. Baskervimme Chem. J., City Coll. N.Y. 4, 8-9.
Senju R.: Investigations on lignin and pulp. 1. A new method of determining lignosulfonic acid in sulfite waste liquor by colloidal titration. Bull. Chem. Soc. Japan 26, 143-147.
Smithies W.R.: Lysing action of enzymes on a sample of mycelium of *Penicillium griseofulvum*. Biochem. J. 55, 346-350.
Warbury O.: Cell chemistry. Bioch. Biophys. Acta 12, 15-22.
Wolfrom M.L., Shen-Han T.M. & Summers C.G.: Sulfated nitrogenous polysaccharides and their anticoagulant activity. J. Am. Chem. Soc. 75, 1519.
Yasuda K.: Histochemical staining of hyaluronic acid with chitosan. Okajimas Folia Anat. Japan 25, 55-60.

1954

Akiya S. & Osawa T.: Nitrogen-containing sugars: configuration of the so-called chitose. J. Pharm. Soc. Japan 74, 1259-1262.
Blank F.: The cell walls of dimorphic fungi causing systemic infections. Can. J. Microbiol. 1, 1-5.
Blank F. & Burke R.C.: Composition of cell wall of *Coccidioides imitis*. Nature 173, 829.
Cushing I.B., Davis R.V., Kratovil E.J. & McCorquodale D.W.: The sulfation of chitin in chlorosulfonic and dichloroethane. J. Am. Chem. Soc. 76, 4590-4591.
Danilov S.N. & Plisko E.A.: Chitin. Action of acids and alkalis on chitin. Zhur. Obs. Khim. 24, 1761-1769.
Danilov S.N. & Plisko E.A.: Chitin. Glyceryl ethers of chitin. Zhur. Obs. Khim. 24, 2071-2075.
Fodor G. & Otvos L.: The conformation of D-glucosamine. Acta Chim. Acad. Sci. Hung. 5, 205-207.
Gehring F.: The destruction of chitin by microorganisms. Zentr. Bakt. Parasitenk. Abt. II, 108, 232-242.
Gyorgy P., Kuhn R. & Zilliken F.: Infant feeding formula. U.S. 2,694,640.

1954 CNTD

Hackman R.H.: Chitin. 1. Enzymic degradation of chitin and chitin esters. Austr. J. Biol. Sci. 7, 168-178.
Hopff H., Spanig H. & Steimmig A.: N-acylated amino sugars. Ger. 915,566.
Jeuniaux Ch.: Chitinase and the intestinal bacterial flora of gastropods. Acad. Roy. Belg. Class. Sci. Mém. Col. 8, 28, 5-45.
Jones R.V.: Drilling muds. U.S. 2,670,329.
Jones R.V.: Chitin sulfate. U.S. 2,689,244.
Kreger D.R.: Observations on cell walls of yeast and some other fungi by x-ray diffraction and solubility tests. Bioch. Biophys. Acta 13, 1-9.
Manskaya S.M., Drozdova T.V. & Tobelko K.I.: Formation of melanoidins from chitin. Dokl. Akad. Nauk. SSSR 96, 569-572.
Reynolds D.M.: Exocellular chitinase from a *Streptomyces* sp. J. Gen. Microbiol. 11, 150-159.
Roth L.W., Shepperd I.M. & Richards R.K.: Anticoagulatns and other pharmacologic effects of sulfated chitin in animals. Proc. Soc. Exp. Biol. Med. 86, 315-318.
Rust J.B.: Shrinkproofing of wool. U.S. 2,669,529.
Sadov F.I. & Markova G.B.: Production of chitosan and its use. Nauk. Issled. Trudy Moskow, Tekst. INst. 13, 70-74.
Yashuda K.: Histochemical staining of hyaluronic acid with chitosan. Excerpta Med. V, 7, 688.

1955

Abbott Lab.: Sulfate esters of partially degraded chitin. Brit. 740,152.
Bruns G.: The membrane envelope of *Pneumocystis carinii*. Naturwissens. 42, 610.
Giles C.H., Jain S.K.& Hassan A.S.A.: Nature of adsorption on cellulose from aqueous solutions. Chem. Ind. 1955, 629-630.
Hackman R.H.: Chitin. 2. Reaction of N-acetyl-D-glucosamine with α-aminoacids, peptides and proteins. Austr. J. Biol. Sci. 8, 83-96.
Hackman R.H.: Chitin. 3. Adsorption of proteins to chitin. Austr. J. Biol. Sci. 8, 530-536.
Han T.S.: Sulfation of chitosan. Diss. Abs. 15, 957.
Jeuniaux Ch.: Chitinolytic properties of aqueous extracts of exuvates of the larvae, prenymphs and nymphs of *Tenebrio molitor*. Arch. Int. Physiol. Bioch. 63, 114-120.
Jeuniaux Ch.: Chitinolytic properties of the exuvial fluids of the silkworm. Experientia 11, 195-196.
Jeuniaux Ch.: The intestinal chitinolytic bacterial flora of the roman snail *Helix pomatia*: quantitative and qualitative analysis. Bull. Soc. Roy. Sci. Liège 24, 254-270.
Kent P.W. & Whitehouse M.W.: Biochemistry of Aminosugars. Butterworths, London.
Krishnan G., Ramachandran G.N. & Santanam M.S.: Occurrence of chitin in the epicuticle of an arachnid, *Palamnaeus swammerdami*. Nature 176, 557-558.
Mataga N. & Koizumi M.: Influence of the addition of high-molecular electrolytes on the absorption spectra and fluorescence of organic dyes. Bull. Chem. Soc. Japan 28, 51-54.
Naito N. & Tani T.: Effects of sodium 2,4-dichlorophenoxyacetate on the nutritional absorption in *Gloeosporium olivarum*. Nippon Shokubutsu Byori Gakkaiko 20, 89-92.
Reissig A.R., Strominger R.S. & Leloir L.F.: A modified colorimetric method for the estimation of N-acetylamino sugars. J. Biol. Chem. 217, 959-966.
Rudall K.M.: The distribution of collagen and chitin. Symp. Soc. Exp. Biol. N° 9, Fibrous Proteins and their Biological Significance 49-71.
Sussman A.S. & Lowri R.J.: Physiology of the cell surface of *Neurospora ascopores*. Cation binding properties of the cell surface. J. Bacteriol. 70, 675-685.
Tracey M.V.: Chitin, in Moderne Methoden der Pflanzenanalyse, vol. 2, Paech K. & Tracey M.V., Eds. Springer-Verlag, Berlin.
Tracey M.V.: Cellulase and chitinase in soil amebas. Nature 175, 815.
Tracey M.V.: Chitinases in some *Basidiomycetes*. Biochem. J. 61, 579-586.
Travis D.F.: The molting cycle of the spiny lobster *Palinurus argus*. Pre-ecdysial histological and histochemical changes in the hepatopancreas and integumental tissues. Biol. Bull. 108, 88-112.
Zilliken F., Braun G.A., Rose C.S. & Gyorgy P.: Crystallized N,N'-diacetylchitobiose. J. Am. Chem. Soc. 77, 1296-1297.

1956

Cushing I.B. & Kratovil E.J.: Sulfation of polysaccharides. U.S. 2,755,275.
Hiyama N., Maki M. & Miyazawa Y.: Anticoagulant activity of heparin and its relation to the chemical structure. Hirosaki Med. J. 7, 284-291.
Jeuniaux Ch.: Purification of microbial chitinase. Arch. Int. Physiol. Bioch. 64, 522-524.
Manskaya S.M., Drozdova T.V. & Emel'yanova M.P.: The combining of uranium with humic acids and melanoids. Geokhimiya 4, 10-23.
Mataga N.: Influence of high-molecular cations on the fluorescence and absorption spectra of dye anions. J. Inst. Polytech. Osaka City Univ. 5-C, 74-84.
Okimasu S. A new method for the determination of pectin in plant materials by colloid titration. Bull. Agr. Chem. Soc. Japan 20, 29-35.
Upjohn Co.: Sulfated chitosan. Brit. 746,870.
Wigglesworth V.B.: The hemocytes and connective tissue formation in an insect, *Rhodnius prolixus*. Quart. J. Micr. Sci. 97, 89-98.

1957

Barker S.A., Foster A.B., Stacey M. & Webber J.M.: Isolation of a homologous series of oligo-
saccharides from chitin. Chem. & Ind. 1957, 208-209.
Blumenthal H.J. & Roseman S.: Quantitative estimation of chitin in fungi. J. Bacteriol. 74,
222-224.
Cantino E.C., Lovett J. & Horenstein E.H.: Chitin synthesis and nitrogen metabolism during
differentiation in *Blastocladiella emersonii*. Am. J. Bot. 44, 498-505.
Carlstrom D.: Crystal structure of α-chitin. J. Biophys. Bioch. Cytol. 3, 669-683.
Doczi J.: Purification of chitosan by means of the salicylic acid salt thereof. U.S. 2,795,579.
Florkin M.: The origin of animal cellulases and chitinases. Proc. Intl. Symp. Enzyme Chem.
Tokyo and Kyoto, 2, 390-392.
Foster A.B. & Hackman R.H.: Application of ethylenediamine tetraacetic acid in the isolation
of crustacean chitin. Nature 180, 40-41.
Glaser L. & Brown D.H.: Enzymic synthesis of chitin by extracts of *Neurospora crassa*. Bioch.
Biophys. Acta 23, 449-450.
Glaser L. & Brown D.H.: Synthesis of chitin in cell-free extracts of *Neurospora crassa*. J.
Biol. Chem. 228, 729-742.
Gyorgy P., Kuhn R. & Zilliken F.: Alcoholized chitin and food product containing it. U.S.
2,783,148.
Heyns K.: Amino sugars and derivations. Stärke 9, 85-98.
Hoffmann-La Roche F. & Co.: N-formulated chitosan sulfuric acid polyesters. Brit. 777,204.
Horowitz S.T., Roseman S. & Blumenthal H.J.: Preparation of glucosamine oligosaccharides. 1.
Separation. J. am. Chem. Soc. 79, 5046-5049.
Inoue Y.: N-acylglucosamine. Japan 5,706.
Jones R.V.: Well-cementing materials and their applications. U.S. 2,782,858.
Klenkova N.I. & Plisko E.A.: Hydrophilic properties and heat of swelling of chitin. Zhur. Obs.
Khim. 27, 399-402.
Matsushima Y. & Fujii N.: Studies on aminohexoses. 4. N-deacylation with hydrazine and de-
amination with nitrous acid, a clue to the structure of aminopolysaccharides. Bull. Chem.
Soc. Japan 30, 48-50.
McMahon P., Von Brand T. & Nolan M.O.: Observations on the polysaccharides of aquatic snails.
J. Cell. Comp. Phys. 50, 219-240.
Robinette H.: Dyeing and finishing. Modern Text. Mag. 38, 50-52.
Thiele H. & Langmaack L.: Structure formation by ion-diffusion simplex-ionotropism. Z. Natur-
forsch. 12-B, 14-23.
Thiele H. & Langmaack L.: Some polyelectrolytes as anions, cations and amphoteric ions. Z.
Physik. Chem. 207, 118-137.
Tracey M.V.: Chitin. Rev. Pure Appl. Chem. 7, 1-14.
Veldkamp H.: Aerobic decomposition of chitin by microorganisms. Mededel. Landbouwhogeschool
Wageningen 55, 127-174.
Wigglesworth V.B.: The physiology of insect cuticle. A. Rev. Entomol. 2, 37-54.
Willems J.: Oriented overgrowth of organic high polymers. Experientia 13, 276-277.

1958

Barker S.A., Foster A.B., Stacey M. & Webber J.M.: Amino sugars and related compounds. 4.
Isolation and properties of oligosaccharides obtained by controlled fragmentation of
chitin. J. Chem. Soc. 1958, 2218-2227.
Berger L.R. & Reinholds D.M.: Chitinase system of a strain of *Streptomyces griseus*. Bioch.
Biophys. Acta 29, 522-534.
Brunet P.C.J. & Carlisle D.B.: Chitin in *Pogonophora*. Nature 182, 1689.
Danilov S.N. & Plisko E.A.: The study of chitin. 3. Hydroxyethyl and ethyl ethers of chitin.
Zhur. Obs. Khim. 28, 2255-2259.
De Chirico A. & Gallo E.: Electrochemical properties of selective membranes of polyglucosamine
and polyhexamethyleneadipamide. Ric. Sci. 28, 1856-1862.
Delangre J.P.: Deacetylated chitin mordant. U.S. 2,842,049.
Giles C.H. & Hassan A.S.A.: Adsorption at organic surfaces. 5. A study of the adsorption of
dyes and other organic solutes by cellulose and chitin. J. Soc. Dyers Color. 74, 846-857.
Giles C.H., Hassan A.S.A., Laidlaw M. & Subramanian R.V.R.: Adsorption at organic surfaces. 3.
Some observations on the constitution of chitin and its adsorption of inorganic and organic
acids from aqueous solutions. J. Soc. Dyers Color. 74, 647-654.
Giles C.H., Hassan A.S.A. & Subramanina R.V.R.: Adsorption at organic surfaces. 4. Adsorption
of sulfonated azo dyes by chitin from aqueous solution. J. Soc. Dyers Color. 74, 682-688.
Jeuniaux Ch.: Recherches sur les chitinases. 1. Dosage néphélométrique et production de chitinase par des Streptomycètes. Arch. Int. Physiol. Bioch. 66, 408-427.
Okimasu S.: Studies of chitin from the standpoint of polymer chemistry. 4. Relation of the ef-
fective charges of glycolchitosan and its N-methyl derivative on interaction with negative
colloids. Nippon Nogei Kagaku 32, 298-302 and 32, 303-308.
Okimasu S.: Physicochemical behavior of carboxymehtylchitosan as polyampholyte. 1. Molecular
weight by light-scattering. Nippon Nogei Kagaku 32, 383-386; 32, 387-389 and 32, 471-478.
Ryan W.H.: Photographic products. U.S. 3,003,875.
Sadov F.I. & Vil'dt E.O.: Chitosan in pigment printing. Tekst. Prom. 18, 38-40.

1958 CNTD

Srinivasan S. & Quastel J.H.: Enzymic syntheses of oligo and polysaccharides containing D-
    -glucosamine. Science 127, 143-144.
Tomita G. & Takeyama h.: Photooxidation and reduction and the catalytic action of natural high
    polymers. Kagaku 28, 308-310.
Tomita G. & Takeyama H.: Activation of the fluorescence by polymers and photooxidation-
    -reduction. Kagaku 28, 528.
Tracey M.V. & Yonatt G.: Cellulase and chitinase in two species of australian termites.
    Enzymol. 19, 70-72.
Vogler K.: Polysulfuric acid esters of N-formylchitosan. U.S. 2,831,851.
Vogler K.: Polysulfate esters of chitin. Swit. 326,791
Vogler K.: Polysaccharide polysulfonic acid esters. Swit. 326,792.
Vogler K.: Polysaccharide esters. Swit. 329,205.
Warner D.T. & Coleman L.L.: Selective sulfonation of amino groups in amino alcohols. J. Org.
    Chem. 23, 1133-1135.
Wolfrom M.L.: Sulfated aminopolysaccharides. U.S. 2,832,766.
Wolfrom M.L., Maher G.G. & Chaney A.: Chitosan nitrate. J. Org. Chem. 23, 1990-1991.
Zimmerman H.K.: Studies on glucosammonium chloride. Ultraviolet absorption behavior in borate
    buffers. Arch. Biochem. Biophys. 75, 520-527.

1959

Aronson J.M. & Machlis L.: The chemical composition of the hyphal walls of the fungus *Allomy-
    ces*. Am. J. Bot. 46, 292-300.
Drozdova T.V.: Chitin and its transformations in natural processes. Biol. 47, 265-276.
Fel'dman R.I., Makarov Y.Y., Reutov O.S. & Goldovskii E.A.: Cardboard. U.S.S.R. 125,127.
Gibbs K.E. & Morrison F.O.: Cuticle of the two-spotted spider mite *Tetranychus telarius*. Can.
    J. Zool. 37, 633-637.
Heller K., Claus L. & Huber J.: Identity of plant and animal chitin. Z. Naturforsch. 14-B,
    476-477.
Horikoshi K. & Ida S.: Effect of lytic enzyme from *Bacillus circulans* and chitinase from
    Streptomyces sp. on *Aspergillus oryzae*. Nature 183, 186-187.
Jeuniaux Ch.: Recherches sur les chitinases. 2. Purification de la chitinase d'un Streptomycè-
    te, et séparation électrophorétique de principes chitinolytiques distincts. Arch. Int.
    Physiol. Bioch. 67, 597-617.
Jeuniaux Ch.: Action consécutive de deux enzymes différents au cours de l'hydrolyse complète
    de la chitine. Arch. Int. Physiol. Bioch. 67, 115-116.
Kurata G. & Sano K.: Investigation of chitin. Seikagaku 31, 153-157.
Leloir L.F.: The role of uridine nucleotides in metabolism. Centennial Lectures Commen. One
    Hundredth Anninv E.R. Squibb & Sons, 101-108.
MacNall E.G.: The chemical structure of the cell wall polysaccharide derived from *Coccidioi-
    des immitis*. Fed. Proc. 18, 585.
Marchessault R.H., Morehead F.F. & Walter N.M.: Liquid crystal systems from fibrillar poly-
    saccharides. Nature 184, 632-633.
Olesen E.S.: Effect of acid polysaccharides on the fibrinolytic system in guinea pig serum.
    Acta Pharmacol. Toxicol. 15, 307-318.
Pope S. & Zilliken F.: N-acetylglucosamine preparation by fermentation. U.S. 2,910,408.
Popowicz J.: Determination of chitin by the turbidimetric method. Chem. Anal. 4, 7-8.
Richards A.G. & Pipa R.L.: The molecular organization of insect cuticle. Smithsonian Inst.
    Publ. Contrib. Knowledge, Misc. Coll. 137, 247-262.
Ryan W.H. & Yankowski E.L.: Photographic image-receiving material. Ger. 1,116,969.
Van Duin P.Y. & Hermans J.J.: Light scattering and viscosity of chitosan in aqueous solutions
    of sodium chloride. J. Polymer Sci. 36, 295-303.
Wolfrom M.L. & Shen-Han T.M.: The sulfonation of chitosan. J. am. Chem. Soc. 81, 1764-1766.

1960

Bernfeld P., Nisselbaum J.S., Berkeley B.J. & Hanson R.W.: The influence of chemical and
    physicochemical nature of macromolecular polyanions and their interactions with β-lipo-
    proteins. J. Biol. Chem. 10, 2852-2859.
Davies D.A.L.: Polysaccharides of gram-negative bacteria. Adv. Carbohydr. Chem. 15, 271-340.
Devigne J. & Jeuniaux Ch.: The origin of intestinal chitinases in *Lombricus* sp. Arch. Int.
    Physiol. Bioch. 68, 833-834.
Dweltz N.E.: Structure of chitin. Bioch. Biophys. Acta 44, 416-435.
Golubchuk M., Cuendet L.S. & Geddes W.F.: Grain storage. 30. Chitin content of wheat as index
    of a mold contamination and wheat deterioration. Cereal Chem. 37, 405-411.
Hackman R.H.: Chitin. 4. The occurrence of complexes in which chitin and protein are covalen-
    tly linked. Aust. J. Biol. Sci. 13, 568-577.
Hamaguchi K., Rokkaku K., Funatsu M. & Hayashi K.: The structure and the enzymic function of
    lysozyme. 1. Enzymic action of lysozyme on glycol chitin. J. Bioch. 48, 351-357.
Harno N.: A specific adsorbent of lysozyme. Osaka Daigku Igaku Zasshi 12, 1531-1532.
Hurd I.S. & Haynes G.M.: Dyeable coatings for glass fibers. U.S. 2,961,344.
Inoue Y., Onodera K., Kitaoka S. & Hirano S.: N-acylation of 2-amino-2-deoxy-D-glucose with
    mixed carboxylic anhydrides. J. Org. Chem. 25, 1265-1267.

1960 CNTD

Jeuniaux Ch.: Chitinase and chitobiase activity of the hepatopancreas and the epidermis of the crustacean *Maia squinado* during shell molting. Arch. Int. Physiol. Bioch. 68, 837-838.

Lenk H.P., Wenzel M. & Schutte E.: Darstellung von Oligosacchariden des Glukosamins und acetylglukosamins und deren Spaltung durch Lysozym. Naturwissens. 22, 1-2.

Locke M.: The cuticular pattern in an insect: the intersegmental membranes. J. Exp. Biol. 37, 398-346.

Lovett J.S. & Cantino E.C.: The relation between bicarbonate, glucosamine synthetase and chitin synthesis in *Blastocladiella*. Mycologia 52, 338-341.

Lunt M.R. & Kent P.W.: Chitinase system from *Carcinus maenas*. Bioch. Biophys. Acta 44, 371-373.

Marchessault R.H., Pearson F.G. & Liang C.Y.: Infrared spectra of crystalline polysaccharides. 4. Effect of orientation on the tilting spectra of chitin films. Bioch. Biophys. Acta 45, 499-507.

Okimasu S. & Ohtakara A.: Decomposition of glycolchitin by exocellular chitinase of the black--koli mold. Nippon Nogeikagaku Kaishi 34, 873-878.

Pearson F.G., Marchessault R.M. & Liang C.Y.: Infrared spectra of crystalline polysaccharides. 5. Chitin. J. Polymer Sci. 43, 101, 116.

Sharan N. & Jeanloz R.W.: The diaminohexose component of a polysaccharide isolated from *Bacillus subtilis*. J. Biol. Chem. 235, 1-5.

Takeda M. & Tagawa S.: Behavior of dilute aqueous solutions of glycolchitin. Norinsho Suisan Koshusho Kenkyu Hokoku 10, 89-95.

Upjohn Co.: N-sulfonic acid derivatives of aminoalcohols. Brit. 838,709.

Wolfrom M.L. & Juliano B.O.: Chitin deacetylation. J. Am. Chem. Soc. 82, 2588-2590.

Zaluska H.: Glycogen and chitin metabolism during development of the silkworm *Bombyx mori*. Acta Biol. Exp. 19, 339-351.

1961

Antonopoulos C.A., Borelius E., Gardell S., Hamnstrom B. & Scott J.E.: Precipitation of poly-anions by long-chain aliphatic ammonium compounds. 4. Elution in salt solutions of mucopolysaccharide quaternary ammonium complexes adsorbed on a support. Bioch. Biophys. Acta 54, 213-226.

Blackith R.E. & Howden G.F.: Food reserves of hatchling locust. Comp. Bioch. Physiol. 3, 108-124.

Claus L.: Chitinase of the insect-killing mold *Beauveria bassiana*. Arch. Mikrobiol. 40, 17-46.

Danilov S.N. & Plisko E.A.: The study of chitin. 4. Preparations and properties of carboxy-methylchitin. Zhur. Obs. Khim. 31, 426-430.

Danilov S.N., Plisko E.A. & Pyaivinen E.A.: Esters and reactivity of cellulose and chitin. Izv. Akad. Nauk. SSSR, Otdel. Khim. Nauk. 1961, 1500-1506.

DeTerra N. & Tatum E.L.: Colonial growth of *Neurospora*. Science 134, 1066-1068.

Devigne J. & Jeuniaux Ch.: Chitinases of the intestinal tissues of earthworms. Arch. Int. Physiol. Bioch. 69, 223-234.

Dweltz N.E.: Structure of β-chitin. Bioch. Biophys. Acta 51, 283-289.

Dweltz N.E. & Awaud N.: Nature of aminosugar in β-chitin. Bioch. Biophys. Acta 50, 357.

Ellenbogen L. & Highley D.R.: Measurement of iron absorption in rats with a plastic scintillator. J. Appl. Physiol. 16, 1126-1128.

Fairbairn D.: The in vitro hatching of *Ascaris lombricoides* eggs. Can. J. Zool. 39, 153-162.

Foster A.B. & Webber J.M.: Chitin. Adv. Carbohydr. Chem. 15, 371-393.

Foster A.B., Martlew E.F., Stacey M., Taylor P.J.M. & Webber J.M.: Aminosugars and related compounds. 8. Some properties of 2-deoxy-2-sulfonamido-D-glucose, heparin and related substances. J. Chem. Soc. 1961, 1204-1208.

Hasz W.: Structure of the chitin of the arthropods. Arch. Physiol. 1961, 295-339.

Jeuniaux Ch.: Chitinase: an addition to the list of hydrolases in the digestive tract of Vertebrates. Nature 192, 135-136.

Kamasastri P.V. & Prabhu P.V.: Preparation of chitin and glucosamine from prawn shell waste. J. Sci. Ind. Res. 20-D, 466.

Krichevskii G.E. & Sadov F.I.: Dyeing of cellulose with fiber-reactive dyes. Izv. Vyss. Ucheb. Zavedenii, Tekknol. Tekstil. Prom. 3, 102-109.

Krishnakumaran A.: Comparative study of the arachnid cuticle. 2. Chemical nature. Z. Vergleich physiol. 44, 478-486.

Lenk H.P., Wenzel M. & Schuette E.: Low-molecular weight decomposition products of chitin. Z. Physiol. Chem. 326, 116-120.

Lingappa Y. & Lockwood J.L.: A chitin medium for isolating, growing and storing actinomycets. Nature 189, 158-159.

Lunt M.R. & Kent P.W.: Evidence of the occurrence of uridine-diphosphate N-acetylglucosamine in crustacean tissues. Bioch. J. 78, 128-134.

Malek S.R.A.: Polyphenols and their quinone derivatives in the cuticle of the desert locust *Schistocerca gregaris*. Comp. Bioch. Physiol. 2, 35-50.

McGarrahan J.F. & Maley F.: Metabolism of glucosamine and N-acetylglucosamine. N.Y. State Dept. Health, Ann. Rep. Div. Lab. Res. 1961, 66-68.

Meenakshi V.R. & Sheer B.T.: Metabolism of glucose in the crabs *Cancer magister* and *Hemigrapsus nudus*. Comp. Biochem. Physiol. 3, 30-41.

1961 CNTD

Micheel F. & Mempel D.: Synthesis of polysaccharides from 2-acetamido-2-deoxy-D-glucose. Makromol. Chem. 48, 24-32.

Mihara S.: Change in glucosamine content of *Chlorella* cells during their life cycle. Plant Cell Physiol. 2, 25-29.

Mitchell R. & Alexander M.: Chitin and the biological control of *Fusarium* diseases. Plant Disease Rep. 45, 487-490.

Murata K.: Further studies on the inhibitory effects of synthetic sulfated monosaccharides on experimental atherosclerosis. Japan Heart J. 2, 493-505.

Noguchi J.: Sulfates. Japan 23,324.

Ohtakara A.: Studies on the chitinolytic enzymes of black-koji mold. 1. Viscometric determination of chitinase activity by application of chitin glycol as a new substrate. Agr. Biol. Chem. 25, 50-54.

Ohtakara A.: Studies on the chitinolytic enzymes of black-koji mold. 3. Liquefying activity and saccharifying activity of the chitinase preparation. Agr. Biol. Chem. 25, 494-499.

Person P. & Fine A.: Reversible inhibitions of cytocrome system components by macromolecular polyions. Arch. Bioch. Biophys. 94, 392-404.

Ryan W.H. & Yankowski E.L.: Photographic image-receiving material. Ger. 1,116,969.

Runham N.W.: The histochemistry of chitin. J. Histoch. Cytoch. 9, 87-92.

Sawicki E., Hauser T.R., Stanley T.W. & Elbert W.: The 3-methyl-2-benzothiazolone hydrazone test. Anal. Chem. 33, 93-96.

Spencer R.: Chitinoclastic activity in the luminous bacteria. Nature 190, 938-939.

Townsley P.M.: Chromatography of tobacco mosaic virus on chitin columns. Nature 191, 626.

Waterhouse D.F. & McKellar J.W.: The distribution of chitinase in the body of the american cockroach. J. Insect Physiol. 6, 185-195.

Waterhouse D.F., Hackman R.H. & McKellar J.W.: Chitinase activity in cockroach and termite extracts. J. Insect Physiol. 6, 96-112.

Wolff R., Brignon J.J. & Schemberg M.: Comparative study of the clearing activity of various intravenous heparin samples and of some heparin-like derivatives in the rabbit. Therapie 15, 78-85.

Yamada K., Kuzuya H. & Noda M.: Studies on some actions of sulfate polysaccharides on arteriosclerosis. 1. Toxicity, lipolytic action and acticoagulant activity. Japan Circul. J. 25, 497-502.

1962

Andreev P.F., Plisko E.A. & Rogozin E.M.: Interactions of dilute solutions of uranyl salts with chitin and some cellulose ethers. Geokhimiya 1962, 536-539.

Bade M.L. & Wyatt G.R.: Metabolic conversions during pupation of the *Cecropia* silkworm. 1. Deposition and utilization of nutrient reserves. Bioch. J. 83, 470-478.

Barat S.K. & Scaria K.Y.: The nature of the cementing substances found in the infested pig bristles. Bull. Centr. Leather Res. Inst. 9, 73-74.

Barker S.A., Lloyd I.R.L. & Stacey M.: Polymerization of glucose induced by gamma radiation. Rad. Res. 16, 224-231.

Bartnicki-Garcia S. & Nickerson W.Y.: Isolation, composition and structure of cell walls of filamentous and yeast-like forms of *Mucor rouxii*. Bioch. Biophys. Acta 58, 102-119.

Bell L.G.E.: Polysaccharides and cell membranes. J. Theoret. Biol. 3, 132-133.

Candy D.J. & Kilby B.A.: Chitin synthesis in the desert locust. J. Exp. Biol. 39, 129-140.

Carlstrom D.: Polysaccharide chain of chitin. Bioch. Piophys. Acta 59, 361-364.

Carmen M.: Radiosensitivity and radioprotection. Action of gamma radiation on the ovomucoid mucoproteins. An. Inst. Invest. Vet. 12, 47-75.

Ciba S.A.: Colorants hautement polymérisés, leur procédé de préparation et leur emploi. Belg. 609,054.

Crook E.M. & Johnson I.R.: Qualitative analysis of the cell walls of elected species of fungi. Bioch. J. 83, 325-331.

Dabrowski J.: Dyeing and/or printing. U.S. 3,023,072.

De Mets R. & Jeuniaux Ch.: Sur les substances organiques constituant la membrane péritrophique des insectes. Arch. Int. Phys. Bioch. 70, 93-96.

Derevitskaya V.A., Likhosherstov L.M., Kara-Murza S.G. & Kochetkov N.K.: Glycopeptides. 2. Synthesis of 6-O-aminoacyl derivatives of glucose. Zhur. Obs. Khim. 32, 2134-2140.

Distler J.J. & Roseman S.: D-glucosamine oligosaccharides. Partial hydrolysis of chitosan. Meth. Carbohydr. Chem. 1, 305-309.

Evans E.E. & Kent S.P.: The use of basic polysaccharides in histochemistry and cytochemistry. 4. Precipitation and agglutination of biological materials by *Aspergillus* polysaccharides and deacetylated chitin. J. Histoch. Cytoch. 10, 24-28.

Frieser E.P.: Some applications of chitin in textile finishing. Melliand Text. 43, 1962-1963.

Hackman R.H.:Chitin. 5. The action of mineral acids on chitin. Aust. J. Biol. Sci. 15, 526-537.

Horikoshi K. & Arima K.: X-ray diffraction patterns of the cell wall of *Aspergillus orizae*. Bioch. Biophys. Acta 57, 392-394.

Kent S.P. & Evans E.E.: The use of basic polysaccharides in hystochemistry and cytochemistry. 2. The demonstration of acidic polysaccharides in tissue sections using fluorescein-labeled deacetylated chitin. J. Histoch. Cytoch. 10, 14-18.

1962 CNTD

Kent S.P. & Evans E.E.: The use of basic polysaccharides in histochemistry and cytochemistry. 3. Altered staining of acidic polysaccharides in tissue sections induced by *Aspergillus* polysaccharides and deacetylated chitin. J. Histoch. Cytoch. 10, 19-24.

Krueger L., Luderitz O., Strominger Y.L. & Westphal O.: Zur Immunochemie der O-Antigen von Enterobacteriaceae. Bioch. 335, 548-558.

Lieflaender M.: Preparation and properties of some N-aminoacylglucosamines. Z. Physiol. Chem. 329, 1-19.

McFarlane J.E.: The cuticles of the egg of the house cricket. Can. J. Zool. 40, 13-21.

Merchants & Manufacturers Inc.: Procédé de pré-traitement de fibres ou de tissus pour en augmenter l'affinité envers les colorants. Fr. 1,284,636.

Mitchell R. & Alexander M.: Microbiological processes. Soil Sci. Soc. Am. Proc. 26, 556-558.

Oeriu S. Lupu E.R., Dimitriu M.A. & Craesco I.: Glucosamine obtained from crustacean shells and its biological and therapeutic importance. Ann. Pharm. Franc. 20, 66-72.

Ohtakara A.: Studies on the chitinolytic enzymes of the black-koji mold. 4. Action of the liquefying chitinase on glycolchitin. Agr. Biol. Chem. 26, 30-35.

Ramachandran G.N. & Ramakrishnan C.: Structure of chitin. Bioch. Biophys. Acta 63, 307-309.

Rehacek Z., Farber G., Sevcik V., Musilek V. & Pilat A.: Submerged cultivation of the *Boletus edulis* var. *reticulatus*. Folia Microbiol. 7, 75-79.

Reisert S.P. & Fuller M.S.: Decomposition of chitin by *Chytriomyces* sp. Mycologia 54, 647-657.

Runham N.W.: Further investigations on the histochemistry of molluscan chitin. J. Histoch. Cytoch. 10, 504.

Salthouse T.N.: Histochemistry and staining of chitin. J. Histoch. Cytoch. 10, 109.

Sayuda T., Yamashina I., Furuhashi T. & Morita K.: Chitosan adducts of polysaccharide sulfuric acid esters. Japan 11,448.

Scott J.E.: Precipitation of polyanions by long chain aliphatic ammonium salts. 4. Affinity of substituted ammonium cations for the anionic groups of some biological polymers. Bioch. J. 84, 270-275.

Sooda Y. & Furuhasi M.: Methods for the preparation of chitosan compounds with polysaccharide sulfuric esters. Japan 11,448.

Stacey M. & Webber J.M.: 2-Amino-2-deoxy-D-glucose from crustaceans. Meth. Carbohydr. Chem. 1, 228-230.

Sundara Rajulu G.: Histochemical observations on chitin in the formative stages of the endo-cuticle of *Cingalobolus bugnioni*, a diplopod. Current Sci. 31, 469-470.

Takeda M. & Abe E.: Isolation of crustacean chitin. Decalcification by disodium ethylenediami-notetraacetate and enzymic hydrolysis of incidental proteins. Norisho Suisan Koshusho Ken-kyu Hokoku 11, 339-406.

Theodorides J.: Auguste Odier (1802-1870) and the discovery of chitin (1821-1823). Coll. Trav. Acad. Int. Histoire Sci. 12, 123-129.

Thiele H., Plohnke K., Brandt E. & Moll G.: Ordering of polyelectrolytes by ion diffusion. Kolloid Z. 182, 24-35.

Wainwright S.A.: An anthozoan chitin. Experientia 18, 18-19.

Wenzel M., Lenk H.P. & Schuette E.: Preparation of tri-N-acetylchitotriose-|^{3}H| and its hydro-lysis by lysozyme. Z. Physiol. Chem. 327, 13-20.

Wigert H.: Chitinase from *Streptomyces violaceus*. Naturwissens. 16, 379-380.

Whistler R.S. & BeMiller J.N.: Chitin. J. Org. Chem. 27, 1161-1163.

1963

Anderson D.M.W. & Garbutt S.: Studies on uronic acid materials. Anal. Chim. Acta 29, 31-38.

Bernfeld P.: Biogenesis of Natural Compounds. Pergamon Press, Oxford.

Bernfeld P. & Kelley T.F.: Inhibitory and activating effects of polyanions on lipoprotein li-pase. J. Biol. Chem. 238, 1236-1241.

Dandrifosse G. & Schoffeniels E.: Inactivation of chitinase by oxygen. Arch. Int. Physiol. Bioch. 71, 788-796.

Farr W.K.: Research on the cytochemistry of microorganisms. AD 423528.

Florkin M. e Stotz E.H., Eds.: Comprehensive Biochemistry, Vol. 5. Elsevier, Amsterdam.

Greig C.G. & Leaback D.H.: Preparation of 2-acyl derivatives of 2-amino-2-deoxy-D-glucose. J. Chem. Soc. 1963, 2644-2647.

Hamiter W.L. & Collins J.M.: Process for coloring glass fabrics. U.S. 3,108,897.

Hayashi K., Imoto T. & Funatsu M.: The enzyme-substrate complex in a muramidase-catalyzed reaction. 1. Difference spectrum of the complex. J. Bioch. 54, 381-387.

Jaworski E., Wang L. & Marco G.: Synthesis of chitin in cell-free extracts of *Prodenia erida-nia*. Nature 198, 790.

Jeuniaux Ch.: Chitine et Chitinolyse. Masson, Paris.

Malek S.R.A.: Chitin in the hyaline exocuticle of the scorpion. Nature 198, 301-302.

Misaki A.: Structural polysaccharide chemistry. Kagaku No Ryoiky 17, 439-448.

Mitchell R.: Addition of fungal cell-call components to soil for biological disease control. Phytopathol. 53, 1068-1071.

Monne L.: Formation and chemistry of the eggshell of *Contracaecum osculatum*. Z. parasitenk. 22, 475-483.

Neville A.C.: Growth and deposition of resilin and chitin in locust rubber-like cuticle. J. Insect Physiol. 9, 265-278.

1963 CNTD

Ohtakara A.: Studies on the chitinolytic enzymes of the black-koji mold. 5. Participation of two different enzymes in the decompostition of glycolchitin to the constituent aminosugar. Agr. Biol. Chem. 27, 454-460.

Oxford Paper Co.: Improvement in coating compositions and planographic printing plates. Brit. 930,765.

Person P., Mora P.T. & Fine A.S.: Macroion interactions involving cytochrome system components. 2. Charge density effects on macrocation inhibition of cytochrome oxidase activity and on block and reversal of such inhibition by macroanions. J. Biol. Chem. 238, 4103-4107.

Plisko E.A. & Danilov S.N.: Properties of chitin and its derivatives. Mat. Vses. Konf. po Probl. "Khim. Obmen. Uglevodov" 3rd, Moscow, 141-145.

Powning R.F. & Irzykiewicz H.: Chitinase from the gut of the cockroach *Periplaneta americana*. Nature 200, 1128.

Quillet M. & Priou M.L.: The presence of chitin in the membranes of some red algae *Ceramiaceae* and *Lomentarieae*. Comp. Rend. 256, 2903-2905.

Ramachandran G.N., Ramakrishnan C. & Sasisekharan V.: Stereochemistry of polypeptide and polysaccharide chain conformations. Aspects Protein Struct. Proc. Symp., Madras 1963, 121-134.

Reese E.T.: Advances in Enzymic Hydrolysis of Cellulose and Related Materials. Pergamon Press, Oxford.

Rudall K.M.: The chitin-protein complexes of insect cuticles. Adv. Insect Physiol. 1, 257-313.

Rogers W.P.: Physiology of infection with nematodes: some effects of the host stimulus on infective stages. Ann. N.Y. Acad. Sci. 113, art. 1; 208-216.

Stegemann H.: Protein (conchagen) and chitin in the supporting tissue of the cuttlefish. Z. Physiol. Chem. 331, 269-279.

Strasdine G.A. & Whitaker D.R.: On the origin of cellulase and chitinase of *Helix pomatia*. Can. J. Bioch. Physiol. 41, 1621-1626.

Sweeley C.C., Bentley R., Makita M. & Wells W.W.: Gas-liquid chromatography of trimethylsilyl derivatives of sugars and related substances. J. Am. Chem. Soc. 85, 2497-2507.

Veyrat R., Manning E.L., Fabre J. & Muller A.F.: Effects of the semi-synthetic adrenostatic heparinoid RO 1-8307 on the secretion of aldosterone. Rév. Franç. Et. Clin. Biol. 8, 667-673.

1964

Ashrafi S.H. & Wafaquani L.: Histochemical localization of polysaccharides in the midgut of the desert locust *Schistocerca gregaria*. Pakistan J. Sci. Ind. Res. 7, 150-151.

Brisou J., Tysset C., De Rautlin De La Roy Y., Curcier R., Moreau R., Moulin M. & Moulin B.: Chitinolysis in sea-water. Ann. Inst. Pasteur 106, 469-478.

Bach M.K.: The inhibition of deoxyribonucleotidyl transferase DNase and RNase by sodium poly-(ethenesulfonic acid). Effect of the molecular weight of the inhibitor. Bioch. Biophys. Acta 91, 619-626.

Brimacombe J.S. & Webber J.M.: Mucopolysaccharides. Elsevier, Amsterdam.

Conrad J.: Chitin. Encycl. Polymer Sci. Technol. 3, 695-705.

Costantino L. & Vitagliano V.: Effect of bivalent anions on the behavior of the polyelectrolyte polyglucosamine. Ric. Sci. 4, 493-450.

Dall W.: The physiology of a shrimp *Metapenaeus mastersii*. 1. Blood constituents. Aust. J. Marine Freshwater Res. 15, 145-161.

Dandrifosse G. & Schoffeniels E.: Action of certain sugars on the secretion of chitinase by isolated gastric mucosa. Arch. Int. Physiol. Bioch. 72, 517-519.

Doczi J., Niger F.C. & Silverman H.I.: Composition for inhibiting pepsin activity and method of preparing same. U.S. 3,155,575.

Fisk F.W. & Rao B.R.: Digestive carbohydrases in the cuban burrowing cockroach. Ann. Entomol. Soc. Am. 57, 40-44.

Hackman R.H. & Goldberg M.: New substrates for use with chitinases. Anal. Bioch. 8, 397-401.

Hayashi K., Imoto T. & Funatsu M.: The enzyme-substrate complex in a muramidase (lysozyme) catalyzed reaction. 2. Evidence for the conformational changes in the enzyme. J. Bioch. 55, 516-521.

Hayashi K., Yamasaki N. & Funatsu M.: Muramidase-catalyzed hydrolysis of glycolchitin. Agr. Biol. Chem. 28, 517-523.

Hillyard I.W., Doczi J. & Kiernan P.B.: Antacid and antiulcer properties of the polysaccharide chitosan in the rat. Proc. Soc. Exp. Biol. Med. 115, 1108-1112.

Jakobi B. & Fredericks J.: Reactions catalyzed by amidases. Acetamidase. J. Biol. Chem. 239, 1978-1982.

Jeuniaux Ch.: Chitine libre et chitine 'masquée' dans les structures squelettiques d'Invertébrés. Soc. Belge Bioch., Réunion Liège, 1964.

Kawasaki C. & Ito Y.: Hydrolysis of chitin by fungal enzyme preparations. Hakko Kogaku Zasshi 42, 212-215.

Kochetkov N.K., Kudryashev L.I. & Senchekova T.M.: Action of gamma radiation on 2-amino-2-deoxy-D-glucose. Dokl. Akad. Nauk SSSR 154, 642-645.

Kooiman P.: Occurrence of carbohydrase in digestive juice and hepatopancreas of *Astacus fluviatilis* and *Homarus vulgaris*. J. Cell. Comp. Physiol. 63, 197-201.

Kravcenko N.A. & Maksimov V.I.: Synthetic activity of hen egg protein lysozyme. Izv. Akad. Nauk SSSR, Ser. Khim. 3, 584.

1964 CNTD

Meyer W.: Transformation of 2-amino-2-deoxy-D-glucose into an aziridine derivative of 2-amino-
-2-deoxy-D-allose. Ber. 97, 325-330.
Neely W.B.: Glucosylated polysaccharides. U.S. 3,133,856.
Okutani K. & Kimata M.: Chitinolytic enzyme in the digestive tracts of japanese sea bass *Lateolabrax japonicus*. Nippon Suisan Gakkaishi 30, 262-266.
Okutani K. & Kimata M.: Chitinolytic enzyme in aquatic animals. 2. The chitinolytic enzyme
present in the liver of japanese sea bass *Lateolabrax japonicus*. Nippon Suisan Gakkaishi
30, 490-494.
Okutani K. & Kimata M.: Chitinolytic enzyme in aquatic animals. 3. Distribution of chitinase
in digestive organs of a few species of aquatic animals. Nippon Suisan Gakkaishi 30, 574-
576.
Ohtakara A.: Chitinolytic enzymes of black-koji mold. 6. Isolation and some properties of
N-acetyl-β-glucosaminidase. Agr. Biol. Chem. 28, 745-751.
Ohtakara A.: Chitinolytic enzymes of black-koji mold. 7. Degradation of glycolchitin and
chitin by the chitinase system of *Aspergillus niger*. Agr. Biol. Chem. 28, 811-818.
Quillet M.: The presence of chitin in the skeletal membranes of two species of brown algae
Dictyota dichotoma and *Dilophus spiralis*. Comp. Rend. 258, 3349-3351.
Ralston G.B., Tracey M.V. & Wrench P.M.: The inhibition of fermentation in baker's yeast by
chitosan. Bioch. Biophys. Acta 93, 652-655.
Raymont J.E.G., Austin J. & Linford E.: Biochemical studies on marine zooplankton. 1. Bioche-
mical composition of *Neomysis integer*. J. Conseil Per. Int. Explor. Mer 28, 354-363.
Rudall K.M.: Protein-polysaccharide interactions in chitinous structures. Struct. Function
Connective Skeletal Tissue: Proc. St. Andrews, Scot. 191-196.
Rupley J.A.: The hydrolysis of chitin by concentrated hydrochloric acid and the preparation of
low-molecular weight substrates for lysozyme. Bioch. Biophys. Acta 83, 245-255.
Rzucidlo L.: Aminosugars and their derivatives in microbiology. 1. Chemistry and analysis.
Postepy Mikrobiol. 3, 3-79.
Takeda M. & Katsuura H.: Purification of king crab chitin. Suisan Daigaku Kenkyu Hokoku 13,
109-116.
Wang P.T.: Chitin substances in *Mucor hiemalis*. Bull. Inst. Chem., Acad. Sinica 9, 37-46.
Wolfrom M.L., Vercellotti J.R. & Horton D.: Methylation studies on carboxyl-reduced heparin:
2-amino-2-deoxy-3,6-di-O-methyl-α-glucopyranose from the methylation of chitosan. J. Org.
Chem. 29, 547-550.

1965

Ali B.S., Heitefuss R. & Fuchs W.H.: The production of chitinase by *Entomophtora coronata*. Z.
Pflanzenkranh. Pflanzenschutz 72, 201-207.
BeMiller J.N.: Chitin. Methods Carbohydr. Chem. 5, 103-106.
Blackwell J., Parker K.D. & Rudall K.M.: Chitin in pogonophore tubes. F. Mar. Biol. Ass. 45,
659-661.
Blake C.C.F., Koening D.F., Mair G.A., North A.C.T., Phillips D.C. & Sarma V.R.: Structure of
hen egg white lysozyme. Nature 206, 757-761.
Boulingand Y.: Sur une architecture torsadée repandue dans de nombreuses cuticules d'arthro-
podes. Comp. Rend. 261, 3665-3668.
Camps J.M. & Arias E.: Microdeterminacion de proteinas, carbohidratos y quitina del plancton.
Reunion Prod. Marina Exploit. Pesquera, 5, Barcelona, 22-27.
Cantino E.C.: Intracellular distribution of ^{14}C during sporogenesis in *Blastocladiella emerso-
nii*: effect of light on hemoprotein. Arch. Mikrobiol. 51, 42-59.
Carey G.F.: Chitin synthesis in vitro by crustacean enzymes. Comp. Bioch. Physiol. 16, 155-158.
Carlstrom D.: Structure of chitin. Funct. Organ. Compound Eye, Proc. Int. Symp. Stockholm,
1965, 15-19.
Dajoz R.: Chitin and chitinolysis in the animal kingdom. Sci. Progr. Nat. 1965, 317-320.
Dall W.: Endocrines and control of molting. Aust. J. Marine Freshwater Res. 16, 1-12.
Dall W.: Composition and structure of the integument. Aust. J. Marine Freshwater Res. 16, 13-
23.
Dandrifosse G. & Schoffeniels E.: Mechanism of chitinase transport through cell membranes.
Bioch. Biophys. Acta 94, 165-174.
Dandrifosse G., Schoffeniels E. & Jeuniaux Ch.: Secretion of chitinase from isolated gastric
mucosa. Bioch. Biophys. Acta 94, 153-164.
Domsch K.H.: Influence of captan on the decomposition of glucose, aesculin, chitin and tannin
in soil. Phytopathol. Z. 52, 1-18.
Durner G., Roemer R. & Schwartz W.: Investigations on the ecological aspects of life in a
sulfuretum. Z. Allgem. Mikrobiol. 5, 206-221.
Frankignoul M. & Jeuniaux Ch.: Effects of chitin on chitinase secretion by rodents. Life Sci.
4, 1669-1672.
Fueller H.: Dichroic staining of chitin with Thiazine Red; histochemical chitin determination.
Zool. Anz. 174, 125-131.
Fields D.L. & Reynolds D.D.: Process for mercaptoethylating amines. U.S. 3,221,013.
Glowacka D. & Popowicz J.: The use of chitin in the chromatographic separation of uranyl from
ferric, calcium and magnesium ions. Rocznik Akad. Med. Bial. 11, 47-51.

1965 CNTD

Hackman R.H. & Goldberg M.: Chitin. 6. Nature of α- and β-chitins. Aust. J. Biol. Sci. 18, 935-946.

Hempel J.: Chemical composition of hegg coverings and cuticle in *Triops cancriformis* and *Lepidurus* products. Zool. Polon. 15, 143-146.

Horton D.& Lineback D.R.: N-deacetylation: chitosan from chitin. Meth. Carbohydr. Chem. 5, 403-406.

Jaworski E.G., Wang L.C. & Carpenter W.D.: Biosynthesis of chitin in cell-free extracts of *Venturia inaequalis*. Phytopathol. 55, 1309-1312.

Jeuniaux Ch.: Chitine et phylogénie: application d'une méthode enzymatique de dosage de la chitine. Bull. Soc. Chim. Biol. 47, 2267-2278.

Khalifa O.: Biological control of *Fusarium* wilt of peas by organic soil amendment. Ann. Appl. Biol. 56, 129-137.

Kimura A.: Studies of the enzyme from the digestive tract of *Helix peliomphala*. Purification and physico-chemical properties of the chitinase from the digestive tract of *Helix peliomphala*. Shikoku Acta Med. 21, 866-879.

Kojima M. & Koaze Y.: Chitinase produced by *Aspergillus fumigatus*. Meiji Seika Kenkyu Nempo 1965, 12-18.

Lemieux R.U. & Nagabhushan T.L.: Synthesis of 2-amino-2-deoxy sugars from acetylated glycals. Tetr. Lett. 26, 2143-2148.

Lipke H., Grainger M.M. & Siakotos A.N.: Polysaccharide and glycoprotein formation in the cockroach. 1. Identity and titer of bound monosaccharides. J. Biol. Chem. 240, 594-600.

Lipke H., Graves B. & Leto S.: Incorporation of D-glucose-^{14}C into bound carbohydrates. J. Biol. Chem. 240, 601-608.

Lloyd A.B., Noveroske R.L. & Lockwood J.L.: Lysis of fungal mycelium by *Streptomyces* species and the chitinase system. Phytopathol. 55, 871-875.

Mahadevan P.R. & Tatum E.L.: Relation of the major constituents of the *Neurospora crassa* cell--wall to wild-type and colonial morphology. J. Bacteriol. 90, 1073-1081.

Maksimov V.I., Kaverzeneva E.D. & Kravchenko N.A.: The action of lysozyme on oligosaccharide fragments of chitin. Biokhimiya 30, 1007-1014.

McGarrahan J.F. & Maley F.: Hexosamine metabolism. 3. The utilization of D-glucosamine-1-^{14}C and N-acetyl-D-glucosamine-1-^{14}C by *Aspergillus parasiticus*. J. Biol. Chem. 240, 2322-2327.

McLachlan J., McInnes A.G. & Falk M.: Chitan (chitin: poly-N-acetylglucosamine) fibers of the diatom *Thalassiosira fluviatilis*. 1. Production and isolation of chitan fibers. Can. J. Bot. 43, 707-713.

Noguchi J.: Anion-exchange resins. Japan 24,400.

Noguchi J., Tokura S., Inomata M. & Asano C.: Anion-exchange resins from chitosan derivatives. Kogyo Kagaku Zasshi 68, 904-908.

Ogawa H. & Ito T.: 6-Deoxy-6-mercapto-D-glucosamine and a process for the preparation of this compound. U.S. 3,176,004.

Okafor N.: Isolation of chitin from the shell of the cuttlefish *Sepia officinalis*. Bioch. Biophys. Acta 101, 193-200.

Okutani K. & Kimata M.: The chitinolytic enzyme present in aquatic animals. 4. Properties of the chitinolytic enzyme present in the stomach of japanese sea bass *Lateolabrax japonicus*. Nippon Suisan Gakkaishi 31, 232-237.

Polivanova E.N.: The functional role of embryonic membranes in insects. Dokl. Akad. Nauk SSSR 180, 243-245.

Porter C.A. & Jaworski E.G.: Biosynthesis of chitin during various stages in the metamorphosis of *Prodenia eridania*. J. Insect Physiol. 11, 1151-1160.

Potgieter H.Y. & Alexander M.: Polysaccharide components of *Neurospora crassa* hyphal walls. Can. J. Microbiol. 11, 122-125.

Powning R.F. & Irzykiewicz H.: Detection of chitin oligosaccharides on paper chromatograms. J. Chromatogr. 17, 621-623.

Powning R.E. & Irzykievicz H.: Chitinase system in bean and other seeds. Comp. Bioch. Physiol. 14, 127-133.

Pruglo N.V., Spasokukotskii N.S. & Bongard S.A.: Effect of introduction of polymeric bases on the fixing of acidic dyes in gelatin layers. 2. A microscopic study of prints with matrix on a layer containing various polymeric bases. Zhur. Nauk. Prikl. Fotogr. Kinematogr. 10, 360-365.

Rudall K.M.: Skeletal structure in insects. Bioch. J. 95, 38-39.

Rzucidlo L.: Amino sugar and their derivatives in microbiology. 2. Chemistry and analysis. Postepy Mikrobiol. 4, 57-96.

Sassen M.M.A.: Breakdown of the plant cell wall during the cell-fusion process. Acta Bot. Neerl. 14, 165-196.

Schmidt E.: The effect of chlorine dioxide on vegetable cell walls and natural albuminoids. Papier 19, 728-735.

Skujins J.J., Potgieter H.Y. & Alexander M.: Dissolution of fungal cell walls by a streptomycete chitinase and β-(1→3) glucanase. Arch. Bioch. Piophys. 111, 358-364.

Tamura Z., Miyazaki M. & Suzuki T.: Metal complexes of 2-amino-2-deoxy-D-glucose and its derivatives. 3. Determination of the acid dissociation constants of 2-amino-2-deoxy-D-glucose and its three O-methyl derivatives. Chem. Pharm. Bull. 13, 330-332.

1965 CNTD

Tamura Z., Miyazaki M. & Suzuki T.: Metal complexes of 2-amino-2-deoxy-D-glucose and its derivatives. 4. Determination of stability and equilibrium constants of metal complexes of 2-amino-2-deoxy-D-glucose by pH titration method. Chem. Pharm. Bull. 13, 333-344.

Tamura Z., Miyazaki M. & Suzuki T.: Metal complexes of 2-amino-2-deoxy-D-glucose and its derivatives. 5. Spectrophotometric investigations on copper, nickel and cobalt complexes of 2-amino-2-deoxy-D-glucose and the preparation of the 2-amino-2-deoxy-D-glucose-copper complex. Chem. Pharm. Bull. 13, 345-351.

Travis D.F.: The deposition of skeletal structures in the Crustacea. 5. The histomorphological and histochemical changes associated with the development and calcification of the branchial exoskeleton in the crayfish Orconectes viridis. Acta Histoch. 20, 193-222.

Weliki N. & Weetall H.H.: The chemistry and use of cellulose derivatives for the study of biological systems. Imm. Chem. 2, 293-322.

1966

Abbott E.C., Gornall A.G., Sutherland D.J.A., Stiefel M. & Laidlaw J.C.: The influence of heparin-like compounds on hypertension, electrolytes and aldosterone in man. Can. Med. Ass. J. 94, 1155-1164.

Abbott E.C., Monkhouse F.C., Steiner J.W. & Laidlaw J.C.: Effect of sulfated mucopolysaccharide (RO1-8307) on the zona glomerulosa of the rat adrenal gland. Endocrinol. 78, 651-654.

Bacon J.S.D., Davidson E.D., Jones D. & Taylor I.F.: Location of chitin in the yeast cell wall. Bioch. J. 101, 36C-38C.

Battista O.A.: Method of preparing stable aqueous dispersion-forming cellulose aggregates. U.S. 3,251,824.

Benjaminson M.A., Bazil S.L., D'Oria F.J. & Carito S.L.: Ferritin-labeled enzyme, a tool for electron microscopy. Nature 210, 1275-1276.

Berg J.G.F.: Penicillin derivatives. Brit. 1,038,367.

Brown M.F.: Localization of cellulose and chitin in the cell walls of Ceratocystis ulmi. Univ. Microfilms, Ann Arbor, N° 66-11, 647.

Buddecke E. & Sittner B.: Mucopolysaccharides and mucopolysaccharide-hydrolyzeing enzymes in transplant tumors. Z. Klin. Chem. 4, 157-163.

Chapman J.A.: Discussion of principles of biomolecular organization. Ciba Found. Symp. 129-130.

Coles C.G.: Studies on resilin biosynthesis. J. Insect Physiol. 12, 679-691.

Condoulis W.V. & Locke M.: The deposition of endocuticle in an insect, Calpodes ethlius. J. Insect Physiol. 12, 311-323.

Conn J.W., Rovner D.R., Cohen E.L., Anderson J.E., Kleinbergs S. & Bough W.M.: Inhibition by a heparinoid of aldosterone biosynthesis in man. J. Clin. Endocrinol. Metab. 26, 527-532.

Dahlquist F.W., Jao L. & Raftery M.: The binding of chitin oligosaccharides to lysozyme. Proc. Natl. Acad. Sci. U.S. 56, 26-30.

Dandrifosse G. & Schoffeniels E.: Secretion of chitinase and the free-flow of water through the gastric mucous membrane of the green lizard. Ann. Endocrinol. 27, 517-520.

Depoortere H. & Magis N.: Mise en évidence, localisation et dosage de la chitine dans la coque des oeuf de Brachionus leydigii Cohn et d'autres rotifères. Ann. Soc. Roy. Zool. Bel. 97, 187-195.

Elorza M.V., Munoz Ruiz E. & Villanueva J.R.: Production of yeast cell wall lytic enzymes on a semi-defined medium by a Streptomyces. Nature 210, 442-443.

Falck M., Smith D.G., McLachlan J. & McInnes A.G.: Studies on the chitin fibres of the diatom Thalossiosira fluviatilis Hustedt. 2. Proton magnetic resonance, infrared and x-ray study. Cna. J. Chem. 44, 2269-2281.

Gaillard B.D.E. & Bailey R.W.: Reaction with iodine of polysaccharides dissolved in strong calcium chloride solution. Nature 212, 202-203.

Gibian H.: Enzyme degrading glycosaminoglycans. Amino Sugar 1966, 181-200.

Jeuniaux Ch.: Chitinases. Methods Enzymol. 8, 644-650.

Kimura A.: Studies of the enzyme from the digestive tract of Helix peliomphala. 2. Substrate specificity of chitinase from snail digestive tract. Shikoku Acta Med. 22, 679-683.

Kimura A.: Studies of the enzyme from the digestive tract of Helix peliomphala. 3. Effect of metal ions on chitinase from snail digestive tract. Shikoku Acta Med. 22, 684-687.

King. M.V. : The precession method in fiber diffraction photography. Acta Cryst. 21, 629-635.

Kochetkov N.K., Kudryashov L.I., Senchenkova T.M. & Nedoborova L.I.: Radiation chemistry of carbohydrates. 7. Radiolysis of aqueous solutions of some glucosamine derivatives. Zhur. Obs. Khim. 36, 1020-1025.

Krueger H.R. & Jaworski E.G.: The fractionation and solubilization of Prodenia eridania chitin synthetase. J. Econ. Entomol. 59, 229-230.

Linker A.: Bacterial mucopolysacccharidase. Methods Enzymol. 8, 650-654.

McCormick J.E.: Properties of periodate-oxidized polysaccharides. 7. The structure of nitrogen containing derivatives as deduced from a study of monosaccharide analogs. J. Chem. Soc. 1966, 2121-2127.

McLachlan J. & Craigie J.S.: Chitan fibres in Cyclotella cryptica and growth of Cyclotella cryptica and Thalassiosira fluviatilis. Some Contemp. Stud. Mar. Sci. 1966, 511-517.

Marchant R.: Wall structure and spore germination in Fusarium culmorum. Ann. Bot. 30, 821-830.

Mehta P.D. & Maisel M.: Chitin in Tunicata. Experientia 22, 820.

1966 CNTD

Monkhouse F.C. & Abbott E.C.: The clearing factor lipase response to synthetic sulfonated polysaccharides in the rat. Can. J. Physiol. Pharmacol. 44, 635-640.

Moshy R.J. & Germino F.J.: Reconstituted tobacco sheet. U.S. 3,421,519.

Murray C.L. & Lovett J.S.: Nutritional requirements of the chytrid *Karlinga asterocysta,* an obligate chitinophile. Am. J. Bot. 53, 469-476.

Nakagami T., Tonomura K. & Tanabe O.: Chitin-decomposing microorganisms and culture conditions for production of the enzymes. Kogyo Gijutsuin, Hakko Kenkyusho Kenkyu Hokoku 30, 19-26.

O'Brien R.W. & Ralph B.J.: The cell wall composition and taxonomy of some Basidiomycetes and Ascomycetes. Ann. Bot. 30, 831-843.

Okafor N.: Estimation of the decomposition of chitin in soil by the method of carbon dioxide release. Soil Sic. 102, 140-142.

Porter C.A. & Jaworski E.G.: The synthesis of chitin by particulate preparations of *Allomyces macrogynus.* Bioch. 5, 1149-1154.

Potgieter H.J. & Alexander M.: Susceptibility and resistance of several fungi to microbial lysis. Bacteriol. 91, 1526-1532.

Powning R.F. & Irzykiewicz H.: Effect of lysozyme on chitin oligosaccharides. Bioch. Biophys. Acta 124, 218-220.

Pruglo N.V., Bongard S.A. & Spasokukotskii N.S.: Effect of introduction of polymeric bases on fixing of acidic dyes in gelatin layers. 4. Effect of addition of polymeric bases on color intensity of gelatin layer. Zhur. Nauk. Prikl. Fotogr. Kinematogr. 11, 194-201.

Reynolds D.D.: Mercaptoethylation of amines with carbamates. U.S. 3,232,936.

Swanson R. & S tock J.J.: Biochemical alterations of dermatophytes during growth. Appl. Microbiol. 14, 438-444.

Takeda H., Strasdine G.A., Whitaker D.R. & Roy C.: Lytic enzymes in the digestive juice of *Helix pomatia;* chitinases and muramidases. Can. J. Bioch. 44, 509-518.

Thomas E.W.: Interaction between lysozyme and acetamido sugars as detected by proton magnetic resonance spectroscopy. Bioch. Biophys. Res. Comm. 24, 611-615.

Tipper D.J. & Strominger J.L.: Isolation of 4-O-β-N-acetylmuramyl-N-acetylglucosamine and 4-O-β-N-6-O-diacetylmuramyl-N-acetylglucosamine and the structure of the walll polysaccharide of *Staphilococcus aureus.* Bioch. Biophys. Res. Comm. 22, 48-56.

Unestam T.: Chitinolytic, cellulolytic and pectinolytic activity in vitro of some parasitic and saprophytic *Oomycetes.* Physiol. Plant. 19, 15-30.

Wang C.S. & MIles P.G.: Cell walls of *Schizophyllum commune.* Am. J. Bot. 53, 792-800.

Weisberg M. & Gubner R.S.: Chitosan-containing antacid composition. U.S. 3,257,275.

Willingham C.A., Roach A.W. & Silvey J.K.G.: Comparative studies of substrate degradation by marine-occurring *Actinomycetes.* Am. Midland Naturalist 75, 232-241.

Yabuki M.: *Aspergillus amstelodami.* 6. The possible role of chitin in binding with β-fructofuranosidase within the cell walls of ascospores. Nippon Nogei Kagaku Kaishi 40, 201-208.

1967

Applegarth D.A.: The cell wall of *Penicillium notatum.* Arch. Bioch. Biophys. 120, 471-478.

Applegarth D.A. & Bozoian G.: Presence of chitin in the cell wall of a griseofulvin-producing species of *Penicillium.* J. Bacteriol. 94, 1787-1788.

Attwood M.M. & Zola H.: The association between chitin and protein in some chitinous tissues. Comp. Bioch. Physiol. 20, 993-998.

Balazs E.A., Davies J.V., Phillips G.O. & Scheufele D.S.: Polyanions and their complexes. 4. A pulse radiolysis study of the interaction between methylene blue and heparin in aqueous solutions. J. Chem. Soc. 1967, 1424-1429.

Balazs E.A., Davies J.V., Phillips G.O. & Scheufele D.S.: Polyanions and their complexes. 5. examination of polyanion-cation interaction using pulse radiolysis. J. Chem. Soc. 1967, 1429-1433.

Balazs E.A., Phillips G.O. & Young M.D.: Polyanions and their complexes. 2. Light-induced paramagnetism in solid glycosaminoglycan-dye complexes. Bioch. Biophys. Acta 141, 382-390.

Beuzekom W.J.C. & Douma A.D.: Preparations for the destruction of Nematodes. Neth. A. 6710529.

Blackwell J., Parker K.D. & Rudall K.M.: Chitin fibers of the diatoms *Thalassiosira fluviatilis* and *Cyclotella cryptica.* J. Mol. Biol. 28, 383-385.

Brauns F.E.: Recovery of lignosulfonates. U.S. 3,297,676.

Charlemagne D. & Jolles P.: Specificity of various lysozymes towards low-molecular weight substrates from chitin. Bull. Soc. Chim. Biol. 49, 1103-1113.

Cherkasov I.A. & Kravchenko N.A.: Enzyme-substrate complex formation between lysozyme and chitin. Mol. Biol. 1, 381-390.

Christias C. & Baker K.F.: Chitinase as a factor in the germination of chlamydospores of *Thielaviopsis basicola.* Phytopathol. 57, 1363-1367.

Clarke A.J., Cox P.M. & Shepherd A.M.: The chemical composition of the egg shells of the potato cyst nematode *Heterodera rostochiensis.* Bioch. J. 104, 1056-1060.

Crompton M. & Birt M.: Changes in the amounts of carbohydrates phosphagen and related compounds during the metamorphosis of the brownfly *Lucilia cuprina.* J. Insect Physiol. 13, 1575-1592.

Dabrowski J.: Nonwoven fabrics of resin-bonded regenerated cellulose. U.S. 3,304,194.

Dandrifosse G. & Schoffeniels E.: Mechanism of chitinase secretion in the gastric mucosa. Bioch. Biophys. Acta 148, 741-748.

1967 CNTD

Domer J.E., Hamilton J.G. & Harkin J.C.: Comparative study of the cell walls of the yeast-like and mycelial phases of *Histoplasma capsulatum*. J. Bacteriol. 94, 466-474.

Imperial Chemical Industries: Sultones. Brit. 1,059,568.

Kravchencko N.A. & Kuznetsov Y.D.: Mechanism of lysozyme action on chitin oligosaccharides. Khim. Biokhim. Uglevodov, Mater. Vses. Konf. 4, 187-191.

Kuo M.J. & Alexander M.: Inhibition of the lysis in fungi by melanins. J. Bacteriol. 94, 624-629.

Leaback D.H.: The application of continuous, fluorogenic assay to the determination of enzyme steady-state characteristics. Bioch. J. 102, 42-P.

Leopold J. & Seichertova O.: Parasitism of the mold *Penicillium purpurogenum* on *Aspergillus niger*. 3. Determination of some enzymes of *Penicillium purpurogenum* in the mycelium lysis of the mold *Aspergillus niger*. Folia Microbiol. 12, 458-465.

Maeda H. & Ishida N.: Specificity of binding of hexopyranosyl polysaccharides with a fluorescent brightener. J. Bioch. 62, 276-278.

Maksimov V.I., Osipov V.I. & Kaverzneva E.D.: The use of chitin low-molecular weight fractions for characterizing the activity of gamma irradiated lysozyme. Biokhimiya 32, 1169-1174.

Manocha M.S. & Colvin J.R.: Structure and composition of the cell walls of *Neurospora crassa*. J. Bacteriol. 94, 202-212.

Michell A.J. & Scurfield G.: Composition of extracted fungal cell walls as indicated by infrared spectroscopy. Arch. Bioch. Biphys. 120, 628-637.

Neville A.C.: Chitin orientation in cuticle and its control. Adv. Insect Physiol. 4, 213-286.

Okutani K., Sawada I. & Kimata M.: Chitinolytic enzyme in aquatic animals. 5. Chitinolytic enzyme in the digestive tract of yellowtail. Nippon Susisan Gakkaishi, 33, 848-852.

Okutani K., Sawada I. & Kimata M.: Chitinolytic enzyme in aquatic animals. 6. Chitinolytic enzyme in rainbouw trout. Nippon Suisan Gakkaishi 33, 952-955.

Owen J.B. & Sagan M.: Treatment of shaped articles. Brit. 1,059,568.

Plessmann-Camargo E., Dietrich C.P., Sonneborn D. & Strominger J.L.: Biosynthesis of chitin in spores and growing cells of *Blastocladiella emersonii*. J. Biol. Chem. 242, 3121-3128.

Powning R.F. & Irzykiewicz H.: Separation of chitin oligosaccharides by thin-layer chromatography. J. Chromatogr. 29, 115-119.

Powning R.F. & Irzykievicz H.: Lysozyme-like action of enzymes from the cockroach *Periplaneta americana* and from some other sources. J. Insect Physiol. 13, 1293-1299.

Rudall K.M.: Conformation in chitin-protein complexes. Conformation of Biopolymers, Vol. 1, Ramachandran G.N., Ed., Academic Press, London.

Ruiz-Herrera J.: Chemical components of the cell wall of *Aspergillus* species. Arch. Bioch. Biophys. 122, 118-125.

Sauer F.: Chitin, ideal building material of exoskeletons. Mikrokosmos 56, 8, 20-22.

Sundara Rajulu G. & Gowri N.: Nature of chitin of the coenosteum of Millipora. Indian J. Exp. Biol. 5, 180-181.

Thomas E.W.: Interaction between diacetylchitobiose methyl glycoside and lysozyme as studied by NMR spectroscopy. Bioch. Biophys. Res. Comm. 29, 635.

Toyama N.: Mycolytic and cellulolytic enzymes of *Trichoderma viride*. Hakko Kogaku Zasshi 45, 663-670.

1968

Arai K., Kinumaki T. & Fujita T.: Toxicity of chitosan. Tokaiku Suisan Kenkyusho Kenkyu Hokoku 56, 89-94.

Balazs E.A., Davies J.V., Phillips G.O. & Sceufele D.S.: A study of polyanion-cation interactions using hydrated electrons. Bioch. Biophys. REs. Comm. 30, 386-392.

Bartnicki-Garcia S.: Cell-wall chemistry, morphogenesis and taxonomy of fungi. Ann. Rev. Microbiol. 22, 87-108.

Beran K., Rehacek J. & Seichertova O.: Infrared spectral analysis of the cell wall. The application of the method in the study of chitin in the cell wall of yeast *Saccharomyces cerevisiae*. Acta Fac. Med., Proc. 2nd Int. Symp. Yeast Protoplast.

Billy C.: A chitinolytic anaerobic bacterium *Clostridium chitinoilum*. Comp. Rend., Ser. D, 266, 1535-1536.

Boissiere J.C.: Demonstration and localization of chitin in some lichens. Mém. Soc. Bot. Fr. 1968, 141-150.

Broussignac P.: Chitosan, a natural polymer not well known by the industry. Chim. Ind. Génie Chim. 99, 1241-1247.

Cherkasov I.A. & Kravchenko N.A.: Sorption of lysozyme on chitin. Biokhimiya 33, 761-765.

Chu S.S.C. & Jeffrey G.A.: The refinement of the crystal structure of β-D-glucose and cellobiose. Acta Crystall. B-24, 830-838.

Domanski R.E. & Miller R.E.: Use of chitinase complex and β-(1→3)-glucanase for spheroplast production from *Candida albicans*. J. Bacteriol. 96, 270-271.

Dweltz N.E., Colvin J.R. & McInnes A.G.: Chitan {β-(1→4)-liked 2-acetamido-2-deoxy-D-glucan} fibers of the diatom *Thalassiosira fluviatilis*. 3. The structure of chitan from x-ray diffraction and electron microscope observations. Can. J. Chem. 46, 1513-1521.

1968 CNTD

Furuhashi T.: Preservative for eye graft material. U.S. 3,371,012.

Garcia Mendoza C. & Novaes Ledieu M.: Chitin in the new wall of regenerating protoplasts of *Candida utilis (Torulopsis utilis)*. Nature 220, 1035.

Harnden D.G.: Digestive carbohydrases of *Balanus nubilis*. Comp. Bioch. Physiol. 25, 303-309.

Hayashi K., Imoto T. & Funatsu M.: Crystallization of the lysozyme-product complex. J. Bioch. 63, 550-552.

Karkocha I.: Determination of nitrogen compounds and their digestibility in *Lactarius vellereus* mushrooms. Rocz. Panstw. Zakl. Hig. 19(3), 307-310.

Lovell R.T., Lafleur J.R. & Hoskins F.H.: Nutritional value of freshwater crayfish waste meal. J. Agr. Food. Chem. 16, 204-207.

Mendoza C.G. e Ledieu M.N.: Chitin in the new wall of regenerating protoplast of *Candida utilis*. Nature 220, 1035.

Miyazaki T. & Matsushima Y.: Amino hexoses. 14. Preparation of partially O-carboxymethylated chitin and its component 3-O- and 6-O-carboxymethyl-2-amino-2-deoxy-D-glucose and the corresponding alditols. Bull. Chem. Soc. Japan 41, 2723-2726.

Miyazaki T. & Matsushima Y.: Mode of the enzymic action on partially O-carboxymethylated chitin. Bull. Chem. Soc. Japan 41, 2754-2757.

Pramer D.: Significant synthesis of heterotrophic soil bacteria. Ecol. Soil Bact. Int. Symp. (1966), 220-223.

Preston R.D. & Goodman R.N.: Structural aspects of cellulose microfibril biosynthesis. J. R. Micr. Soc. 88, 513-527.

Pruglo N.V., Bongard S.A. & Veitsman A.I.: Effect of polymeric bases on the fixation of acid dyes in gelatin layers; 5. Comparative evaluation of the effect of polymeric bases on lateral diffusion of dyes in the receiving layer of blank film. Zhur. Nauk. Prikl. Fotogr. Kinematogr. 13, 96-99.

Raftery M.A. & Rand-Meir T.: On distinguishing between possible mechanistic pathways during lysozyme-catalyzed cleavage of glycosidic bonds. Bioch. 7, 3281-3289.

Ritter H. & Bray M.: Chitin synthesis in cultivated cockroach blood. J. Insect Physiol. 14, 361-366.

Sera H. & Okutani K.: Chitinolytic enzyme present in aquatic animals. 7. Chitinolytic enzyme present in a sea brean, *Acanthopagrus schlegedi*. Nippon Suisan Gakkaishi 34, 920-924.

Toyama N. & Ogawa K.: Purification and properties of *Trichoderma viride* mycolytic enzyme. Hakko Kogaku Zasshi 46, 626-633.

Trujillo R.: Preparation of carboxymethylchitin. Carbohydr. Res. 7, 483-485.

Turner J.F. & Doczi J.: Isolation and purification of thyroglobulin. U.S. 3,368,940.

Unestam T.: Some properties of unpurified chitinase from the crayfish plague fungus *Aphanomyces astaci*. Physiol. Plant. 21, 137-147.

Weis-Fogh T.: Properties of resilin and elastomers in insects. U.S. Clearinghouse Fed. Sci. Techn. Inf. AD 666088.

Wieckowska E.: Determination of chitin on the basis of glucosamine content. Chem. Anal. 13, 1311-1317.

1969

Balassa L.L.: Wound healing accelerators. Ger. 1,906,155.

Banno T. & Yamano T.: Chitinase. Japan 01,180.

Bartnicki Garcia S.: & Lippman E.: Fungal morphogenesis: cell wall construction in *Mucor rouxii*. Science 165, 302-304.

Beaulaton J.: Cytochemical electron-microscopic study of the cuticular intima of the insect trachea. Action of chitinases and proteases on ultrafine tissue sections. Comp. Rend., Ser. D, 269, 2388-2391.

Benjaminson A.A.: Conjugates of chitinase with fluorescein isothiocyanate or lissamine rhodamine as specific stains for chitin in situ. Stain Technol. 44, 27-31.

Bernadet M.F.: Skin cosmetics from chitin. Fr. 1,552,076.

Beuzekom W.J.C. & Douma A.D.: Preparations for the destruction of Nematodes. Neth. 6,710,529.

Bittiger H., Huseman E. & Kuppel A.: Electron-microscope investigations on fibril formation. J. Polymer Sci. C-28, 45-56.

Blackwell J.: Structure of β-chitin or parallel chain systems of poly-β-(1→4)-N-acetyl-D--glucosamine. Biopolymers 7, 281-289.

Chet I. & Henis Y.: Effect of catechol and disodium EDTA on melanin content of hyphal and sclerotial walls of *Sclerotium rolfsii* and the role of melanin in the susceptibility of these walls to β-(1→3)-glucanase and chitinase. Soil Biol. Bioch. 1, 131-138.

Cohen J., Katz D. & Rosenberger R.F.: Temperature sensitive mutant of *Aspergillus nidulans* lacking amino sugars in its cell wall. Nature 224, 713-715.

Crompton M. & Polakis S.E.: The labeling of tissue components during the pharate adutl life of *Lucilia cuprina* after injection of ^{14}C-glucose. J. Insect Physiol. 15, 1323-1329.

Dahlquist F.W. & Raftery M.A.: Properties of contiguous binding subsites on lysozyme as determined by proton magnetic resonance spectroscopy. Bioch. 8, 713-725.

Dandrifosse G.: Action of some sugars on the secretion of chitinase. Arch. Int. Physiol. Bioch. 77, 639-648.

Dandrifosse G.: Metabolically dependent movement of the gastric chitinase across the cellular membrane. Bull. Cl. Sci. Acad. Roy. Belg. 55, 701-707.

1969 CNTD

Delachambre J.: Periodic acid Schiff positive reaction of chitin. Histochemie 20, 58–67.

Erickson B.W. & Blanquet R.S.: Occurrence of chitin in the spore wall of *Glugea weissembergi*. J. Invert. Pathol. 14, 358–364.

Fugono Y. & Yamano F.: Method for the preparation of chitinase. Japan 44–1180.

Furlan M. & Moroson H.: Effect of polycation on tumor cells in vivo. Rad. Prot. Sensitization Proc. Int. Symp. 2, 233–242.

Giles C.H. & Aghinotri V.G.: Monolayers of chitin. Chem. Ind. 754–755.

Goffinet G.: Electron microscope study of the organized structures of the constituents of conchiolin from *Nautilus pompilius* mother-of-pearl. Comp. Bioch. Physiol. 29, 277–282.

Hayashi K., Fujimoto N., Kugimiya M. & Funatsu M.: Enzyme–substrate complex of lysozyme with chitin derivatives. J. Bioch. 65, 401–405.

Ikan R.: Natural Products. Academic Press, New York.

Jeanloz R.W. & Balzs E.A.: The Amino Sugars. Academic Press, London.

Justus D.E. & Ivey M.H.: Chitinase activity in developmental stages of *Ascaris suum* and its inhibition by anti-body. J. Parasitol. 55, 472–476.

Kurup P.A., Leelamma S., Vijayammal P.L. & Vijayalekshmi N.R.: Nature of the cuticle of the scorpion *Heterometrus scaber*. Indian J. Exp. Biol. 7, 27–28.

Lestan P.: Chitin and chitinolysis in Ascaridae. Biol. 24, 601–606.

Maksimov V.I. & Mosin V.A.: Simple method for detecting and separating oligosaccharides, fragments of chitosan, after ion-exchange chromatography. Izv. Akad. Nauk SSSR Ser. Khim. 11, 2579–2581.

Marzotto A. & Galzigna L.: Enzymic hydrolysis of carboxymethylchitin by lysozyme. Hoppe–Seiler Z. Physiol. Chem. 350, 427–430.

Monreal J. & Reese E.T.: Chitinase of *Serratia marcescens*. Can. J. Microbiol. 15, 689–696.

Moreno R.E., Kanetsuna F. & Carbonell L.M.: Isolation of chitin and glucan from the cell wall of the yeast for of *Paracoccidioides brasiliensis*. Arch. Bioch. Biophys. 130, 212–217.

Moshi R.J.: Reconstituted tobacco sheet. U.S. 3,421,519.

Muzzarelli R.A.A. & Tubertini O.: Chitin and chitosan as chromatographic supports and adsorbents for collection of metal ions from organic and aqueous solutions and sea-water. Talanta 16, 1571–1579.

Neville A.C. & Luke B.M.: A two-system model for chitin protein complexes in insect cuticles. Tissue Cell 1, 689–707.

Noguchi J., Arato K. & Komai T.: Chitosan epichlorohydrin anion exchange resin with primary amine as absorption site. Kogyo Kagaku Zasshi 72, 796–799.

Peniston Q.P., Johnson E.L., Turrill C.N. & Hayes M.L.: New process for recovery of by-product from shell-fish waste. Eng. Bull. Purdue Univ. Eng. Ext. Serv. 135(1), 402–412.

Pfeiffer E., Jaeger K. & Reisener H.J.: Host-parasite metabolic relation between *Puccinia graminis tritici* and *Triticum vulgare* (wheat). Planta 85, 194–201.

Raftery M.A., Rand-Meir T., Dahlquist F.W., Parson S.M., Borders C.L., Wolcott R.G., Beranek W. & Jao L.: Separation of glycosaminoglycan saccharide and glycoside mixtures by gel filtration. Anal. Bioch. 30, 427–435.

Rudall K.M.: Chitin and its association with other molecules. J. Polymer Sci. C-28, 83–102.

Sannasi A.: Resilin in the cuticle of the click beetles. J. Ga. Entomol. Soc. 4, 31–32.

Sarko A. & Marchessault R.H.: Supermolecular structure of polysaccharides. J. Polymer Sci. C-28, 317–331.

Speck U. & Urich K.: Metabolic fate of nutrients in the crayfish *Orconectes limosus*. Incorporation of glucose-U-^{14}C and palmitate-U-^{14}C into body constituents. Z. Vergl. Physiol. 63, 395–404.

Spencer D.W. & Brewer P.G.: The distribution of copper, zinc and nickel in sea-water of the Gulf of Maine and the Sargasso Sea. Geoch. Cosmoch. Acta 33, 325–339.

Sundararajan P.R. & Rao V.S.R.: The conformation of α-chitin. Biopolymers 8, 305–312.

Takeda M. & Tomida T.: Chitin adsorbent in thin layer chromatography for separation of phenols and aminoacids. J. Shim. Univ. Fish. 18, 36–44.

Takeda M. & Tomida T.: Thin layer chromatography of saccharides obtained by degradation of chitin. 2. Separation of chitin oligosaccharides. Suisan Daigakku Kenkyu Hokoku, 143–153.

Tono T.: Determination of cow milk proteins by colloidal titration. Tottori Nogakkaiho 21, 81–86.

Toyama N.: Mycolytic enzyme of *Trichoderma viride*. 2. Hakko Kyokaishi 27, 24–29.

Troy F.A. & Koffler H.: The chemistry and molecular architecture of the cell walls of *Penicillium chrysogenum*. J. Biol. Chem. 244, 5563–5576.

Tsuji A., Kinoshita T. & Hoshino M.: Analytical chemical studies of amino sugars. 2. Determination of hexosamines using 3-methyl-2-benzothiazolone hydrazone hydrochloride. Chem. Pharm. Bull. 17, 1505–1510.

Ulmer D.D.: Effect of metal binding on the hydrogen-tritium exchange of canalbumin. Bioch. Biophys. Acta 181, 305–310.

Zabin B.A.: Method of making agarose. U.S. 3,423,396.

1970

Adachi S. & Fujita T.: A batik dyeing method. Japan 02,799.

Balassa L.L.: Chitin effect on healing of wounds. Ger. 1,906,159.

Benson D.M. & Baker R.: Rhizosphere competition in model soil system. Phytopathol. 60, 1058-1061.

Beran K., Rehacek J. & Seichertova O.: Infrared spectral analysis of the cell wall. The application of the method in the study of chitin in the cell wall of the yeast *Saccharomyces cerevisiae*. Acta Fac. Med. Univ. Brun. 37, 171-182.

Bull A.T.: Chemical composition of wild-type and mutant *Aspergillus nidulans* cell walls. Nature of polysaccharide and melanin constituents. J. Gen. Microbiol. 63, 75-94.

Bussers J.C. & Jeuniaux Ch.: Recherches préliminaires sur la présence de chitine dans les kystes de quelques Ciliés. J. Protozool. 17, suppl. 119, 32.

Capon B. & Foster R.L.: Preparation of chitin oligosaccharides. J. Chem. Soc. 1970-C, 1654.

Carbonell L.M., Kanetsuna F. & Gill F.: Chemical morphology of glucan and chitin in the cell wall of the yeast phase of *Paracoccidioides brasiliensis*. J. Bacteriol. 101, 636-642.

Charlemagne D. & Jolles P.: Inhibition by N-acetylglucosamine polymers of the lysing activity of lysozyme of various origins against *Micrococcus lysodeikticus* at pH 6.2. Comp. Rend. D-270, 2721-2723.

Cherkasov I.A. & Kravchenko N.A.: Chromatographic approach to the study of the binding of lysozyme to substrate. Bioch. Biophys. Acta 206, 289-294.

Cherkasov I.A. & Kravchenko N.A.: Determination of the heat of the lysozyme-chitin enzyme-substrate complex by a chromatographic method. Biokhimiya 35, 182-186.

Dandrifosse G.: Effect of certain inorganic ions on gastric chitinase secretion. Arch. Intl. Physiol. Bioch. 78, 339-346.

Deutsch V.: Ultrastructural characteristics of insect cells producing chitinase in vitro. Comp. Rend. D-270, 1491-1494.

Eka O.V.: *Epicoccum nigrumi*. 2. Influence of mineral salts on production of chitin and lipid. West Afr. J. Biol. Appl. Chem. 13, 13-17.

Friedman S.: Metabolism of carbohydrates in insects, in Chemical Zoology, Vol. 5, Florkin M. & Scheer B.T., Eds., Academic Press, New YorK.

Fujita T.: Chitosan. Japan 13,599.

Gilby A.R. & MacKellar J.W.: Compositions of the empty puparia of a blowfly. J. Insect Physiol. 16, 1517-1529.

Grove S.N. & Bracker C.E.: Protoplasmic organization of hyphal tips among fungi. J. Bacteriol. 104, 989-1009.

Ham J.T. & Williams D.G.: Cellobiose structure. Acta Crystall. B-26, 1373.

Hara S., Nakagawa Y. & Matsushima Y.: Substrate specificity of egg white lysozyme. Mode of the enzymic attack on partially O-hydroxyethylated chitin. J. Bioch. 68, 53-62.

Hohnke L. & Scheer B.T.: Carbohydrate metabolism in crustacea, in Chemical Zoology, Vol. 5, Florkin M. & Scheer B.T., Eds., Academic Press, New York.

Hunt. S.: Polysaccharide Protein Complexes in Invertebrates. Academic Press, New York.

Iten W.: Function of hydrolytic enzymes in the autolysis of *Coprinus*. Ber. Schweitz. Bot. Ges. 79, 175-198.

Iten W. & Matile P.: Role of chitinase and other lysosomal enzymes of *Coprinus lagopus* in the autolysis of fruiting bodies. J. Gen. Microbiol. 61, 301-309.

Janda V. & Socha R.: Decomposition and transport of nutrients in *Dixippus morosus* in relation to growth and metamorphosis. J. Insect Physiol. 16, 1851-1862.

Jansons V.K. & Nickerson W.Y.: Chemical composition of chlamidospores of *Candida albicans*. J. Bacteriol. 104, 922-932.

Kakiki K., Hori M., Abe H. & Misato T.: A simple procedure for the preparation of UDP-N-acetylglucosamine-^{14}C. Bioch. Biophys. Res. Commun. 39, 718-722.

Kanetsuna F. & Carbonell L.M.: Cell wall glucans of the yeast and mycelial forms of *Paracoccidioides brasiliensis*. J. Bacteriol. 101, 675-680.

Katz D. & Rosenberger R.F.: A mutation in *Aspergillus nidulans* producing hyphal walls which lack chitin. Bioch. Biophys. Acta 208, 452-460.

Krishnan G.: Chemical nature of the cuticle and its mode of hardening in *Eoperipatus weldoni*. Acta Histochem. 37, 1-17.

Kuznetsov V.D. & Yangulova I.V.: Use of a medium containing chitin for isolation of Actinomycetes from soil, and their quantitative enumeration. Mikrobiol. 39, 902-906.

Lagunov L.L., Kryuchkova M.I. & Nikolaeva N.E.: Food product based on planktonic crustacean protein, especially from krill. Fr. 2,024,710.

Landureau J.C. & Jolles P.: Lytic enzyme produced in vitro by insect cells: lysozyme or chitinase. Nature 225, 968-969.

Leopold J. & Samsinakova A.: Quantitative estimation of chitinase and several other enzymes in the fungus *Beauveria bassiana*. J. Invert. Pathol. 15, 34-42.

Lin C.C. & Aronson J.M.: Chitin and cellulose in the cell walls of the oomycete *Apodachlya* sp. Arch. Mikrobiol. 72, 111-114.

Linskens H.F., Linder R., Salden M., Havez R., Randoux A., Laniez D. & Coustaut D.: Glycoproteins and glycan hydrolases during pollination of autoincompatible hydrid petunias. Bull. Soc. Pharm. Lille 1, 7-22.

1970 CNTD

Liston J. & Shenouda S.Y.K.: Decomposition of organic materials by marine benthic bacteria. U.S. Nat. Tech. Inf. Serv. 728468.

Mahadevan P.R. & Mahadkar U.R.: Major constituents of the conidial wall of *Neurospora crassa*. Indian J. Exp. Biol. 8, 207-210.

Marks E.P. & Leopold R.A.: Cockroach leg regeneration: effects of ecdysterone in vitro. Science 167, 61-62.

Martin J.H.: The possible transport of trace metals via molted copepods exoskeletons. Limnol. Oceanogr. 15, 756-761.

McMurrough I. & Bartnicki-Garcia S.: Chitin synthesis in *Mucor rouxii*. Bioch. J. 1970, 119.

Moskowitz R.W., Schwartz H.J., Michel B., Katnoff O.D. & Astrup T.: Generation of kinin-like agents by chondroitin sulfate, heparin, chitin sulfate and human articular cartilage. Possible pathophysiologic implications. J. Lab. Clin. Med. 76, 790-798.

Muzzarelli R.A.A.: Uptake of nitrosyl ruthenium-106 on chitin and chitosan from waste solutions and polluted sea-water. Water Res. 4, 451-455.

Muzzarelli R.A.A., Raith G. & Tubertini O.: Separation of trace elements from sea-water, brine and sodium and magnesium salt solutions by chromatography on chitosan. J. Chromatogr. 47, 414-420.

Muzzarelli R.A.A. & Tubertini O.: Purification of thallium(I) nitrate by column chromatography on chitosan. Microch. Acta 5, 892-899.

Nagasawa K., Watanabe H. & Ogamo A.: Ion-exchange chromatography of nucleic acid constituents on chitosan-impregnated cellulose thin layers. J. CHromatogr. 47, 408-413.

Nagasawa K., Kaneko F. & Ogamo A.: Ion-exchange chromatography of nucleic acid constituents on chitosan formate impregnated papers. Seikagaku 42, 255-259.

Nud'ga L.A., Plisko E.A. & Danilov S.N.: Production of chitosan and study of its fractionation. Zhur. Obs. Khim. 41, 2555-2558.

Ohta N., Kakiki K. & Misato T.: Mode of action of polyoxin D. 2. Effect of polyoxin D on the synthesis of fungal cell wall. Agr. Biol. Chem. 34, 1224-1234.

Okuyama T.: Chitin derivatives. Tampakushitsu Kakusan Koso 15, 47-53.

Parvathy K.: Blood sugars in relation to chitin synthesis during cuticle formation in *Emerita asiatica*. Marine Biol. 5, 108-112.

Peniston Q.P. & Johnson E.L.: Treating an aqueous medium with chitosan and derivatives of chitin to remove an impurity. U.S. 3,533,940.

Piavaux A. & Magis N.: Données complémentaires sur la localisation de la chitine dans les enveloppes des oeufs de rotifères. Ann. Soc. Roy. Zool. Belg. 100, 49-59.

Prudden J.F., Migel P., Hanson P., Friedrich L. & Balassa L.: Discovery of a potent pure chemical wound healing accelerator. Am. J. Surg. 119, 560-564.

Rees D.A. & Skerret R.J.: Conformational analysis of polysaccharides. 4. Long-range contacts in some β-glucans by model building in the compute: and the influence of oligosaccharide conformation on optical rotation. J. Chem. Soc. B-1, 189-193.

Sannasi A. & Hermann H.R.: Chitin in the Cephalochordata *Branchiostoma floridae (amphioxus)*. Experientia, 26, 351-352.

Sawasaki T. & Shimokawa H.: Enzymic lysis of native *Candida* cells in the presence of chemical agents and antibiotics. Agr. Biol. Chem. 34, 243-247.

Shrivastava S.C.: Cuticular components of common indian arachnids and myriapods. Experientia 26, 1028-1029.

Skujins J.J., Pukite A.H. & McLaren A.D.: Chitinase of *Streptomyces* sp. purification and properties. Enzymologia 39, 353-370.

Smith M.J. e Dehnel P.A.: The chemical and enzymatic analysis of the tunic of the ascidian *Halocynthia aurantium*. Comp. Bioch. Phys. 35, 17-30.

Soeder P.O., Nord C.E., Lundblad G. & Kjellman O.: Investigation on lysozyme, protease and hyaluronidase activity in extracts from human leucocytes. Acta. Chem. Scand. 24, 129-136.

Sundara Rajulu G.: Nature of the carbohydrases in a centipede, *Scolopendra heros*, together with observations on hydrogen ion concentration of the alimentary tract. J. Anim. Morph. Physiol. 17, 56-64.

Tanaka H., Ogasawara N., Nakajima T. & Tamari K.: Cell walls of *Piricularia oryzae*. 1. Selective enzymolysis by wall-lytic enzymes of *Bacillus circulans* WL 12. J. Gen. Appl. Microbiol. 16, 39-60.

Takeda M.: Isomerization of D-glucose to D-fructose by a strongly polysaccharide 6-O-(2--hydroxyethyl)-chitosan (trimethylammonium hydroxide). Suisan Daigakko Kenkyu Hokoku 18, 277-286.

Tsai C.S.: Determination of the degree of polymerization of N-acetylchitooligoses by chromatographic methods. Anal. Bioch. 36, 114-122.

Tsai C.S. & Matsumoto K.: Acetamido groups in the binding and catalysis of synthetic substances by lysozyme. Bioch. Biophys. Res. Commun. 39, 684-689.

Wang M.C. & Bartnicki-Garica S.: Structure and composition of walls of the yeast form of *Verticillium albo-atrum*. J. Gen. Microbiol. 64, 41-54.

Weis-Fogh T.: Structure and formation of insect cuticle. Symp. Roy. Ent. Soc. 5, 165-185.

Zvyaginstsev D.G.: Selective adsorption of soil microorganisms by ion-exchange resins. Biol. Nauk. 1, 81-86.

1971

Abeles F.B., Bosshart R.P., Forrence L.E. & Habig W.: Preparation and purification of glucanase and chitinase from bean leaves. Plant Physiol. 47, 129-134.

Allan G.G., Johnson P.G., Lay Y.Z. & Sarkanen K.V.: Marine polymers. 1. New procedures for the fractionation of agar. Carbohydr. Res. 17, 234-236.

Allan G.G., Johnson P.G., Lay Y.Z. & Sarkanen K.V.: Solubility and reactivity of marine polymers in dimethylformamide + dinitrogen tetroxide. Chem. Ind. 1971, 127.

Balassa L.L.: Chitin-containing wound healing compositions. Brit. 1,252,373.

Bernier I., Van Leemputten E., Horisberger M., Bush D.A. & Jolles P.: Turnip lysozyme. FEBS Lett. 14, 100-104.

Bradleigh S. & Law P.K.: Cuticular composition and DDT resistance in the tobacco budworm. J. Econ. Entomol. 64, 1387-1390.

Bryan W.P. & Linder P.W.: Chitotriose, a tritium exchange probe of the active cleft of lysozyme.. J. Am. Chem. Soc. 93, 3061-3062.

Cabib E. & Bowers B.: Chitin and yeast budding. Localization of chitin in yeast bud scars. J. Biol. Chem. 246, 152-159.

Cabib E. & Farkas V.: The control of morphogenesis: an enzymatic mechanism for the initiation of septum formation in yeast. Proc. Nat. Acad. Sci. 68, 2052-2056.

Cabib E. & Keller F.A.: Chitin and yeast budding. Allosteric inhibition of chitin synthetase by a heat stable protein from yeast. J. Biol. Chem. 246, 167-173.

Carlson D.M., Swenson A. & Roseman S.: 2-amino-2-deoxy-α-D-galactopyranosyl dihydrogen phosphate. Bioch. Prep. 13, 3-7.

Cerezo A.S.: Proton absorption in the NMR spectra of N-acylglycosylamines, N-acetyl-2-amino--2-deoxy-D-glycoses and 1,1-bis(acylamido)-1-deoxy-alditols. Chem. Ind. 1971, 96-97.

Cherniak R.: 2-amino-2-deoxy-D-glucopyranoside phosphate. Bioch. Prep. 13, 7-13.

Colin D.A. & Perez G.: Comparative studies on the chitinolytic enzymic activity in the digestive mucosa of marine telosts. Ann. Inst. Michel Pacha 4, 1-9.

Cowey C.B. & Foster J.R.M.: Essential aminoacid requirements of the prawn *Palaemon serratus*. Growth of prawns on diets containing proteins of different aminoacid composition. Marine Biol. 10, 77-81.

Crowe J.H., Newell I.M. & Thomson W.W.: Fine structure and chemical composition of the cuticle of the tardigrade *Macrobiotus areolatus*. J. Microsc. 11, 107-120.

Domer J.E.: Monosaccharide and chitin content of cell walls of *Histoplasma capsulatum* and *Blastomyces dermatitidis*. J. Bacteriol. 107, 870-877.

Embery G., Lloyd A.G. & Fowler L.J.: Heparin degradation. 2. Metabolic fate of the potassium salts of $|^{35}S|$-fulfoamino heparin, $|^{35}S|$-sulfoamino chitosan 2-deoxy-2-$|^{35}S|$-sulfoamino-D--glucose and $|^{35}S|$-sulfoamino-L-serine in the rat. Bioch. Pharmacol. 20, 649-658.

Fossato V.U.: Chemical composition of the zooplancton collected in the Gulf of Venice. Arch. Oceanogr. Limnol. 7, 19-26.

Gardner K.H. & Blackwell J.: Substructure of crystalline cellulose and chitin microfibrils. J. Polymer Sci. C-36, 327-340.

Guire P.E.: Fractionation of oligosaccharides by polyacrylamide gel filtration. Proc. Oklahoma Acad. Sci. 51, 63-65.

Haga H.: Method to prepare an anion-exchanger. Japan 46-39,322.

Hohnke L.A.: Enzymes of chitin metabolism in the decapod *Hemigrapsus nudus*. Comp. Bioch. Physiol. B-40, 757-779.

Holan Z., Beran K., Prochazkova V. & Baldrian J.: Gas-liquid chromatographic determination of the content of N-acetylglucosamine and its polymer in the cell wall of yeast. Yeast Models Sci. Tech. Proc. Spec. Int. Symp. 71, 239-260.

Hori M., Kakiki K., Suzuchi S. & Misato T.: Mode of action of polyoxins. 3. Relation of polyoxin structure to chitin synthetase inhibition. Agr. Biol. Chem. 35, 1280-1291.

Hormung D.E. & Stevanson J.R.: Changes in the rate of chitin synthesis during the crayfish molting cycle. Comp. Bioch. Physiol. B-40, 341-346.

Ikeda T.: Changes in respiration rate and in composition of organic matter in *Calanus cristatus* during stravation. Hokkaido Daigaku Suisan Gakubu Kenkyu Iho 21, 280-298.

Jaskoski B.J. & Butler V.L.: *Rhipicephalus sanguineus*. Aminoacid composition of egg shells. Exp. Parasitol. 30, 400-406.

Jeuniaux Ch.: Chitinous structures, in Comprehensive Biochemistry, Vol. 26-C, Florkin M. & Stotz E.H., Eds., Elsevier, Amsterdam.

Jeuniaux Ch.: On some biochemical aspects of regressive evolution in animals, in Biochemical Evolution and the Origin of Life, Shoffeniels E., Ed., North-Holland, Amsterdam.

Kanetsuna F. & Carbonell L.M.: Cell wall composition of the yeast-like and mycelial form of *Blastomyces dermatitidis*. J. Bacteriol. 106, 946-948.

Katz D. & Rosenberger R.F.: Hyphal wall synthesis in *Aspergillus nidulans*. Effect of protein synthesis inhibition and osmotic shock on chitin insertion and morphogenesis. J. Bacteriol. 108, 184-190.

Katz D. & Rosenberger R.F.: Lysis of an *Aspergillus nidulans* mutant blocked in chitin synthesis and its relation to wall assembly and wall metabolism. Arch. Mikrobiol 80, 284-292.

Keller F.A. & Cabib E.: Chitin and yeast budding. Properties of chitin synthetase from *Saccharomyces carlsbergensis*. J. Biol. Chem. 246, 160-166.

1971 CNTD

Lejohn H.B.: Enzyme regulation, lysine pathway and cell wall structure as indicators of major lines of evolution in fungi. Nature 231, 5299.

Lipke H. & Geoghegan T.: Composition of peptidochitodextrins from sarcophagid puparial cases. Bioch. J. 125, 703-716.

Lloyd A.G., Embery G. & Fowler L.J.: Heparin degradation. 1. Preparation of $|^{35}S|$-sulfamate derivatives for studies on heparin-degrading enzymes of mammalian origin. Bioch. Pharmacol. 20, 637-648.

Maksimov V.I. & Mosin V.A.: Separating oligosaccharide fragments of chitosan. USSR 319,607.

McMurrough I. & Bartnicki-Garcia S.: Properties of a particulate chitin synthetase from *Mucor rouxii*. J. Biol. Chem. 246, 4008-4016.

McMurrough I., Flores-Carreon A. & Bartnicki-Garcia S.: Pathway of chitin synthesis and cellular localization of chitin synthetase in *Mucor rouxii*. J. Biol. Chem. 246, 3999-4007.

Muzzarelli R.A.A.: Selective collection of trace metal ions by precipitation of chitosan and new derivatives of chitosan. Anal. Chim. Acta 54, 133-142.

Muzzarelli R.A.A.: Applications of natural polymers in marine ecology. Rev. Int. Océanogr. Méd. 21, 93-103.

Muzzarelli R.A.A.: Recovery of transition metal ions from sea-water and from solutions, using chelating polymers. Symp. Anal. Chem. Water, Bormio.

Muzzarelli R.A.A.: Collection of trace metals by precipitation of chitosan. Proc. VI Int. Symp. Microtechn., Graz, B-34, 187-192.

Muzzarelli R.A.A.: Applications of chitosan as a chelating polymer for collection of metal ions from natural waters and from industrial solutions. Proc. KEM-TEK 2 Congress, Copenhagen, 3, V/3.

Muzzarelli R.A.A. & Isolati A.: Methyl mercury acetate removal from waters by chromatography on chelating polymers. Water, Air & Soil Poll. 1, 65-71.

Muzzarelli R.A.A. & Sipos L.: Chitosan for the collection from sea-water of naturally occurring zinc, cadmium, lead and copper. Talanta, 18, 853-858.

Nagasawa K., Tohira Y., Inoue Y. & Tanoura N.: Reactions between carbohydrates and sulfuric acid. 1. Depolymerization and sulfonation of polysaccharides by sulfuric acid. Carbohydr. Res. 18, 95-102.

Nagasawa K., Watanabe H. & Ogamo A.: Ion-exchange chromatography of dinucleoside-3'-5'--phosphates on chitosan-impregnated cellulose thin-layers. J. Chromatogr. 56, 378-381.

Nanba H. & Kuroda H.: Fungicides. 9. Chemical structure of chitin-like substances of cell wall of *Cochliobotus miyabeanus*. Chem. Pharm. Bull. 19, 1402-1408.

Oliver G., Pesce de Ruiz H., Aida A., Pons J.J. & Manca M.C.: Chemical composition of *Aspergillus phoenicis* cell walls. Arch. Bioquim. Quim. Farm. 17, 91-100.

Parvathy K.: Blood sugar metabolism during molting in the isopod crustacean *Ligia exotica*. Marine Biol. 9, 323-327.

Paszkiewicz-Gadek A. & Niedzwiecka J.: Potentiometric determination of chitin. Chem. Anal. 16, 443-444.

Petrushko G.M. & Kalyuzhnyi M.Y.: Chitin of yeat-like organisms of the *Candida* genus. Prikl. Biokhim. Mikrobiol. 7, 637-641.

Petrushko G.M. & Kalyuzhnyi M.Y.: Polysaccharides of *Candida* yeast-like organisms determining the lyophilic and lyophobic properties of cell membranes. Vses. Nauk.-Issled. Inst. Gidraliza Rast. Mater. 21, 110-114.

Porter W.R.: Chitin antigens for rendering infected dogs tick-free. U.S. 3,590,126.

Raymont J.E.G., Srinivasagam R.T. & Raymont K.K.B.: Biochemical studies on marine zooplancton. 8. *Meganyctiphanes norvegica*. Deep-Sea Res. Oceanogr. Abs. 18, 1167-1178.

Raymont J.E.G., Srinivasagam R.T. & Raymont K.K.B.: Biochemical studies on marine zooplancton. 9. The biochemical composition of *Euphausia superba*. J. Mar. Biol. Ass. U.K. 51, 581-588.

Ride J.P. & Drysdale R.B.: A chemical method for estimating *Fusarium oxysporum f. lycopersici* in infected tomato plants. Physiol. Plant Pathol. 1, 409-420.

Sahu C.R. & Mitra A.K.: Fundamental studies on the interaction of transition metal ions with carbohydrate derivatives as ligands. J. Indian Chem. Soc. 48, 795-798.

Scogin R. & Miller C.E.: Chemical studies on the genus *Chytriomyces*. J. Eliska Mitchell Sci. Soc. 87, 31-35.

Sirica A.E. & Woodman R.J.: Selective aggregation of L-1210 leukemia cells by the polycation chitosan. J. Nat. Cancer Inst. 47, 377-388.

Smirnoff W.A.: Effect of chitinase on the action of *Bacillus thuringiensis*. Can. Entomol. 103, 1829-1831.

Sneh B., Katan J. & Henis Y.: Mode of inhibition of *Rhizoctonia solani* in chitin-amended soil. Phytopathol. 61, 1113-1117.

Sundara Rajulu G.: Electron microscopic study of the ultrastructure of the peritrophic embrane of a chilopod, *Ethmostigmus spinosus* together with observations on its chemical composition. Curr. Sci. 41, 134-135.

Tiunova N.A., Feniksova R.V., Kobzeva N.Y., Kuznetsov V.D., Kostycheva L.I., Rodionova N.A., Itomlesnskite I. & Pirieva D.G.: Hydrolytic enzymes of *Actynomycetes*. Prikl. Biokhim. Microbiol. 7, 456-461.

Vinson S.B. & Law P.K.: Cuticular composition and DDT resistance of the tobacco budworm. J. Econ. Entomol. 64, 1387-1390.

1971 CNTD

Wadstrom T.: Chitinase activity and substrate specificity of endo-β-N-acetylglucosaminidase of *Staphilococcus aureus*. Acta Chem. Scand. 25, 1807-1802.

Wahlquist B.T.: Effect of soil moisture and soil temperature on the decomposition of chitin and chitin containing complexes in soil. Univ. Microfilms Ann Harbor 72-2136.

Weed L.J.: Additive color film for the silver halide diffusion process. Ger. 2,052,674.

Whistler R.J. & Kosik M.: Anticoagulant activity of oxidized and N- and O-sulfated chitosan. Arch. Bioch. Biophys. 142, 106-110.

Williams S.T. & Mayfield C.I.: Ecology of Actinomycetes in soil. 3. Bahavior of neutrophilic *Streptomycetes* in acid soil. Soil Biol. Bioch. 3, 197-208.

Winicur S.: Motility and biochemistry of cilia. Chitinase activity during *Drosophila* development. Diss. Abs. Int. B-32, 85.

Yamamoto S., Nagasaki S. & Lampen J.O.: Microbial enzymes active in hydrolyzing yeast cell wall. 2. Lysis of mannan layer by some enzyme preparations and reducing agents. Hakko Kagaku Zasshi 49, 338-342.

Young S.D.: Organic material from scleractinian coral skeletons. 1. Variation in composition between several species. Comp. Bioch. Physiol. B-40, 113-120.

Zehavi U., Jeanloz R.W., Horton D. & Philips K.D.: Oligosaccharides derived from chitin. Bioch. Prep. 13, 14-18.

Zonneveld B.J.M.: Biochemical analysis of the cell wall of *Aspergillus nidulans*. Bioch. Biophys. Acta 249, 506-514.

1972

Allan G.G., Crosby G.D., Lee J.H., Miller M.L. & Reif W.M.: New bonding systems for paper. Proc. Symp. Man-made Polymers Papermaking, Helsinki, 1972, 85-96.

Arnott S. & Scott W.E.: Accurate x-ray diffraction analysis of fibrous polysaccharides containing pyranose rings. 1. Linked-atom approach. J. Chem. Soc. Perkin T. 2, 324-335.

Bader G., Macak J. & Rohde E.: Differences in histochemical oxidation of polysaccharides in pathogenic fungi. Conclusions in zonal structure of the fungal cell wall. Zentr. Bakteriol. Parasitenk. Infektionskr. Hyg., Abt. 2, 127, 13-24.

Balassa L.L.: Use of chitin for promoting wound healing. U.S. 3,632,754.

Beran K., Holan Z. & Baldrian J.: Chitin-glucan complex in *Saccharomyces cerevisiae*. 1. Infra red and x-ray observations. Folia Microbiol. 17, 322-330.

Berkeley R.C.W., Brewer S.J. & Ortiz J.M.: Preparation of 2-acetamido-2-deoxy-β-D-glucose oligosaccharides from acid hydrolyzates of chitin by electrolytic desalting and esclusion chromatography. Anal. Bioch. 46, 687-689.

Blanquet R.S.: Structural and chemical aspects of the podocyst cuticle of the scyphozoan medusa *Chrysaora quinquecirrha*. Biol. Bull. 142, 1-10.

Bondietti E., Martin J.P. & Haider K.: Stabilization of amino sugar units in humic-type polymers. Soil Sci. Soc. Amer., Proc., 36, 597-602.

Bongard S.A., Pruglo N.V. & Spasokukotskii N.S.: Advantages of using polymeric bases as fixatives in hydrotype. usp. Nauk. Fotogr. 16, 51-58.

Bridgeford D.J., Turbak A.F. & Burke N.I.: Antisoil coating-containing shaped articles. U.S. 3,649,346.

Bridgeford D.J., Turbak A.F. & Burke N.I.: Sulfonated polymeric alcohols. U.S. 3,689,466.

Cabib E.: in Methods in Enzymology, Vol. 28, Academic Press, New York.

Clever U. & Bultmann H.: Chromosomal control of foot pad development in *Sarcophaga bullata*. 3. Requirements of RNA and protein syntheis for cuticle formation and tanning. Cell Diff. 1, 37-42.

Colin D.A.: Relation between the type of natural nutrients and chitinolytic activity in the digestive tract of marine telosts. Comp. Rend. Soc. Biol. 166, 95-98.

Cox R.A. & Best G.K.: Cell wall composition of two strains of *Blastomyces dermatitidis*, exhibiting differences in virulence for mice. Infect. Immunity 5, 449-453.

Debruyn F.E.: Lichtempfindliches fotographisches Aufzeichnungsmaterial. Ger. 2,052,648.

Domsch K.H.: Interactions of soil microbes and pesticides. Symp. Biol., Hung. 11, 337-347.

Elyakova L.A.: Distribution of cellulases and chitinases in marine invertebrates. Comp. Bioch. Physiol. B-43, 67-70.

Foster W.A.: Use of water soluble polymers as flocculants in papermaking. Am. Chem. Soc. Div. Org. Coating Plat. Chem. Pap. 32, 210-216.

Fujita T.: Recovery of proteins. Japan 01,633.

Gooday G.W.: The role of chitin synthetase in the elongation of fruit bodies of *Coprinus cinereus*. J. Gen. Microbiol. 73, 31.

Gooday G.W.: Effect of polyoxin D on morphogenesis of *Coprinus cinereus*. Bioch. J. 129, 17-18.

Goodwin T.W. & Hercer E.I.: Introduction to Plant Biochemistry. Pergamon Press, Oxford.

Gwinn J.F.: Role of acetylglucosamine in chitin synthesis in crayfish. Diss. Abs. Int. B-33, 885-886.

Haberlin R.J.: Photographic film units and processing. U.S. 3,674,482.

Hamaguchi K. & Ikeda K.: Applications of circular dichroic spectroscopy to the study of the interaction between lysozyme and small molecules. Ann. Rep. Biol. Works, Fac. Sci Osaka Univ. 19, 1-29.

1972 CNTD

Hara S. & Matsushima Y.: Simultaneous N-acetylation and O-trimethylsilylation of hexosamines for gas-liquid chromatography. J. Bioch. 71, 907-909.

Heneine I.F. & Kimmel J.R.: Sulfhydryl groups of papaya lysozyme and their relation to biological activity. J. Biol. Chem. 247, 6589-6596.

Hylleberg K.J.: Carbohydrases of some marine invertebrates with notes on their food and on the natural occurrence of the carbohydrates studied. Marine Biol. 14, 130-142.

Hunt S.: Scleroprotein and chitin in the exoskeleton of the ectoproct *Flustra foliacea*. Comp. Bioch. Physiol. B-43, 571-577.

Jensen H.B. & Kleppe K.: T-4 lysozyme. Affinity for chitin and use of chitin in the purification of the enzyme. Eur. J. Bioch. 26, 305-312.

Kim J. & Zobell C.E.: Agarase, amylase, cellulase and chitinase activity at deep-sea pressure. Nippon Kaiyo Gakkai-shi 28, 131-137.

Kitazima Y., Banno Y., Noguchi T., Nozawa Y. & Ito Y.: Effects of chemical modification of structural polymer upon the cell wall integrity of *Tricophyton*. Arch. Bioch. Biophys. 152, 811-820.

Kumar S. & Hansen P.M.T.: New reaction mixture for spectrophotometric determination of N-acetylhexosamines. Anal. Chem. 44, 398-400.

Lipke H. & Strout V.: Peptidochitodextrins of sarcophagid cuticle. 2. Linkage region. Is. J. Entomol. 7, 117-128.

Masri M.S. & Friedman M.: Mercury uptake by polyamine-carbohydrates. Environ. Sci. Technol. 6, 745-746.

Millet F. & Raftery M.A.: A ^{19}F nuclear magnetic resonance study of the binding of trifluoro-acetylglucosamine oligomers to lysozyme. Bioch. 11, 1639-1643.

Muzzarelli M.G.: Chitin and chitosan as chromatographic supports and adsorbents for collection of metal ions from organic and aqueous solutions and sea water. U.S. 3,635,818.

Muzzarelli R.A.A., Ferrero A. & Pizzoli M.: Light-scattering, x-ray diffraction, elemental analysis and infrared spectrophotometry characterization of chitosan, a chelating polymer. Talanta, 19, 1222-1226.

Muzzarelli R.A.A. & Marinelli M.: Use of chitosan to monitor and prevent sea water pollution by metals. Inquinamento 14(3), 27-34.

Muzzarelli R.A.A. & Marinelli M.: Use of chitosan to monitor and prevent sea water pollution by metals. Inquinamento 14(4), 29-35.

Muzzarelli R.A.A., Rocchetti R. & Marangio G.: Separation of zirconium, niobium, cerium and ruthenium for the determination of cesium in nuclear fuel solutions. J. Radioanal. Chem. 10, 17-26.

Muzzarelli R.A.A. & Spalla B.: Removal of cyanide and phosphate traces from brines and sea water on metal ion derivatives of chitosan. J. Radioanal. Chem. 10, 27-35.

Muzzarelli R.A.A. & Tubertini O.: Radiation resistance of chitin and chitosan applied in the chromatography of radioactive solutions. J. Radioanal. Chem. 12, 431-440.

Nagasawa K. & Tanoura N.: Reaction between carbohydrates and sulfuric acid. 3. Depolymerization and sulfation of chitosan by sulfuric acid. Chem. Pharm. Bull. 20, 157-162.

Nakabayashi T. & Makita T.: Tannins of fruits and vegetables. 11. Adsorption of tannins on chitin. Nippon Shokuhin Kogyo Gakkai-shi. 19, 111-119.

Nickerson W.J.: Decomposition of naturally occurring organic polymers. Org. Compounds Aquatic Environ., Rudolfs Res. Conf. 5th, 599-609.

Nord C.E. & Wadstrom T.: Chitinase activity and substrate specificity of three bacteriolytic endo-β-N-acetylmuramidases and one endo-β-N-acetylglucosaminidase. Acta Chem. Scand. 26, 653-660.

Otness J.S. & Medcalf D.G.: Chemical and physical characterization of barnacle cement. Comp. Bioch. Physiol. B-43, 443-449.

Pariser E.R. & Boch S.: Chitin and chitin derivatives. An annotated bibliography of selected publications from 1965 through 1971. MIT Rep. MITSG 73-2.

Peters W.: Occurrence of chitin in Mollusca. Comp. Bioch. Physiol. 41, 541-550.

Piavaux A.: β-Polysaccharidases of the digestive tract of *Rana temporaria* during growth and metamorphosis. Arch. Int. Physiol. Bioch. 80, 153-160.

Pigott G.M.: Total utilization concept in fish processing. Meeting Ass. Sea Grant Inst., Houston, 1972.

Plisko E.A., Nud'ga L.A. & Danilov S.N.: Polyampholytes. USSR 325,234.

Prasad N. & Ramakrishnan C.: Structure and conformation of chitins. 1. Mathematics of the rigid body refinement procedure as applied to the structure of α-chitin. Indian J. Pure Appl. Phys. 10, 501-505.

Pruglo N.V., Bongard S.A. & Spasokukotskii N.S.: Determination of the relative fixing strength of acid dyes in gelatin layers with polymer bases. Zhur. Nauk. Prikl. Fotogr. Kinematogr. 17, 222-223.

Ramakrishnan C. & Prasad N.: Rigid body refinement and conformation of α-chitin. Bioch. Biophys. Acta 261, 123-135.

Rao K.K. & Modi V.V.: Biochemical changes in biotin-deficient *Aspergillus nidulans*. Indian J. Exp. Biol. 10, 385-388.

Reisert P.S.: Studies of the chitinase system of *Chytriomyces hyalinus* using a ^{14}C-chitin assay. Mycologia 64, 288-297.

1972 CNTD

Ride J.P. & Drisdale R.B.: A rapid method for the chemical estimation of filamentous fungi in plant tissues. Physiol. Plant Pathol. 2, 7-15.

Saxena S.C. & Sarin K.: Chitinase in the alimentary tract of the lesser mealworm *Alphitobius diaperinus*. Appl. Entomol. zool. 7, 94.

Sneh B. & Henis Y.: Production of antifungal substances active against *Rhizoctonia solani* in chitin-amended soil. Phytopathol. 62, 595-600.

Schippers B. & De Weyer W.M.M.M.: Chlamydospore formation and lysis of macroconidia of *Fusarium solani* in chitin-amended soil. Neth. J. Plant Pathol. 78, 45-54.

Speck U. & Urich K.: Absorption of the old exoskeleton before molt in the crayfish *Orconectes limosus*. Metabolic fate of liberated N-acetylglucosamine. J. Comp. Physiol. 78, 210-220.

Stevenson J.R.: Changing activities of the crustacean epidermis during the molting cycle. Am. Zool. 12, 373-380.

Sundarrai N. & Bhat J.V.: Breakdown of chitin in *Cytophaga johnsonii*. Arch. Mikrobiol. 85, 159-167.

Surholt B. & Zebe E.: In vivo studies on chitin synthesis in the migratory locust. J. Comp. Physiol. 78, 75-82.

Takeda M. & Tomida T.: Chitin adsorbent in thin layer chromatography for separation of nucleic acid derivatives. J. Shim. Univ. Fish. 20, 107-114.

Toei K. & Kawada K.: Fundamental study on colloid titration. 1. Colloid titration reagents. Bunseki Kagaku 21, 1510-1515.

Van Daalen J.J., Meltzes J., Mulder R. & Wellinga K.: A selective insecticide with a novel mode of action. Naturwissens. 7. 312-313.

Vanlerberghe G. & Sebag H.: Chitosan derivatives. Ger. 2,222,733.

Voss-Foucart M.F. & Jeuniaux Ch.: Lack of chitin in a sample of Ordovician chitinozoa. J. Paleontol. 46, 769-770.

Young R.A., Sarkanen K.V., Johnson P.G. & Allan G.G.: Marine plant polymers. 3. A kinetic analysis of the alkaline degradation of polysaccharides with specific reference to (1→3)-β-D-glucans. Carbohydr. Res. 21, 111-112.

Ward K.A. & Fairbairn D.: Chitinase in developing eggs in *Ascaris suum*. J. Parasitol. 58, 546-549.

Wessels J.G.H., Kreger D.R., Marchant R., Regensburg B.A. & De Vries O.M.H.: Chemical and morphological characterization of the hyphal surface of the basidiomycete *Schizophyllum commune*. Bioch. Biophys. Acta 273, 346-358.

1973

Akagane K. & Allan G.G.: Clarification of polluted water from paper mills with combination of cationic polymers and alum. Shikizai Kyokaishi 46, 365-369.

Akagane K. & Allan G.G.: Antifouling polymers for controlling sea-weeds. Shikizai Kyokaishi, 46, 370-375.

Andersen S.O.: Comparison between the sclerotization of adult and larval cuticle in *Schistocerca gregaria*. J. Insect Physiol. 19, 1603-1614.

Balassa L.L.: Process for facilitating wound healing with N-acetylated, partially depolymerized chitin materials. U.S. 3,914,413.

Bongard S.A. & Pruglo S.A.: Effect of fixations on the storage of colored hydrotype images. Zhur. Nauk. Prikl. Fotogr. Kinematogr. 18, 132-134.

Brimacombe J.S.: Carbohydrate Chemistry, Vol. 6, The Chemical Society, London.

Bussers J.C. & Jeuniaux Ch.: Chitinous cuticle and systematic position of Tardigrada. Bioch. Syst. 1, 77-78.

Bussers J.C. & Jeuniaux Ch.: Structure and composition of *Macrobiotus* sp. and *Milnesium tardigradum*. Ann. Soc. Roy. Zool. Belg. 103, 271-279.

Cabib E., Farkas V., Ulane R.E. & Bowers B., in Yeast, Mould and Plant Protoplasts, Villanueva J.R., Garcia-Acha I., Gascon S. & Uruburu F., Eds., Academic Press, New York.

Cabib E. & Ulane R.: Chitin synthetase activating factor from yeast, a protease. Bioch. Biophys. Res. Comm. 50, 186-191.

Campbell C.K.: Method of feeding chitinous meal to crustacea. U.S. 3,733,204.

Chattaway F.W., Bishop R., Holmes M.R., Odds F.C. & Barlow A.J.E.: Enzymes activities associated with carbohydrate synthesis and breakdown of the yeast and mycelia forms of *Candida albicans*. J. Gen. Microbiol. 75, 97-109.

Cho H., Teshirogi T., Sakamoto M. & Tonami M.: Synthesis of aminodeoxycellulose. Chem. Lett. 1973, 595-597.

Collins M. & Pappagianis D.: Effects of lysozyme and chitinase on the spherules of *Coccidioides immitis* in vitro. Infect. Immunity 7, 817-822.

Czeczuga B.: Carotenoids giving a specific color to the body of *Branchinecta paludosa* from the Dwoisty Staw Gasienicowy Lake. Bull. Pol. Sci., Ser. Sci. Biol. 21, 365-368.

Dennel R.: The structure of the cuticle of the shore crab *Carcinus maenas*. Zool. J. Linn. Soc. 52, 159-163.

Devries O.M.H. & Wessels J.G.H.: Release of protoplasts from *Schizophyllum commune* by combined action of purified α-1,3-glucanase and chitinase derived from *Trichoderma viride*. J. Gen. Microbiol. 76, 319-330.

Fox D.L.: Chitin-bound keto carotenoids in a crustacean carapace. Comp. Bioch. Physiol. B-44, 953-962.

1973 CNTD

Friedman M., Harrison C.S., Ward W.H. & Lundgren H.P.: Sorption behavior of mercuric and methylmercuric salts on wool. J. Appl. Polymer Sci. 17, 377-390.

Friedman M. & Masri M.S.: Sorption behavior of mercuric salts on chemically modified wools and polyaminoacids. J. Appl. Polymer Sci. 17, 2183-2190.

Glaser L.: Bacterial cell surface polysaccharides, in Annual Review of Biochemistry, Vol. 42, Snell E.E., Ed., Annual Reviews Inc., Palo Alto.

Gooday G.W.: Activity of chitin synthetase during the development of fruit bodies of the toadstool Coprinus cinereus. Bioch. Soc. Trans. 540, 1105-1106.

Gwinn J.F. & Stevenson J.R.: Role of acetylglucosamine in chitin synthesis in crayfish. 1. Correlations of ^{14}C-acetylglucosamine incorporation with stages of the molting cycle. Comp. Biohc. Physiol. 45-B, 769-776.

Gwinn J.F. & Stevenson J.R.: Role of acetylglucosamine in chitin synthesis in crayfish. 2. Enzymes in the epidermis for incorporation of acetylglucosamine into UDP-acetylglucosamine. Comp. Bioch. Physiol. B-45, 777-784.

Harmon R.E., De K.K. & Gupta S.K.: New procedures for preparing trimethylsilyl derivatives of polysaccharides. Carbohydr. Res. 31, 407-409.

Hayashi K., Hamasu Y., Doi Y. & Funatsu M.: Effect of neutral salts on the formation of lysozyme-substrate complex. J. Fac. Agr. Kyushu Univ. 17, 327-337.

Hayashibe M. & Katohada S.: Initiation of budding and chitin-ring. J. Gen. Appl. Microbiol. 19, 23-39.

Hiraga T., Komatsui K. & Takagi Y.: Bacteriolytic enzyme from Streptomyces. Japan 35,476.

Hood M.A.: Chitin degradation in the salt marsh environment. Univ. Microfilms, Ann Harbor 73, 27,843.

Horton D. & Just E.K.: Preparation from chitin of $(1\rightarrow4)$-2-amino-2-deoxy-β-D-glucopyranuronan and its 2-sulfoamino analog having blood-anticoagulants properties. Carbohydr. Res. 29, 173-179.

Hunt S.: Chemical and physical studies of the chitinous siphon sheath in the lamellibranch Lutraria lutraria and its relation to the periostracum. Comp. Bioch. Physiol. B-45, 311-323.

Ikeda K. & Hamaguchi K.: Interaction of manganese(II), cobalt(II) and nickel(II) ions with hen egg white lysozyme and with its N-acetylchitooligosaccharide complexes. J. Bioch. 73, 307-322.

Imoto T. & Yagishita K.: Chitin coated cellulose as an adsorbent of lysozyme-like enzymes. Appl. Agr. Biol. Chem. 37, 1192-1195.

Inoue Y. & Nagasawa K.: Synthesis and spectral properties of N-sulfated and/or O-sulfated aminoalcohols. J. Org. Chem. 38, 1810-1813.

Joffe I. & Hepburn H.R.: Regenerated chitin films. J. Material Sci. 8, 1751-1754.

Kennedy J.F. & Doyle C.E.: Active, water-insoluble derivatives of D-glucose oxidase and alginic acid, chitin and celite. Carbohydr. Res. 28, 89-92.

Kikuchi Y.: Polyelectrolyte complex of heparin with chitosan. Nippon Kagaku Kaishi 12, 2436--2438.

Kimura S.: Chitinolytic enzymes in the larval development of the silkworm Bombyx mori. Appl. Entomol. Zool. 8, 234-236.

Kimura S.: Control of chitinase activity by ecdysterone in larvae of Bombyx mori. J. Insect Physiol. 19, 115-123.

Kimura S.: Control of chitinase activity by ecdysterone in larvae of Bombyx mori. J. Insect Physiol. 19, 2177-2181.

Kochnev I.G., Molodkin V.M., McHedlishvili B.V. & Plisko E.A.: Sulfonation of chitosan films and their anticoagulant action in vitro. Zhur. Prikl. Khim. 46, 1141-1142.

Koshijima T., Tanaka R., Muraki E., Yamada A. & Yaku F.: Chelating polymers derived from cellulose and chitin. 1. Formation of complexes with metal ions. Cellul. Chem. Technol. 7, 197-208.

Lotan R., Guss A.E.S., Lis H. & Sharon N.: Purification of wheat germ agglutinin by affinity chromatography on a Sepharose-bound N-acetylglucosamine derivative. Bioch. Biophys. Res. Comm. 52, 656-662.

Marks E.P.: Induction of molting in insect organ cultures. Proc. 3rd Int. Coll. Invertebrate Tissue Culture, Slovak Acad. Sci.

Martin M., Michael M., Gieselmann M.J. & Martin J.S.: Rectal enzymes of attine ants. α-Amylase and chitinase. J. Insect Physiol. 19,1409-1416.

McMurrough I. & Bartnicki-Garcia S.: Inhibition and activation of chitin synthesis by Mucor rouxii cell extracts. Arch. Bioch. Biophys. 158, 812-816.

Micha J.C., Dandrifosse G. & Jeuniaux Ch.: Distribution et localisation tissulaire de la synthèse des chitinases chez les Vertébrés inférieurs. Arch. Int. Physiol. Bioch. 81, 439-451.

Micha J.C., Dandrifosse G. & Jeuniaux Ch.: Activité des chitinases gastriques des Reptiles en fonction du pH. Arch. Int. Physiol. Bioch. 81, 629-637.

Miura T. & Takahashi R.M.: Insect development inhibitors. 3. Effects on nontarget aquatic organisms. J. Econ. Entomol. 66, 917-922.

Molise E.M. & Drake C.H.: Chitinolysis by Serratine including Serratia liquefaciens. Int. J. Syst. Bacteriol. 23, 278-280.

Monaghan R.L., Eveleigh D.E., Tewari R.P. & Reese E.T.: Chitosanase, a novel enzyme. Nature New Biol. 245, 78-80.

1973 CNTD

Mulder R., Gijswijt M.Y.: The laboratory evaluation of two promising insecticides which interfere with cuticle deposition. Pestic. Sci. 4, 737-745.

Mulder R. & Swennen A.A.: Small scale field experiments with PH-6038 and PH-6040 insecticides inhibiting chitin synthesis. 7th Br. Insecticide & Fungicide Conf. 1973.

Mustafa M. & Kamat D.N.: Mucopolysaccharide histochemistry of *Musca domestica*. 6. Structure and development of chitin. Acta Histoch. 45, 241-253.

Muzzarelli R.A.A.: Natural Chelating Polymers. Pergamon Press, Oxford.

Muzzarelli R.A.A. & Rocchetti R.: The determination of molybdenum in sea-water by hot graphite atomic absorption spectrometry after concentration on *p*-aminobenzylcellulose or chitosan. Anal. Chim. Acta 64, 371-379.

Noguchi J., Wada O., Seo H., Tokura S. & Nishi N.: Chitin and chitin-cellulose fibers. Kobunshi Kagaku 30, 320-326.

Nozawa Y., Kitajima Y. & Ito Y.: Chemical and ultrastructural studies of isolated cell walls of *Epidermophyton floccosum*. Presence of chitin inferred from x-ray diffraction analysis and electron microscopy. Bioch. Biophys. Acta 307, 92-103.

Nud'ga L.A., Plisko E.A. & Danilov S.N.: O-Alkylation of chitosan. Zhur. Obs. Khim. 43, 2752--2756.

Nud'ga L.A., Plisko E.A. & Danilov S.N.: N-Alkylation of chitosan. Zhur. Obs. Khim. 43, 2756--2760.

Ogasawara N., Tanaka H., Yoshinaga E. & Hirara K.: Agricultural antimicrobial agents containing β-1,3-glucanase and chitinase. Japan 56,819.

Pegg G.F. & Vessy J.C.: Chitinase activity in *Lycopersicon esculentum* and its relationship to the in vitro lysis of *Verticillium albo-atrum* mycelium. Physiol. Plant Pathol. 3, 207-222.

Pellizzari E.D., Liv J., Twine M.E. & Cook C.E.: A novel silver-sulfathyl cellulose column for purification of ethynyl steroids from biological fluids. Anal. Bioch. 56, 178-190.

Pigott G.M.: Total utilization of fishery products. Proc. Gulf Caribbean Fish. Inst. 25, 115--120.

Post L.C. & Vincent W.R.: A new insecticide inhibits chitin synthesis. Naturwissens. 60, 431--432.

Rudall K.M. & Kenchington W.: The chitin system. Biol. Rew. 48, 597-636.

Seichertova O., Beran K., Holan Z. & Pokorny V.: Chitin-glucan complex of *Saccharomyces cerevisiae*. 2. Location of the complex in the encircling region of the bud scar. Folia Microbiol. 18, 207-211.

Sharma S.C. & Pant R.: Larval and pupal cuticular proteins of *Philosamia rieini*. Indian J. Exp. Biol. 11, 349-351.

Skujins J., Pukite A. & McLaren A.D.: Adsorption and reactions of chitinase and lysozyme on chitin. Mol. Cell. Bioch. 2, 221-228.

Slagel R.C. & DiMarco Sinkovitz G.: Process and product for making paper products of improved dry strength. U.S. 3,709,780.

Slagel R.C. & DiMarco Sinkowitz G.: Chitosan graft copolymers for making paper products of improved dry strength. U.S. 3,770,673.

Smirnoff W.A., Randall A.P., Martineau R., Haliburton W. & Juneau A.: Field test on the effectiveness of chitinase additive to *Bacillus thuringiensis* against *Choristonuera fumiferana*. Can. J. Forest Res. 3, 228-236.

Smith L.E.H., Mohr L.H. & Raftery M.A.: Mechanism for lysozyme-catalyzed hydrolysis. J. Am. Chem. Soc. 95, 7497-7500.

Stagg C.M. & Feather M.S.: The characterization of a chitin-associated D-glucan from the cell walls of *Aspergillus niger*. Bioch. Biophys. Acta, 320, 64-72.

Stanely W.L. & Ralter R.: Lactase immobilization on phenolic resins. Biotechnol. Bioengin. 15, 597-602.

Tang J. & Whistler R.L.: Cellulose and chitin. Phytoch. 1, 249-269.

Thomas E.W.: Unsatured disaccharide found in quenched acetolyzates of chitin. Carbohydr. Res. 26, 225-226.

Tiunova N.A., Pirieva D.A. & Epshetein M.G.: Purification and fractionation of chitinase and chitobiase of *Actinomyces* sp. Dokl. Akad. Nauk SSSR 211, 740-743.

Towle G.A. & Whistler R.L.: Hemicellulose and gums. Phytoch. 1, 198-248.

Usov A.I., Krylova R.G., Ryadovskaya S.N., Kadentsev V.I. & Chizhov O.S.: Mass spectrometric determination of the position of amino groups in amino sugars. Izv. Akad. Nauk SSSR, Ser. Khim. 12, 2780-2783.

Vessey J.C. & Pegg G.P.: Autolysis of chitinase production in cultures of *Verticillium albo--atrum*. Trans. Brit. Mycol. Soc. 60, 133-143.

Walton A.G. & Blackwell J.: Biopolymers. Academic Press, New York.

Watanabe T., Iwamoto Y. & Tsunemitsu A.: Hydroclysis of glycol chitin by an endo- and exo-type of β-N-acetylglucosaminidase from human parotid saliva. J. Dent. Res. 52, 846.

Whistler R.L.: Chitin, in Industrial Gums, Whistler R.L., Ed., Academic Press, New York.

Yaku F. & Koshishima T.: Removal and recovery of haevy metal ions from waste water. Japan 16,863.

Yaku T. & Yamashita I.: Chitin films. Japan 19,213.

Yaku T. & Yamashita I.: A method for the production of chitin films. Japan 20,747.

Yamasaki N., Tsujita T. & Takakuwa M.: Simple colorimetric method for the determination of lysozyme activity. Agr. Biol. Chem. 37, 1507-1508.

1974

Aksnes L. & Grow A.: Bacteriolytic activity of *Streptomyces albus* culture filtrates. 3. Affinity for chitin. Acta Chem. Scand. 28, 221-224.

Angyal S.J., Bender V.J. & Ralph B.J.: Structure of polysaccharides from *Polyporus tumulosus* cell wall. Bioch. Biophys. Acta 362, 175-187.

Anno K., Otsuna K. & Seno N.: A chitin sulfate-like polysaccharide from the tunicate *Halocynthia roretzi*. Bioch. Biophys. Acta 362, 215-219.

Araki Y. & Ito E.: Pathway of chitosan formation in *Mucor rouxii*. Enzymic deacetylation of chitin. Bioch. Biophys. Res. Comm. 56, 669-675.

Aso Y., Koga D., Mori F., Hayashi K. & Funatsu M.: Kinetic analysis of lysozyme inhibitor complex formation. J. Fac. Agr. 18, 89-106.

Bade L.M.: Localization of molting chitinase in insect cuticle. Bioch. Biophys. Acta 372, 474--477.

Bernier I., Landereau J.C., Grellet P. & Jolles O.: Lysozymes and chitinases from hemolymph and cell cultures of cockroach *Periplaneta americana*. Comp. Bioch. Physiol. 47, 41-44.

Bloch R. & Burger M.M.: Purification of wheat germ agglutinin using affinity chromatography on chitin. Bioch. Biophys. Res. Comm. 58, 13-19.

Bordereau C. & Bois J.F.: Physogastry in termite queens. Incorporation of aminoacids and glucosamine into the cuticle of the *Cubirtermes fungifaber* queen. Comp. Rend. D-278, 2165--2168.

Boud W.D., Claiborne H.C. & Lenze R.E.: Method for removal of actinides from high-level waste. Nuclear Techn. 24, 362-370.

Bouligand Y.: Crystaux liquides. Encyclopédie Française.

Bowers B., Levin G. & Cabib E.: Effect of polyoxin D on chitin synthesis and septum formation in *Saccharomyces cerevisiae*. J. Bacteriol. 119, 564-575.

Brine C.J. & Austin P.R.: Utilization of chitin, a cellulose derivative from crab and shrimp waste. Rep. COM-75-10195/6GA.

Bussers J.C. & Jeuniaux Ch.: Recherche de la chitine dans les productions métaplasmatiques de quelques Ciliés. Protistologica 10, 43-46.

Campbell K.C.: Farming aquatic animals. Can. 951,562.

Cerf D.C. & Georghiou G.P.: Cross-resistance to an inhibitor of chitin synthesis TH-6040 in insecticide-resistant strains of the house fly. J. Agr. Food Chem. 22, 1145-1146.

Charlemagne D. & Jolles P.: Inhibition par des polymères de la N-acetylglucosamine de l'activité lysante de lysozyme vis-à-vis de *Micrococcus luteus*. Comp. Rend. 279, 299-301.

Costa N.M.M.D.: New species of chitinozoa in trombetas formation, Para, Brazil. Anais Acad. Brazil. Ciencias 46, 287-302.

Cutler B. & Richards A.G.: On the absence of chitin in the endosternite of Chelicerata. Experientia 30, 1393-1394.

Dandrifosse G.: Redox changes in the respiratory chain related to chitinase excretion by the gastric mucosa. J. Physiol. 68, 387-394.

Delmotte F.M. & Monsigny M.L.P.: Synthesis of p-nitrobenzyl-1-thio-β-chitobioside and 1-thio-β-chitotrioside and p-nitrophenyl-β-chitotrioside. Carbohydr. Res. 36, 219-226.

Elbein A.D.: Interactions of polynucleotides and other polyelectrolytes with enzymes and other proteins, in Advances in Enzymology, Vol. 40, Meister A., Ed., John Wiley & Sons, London.

Environment Protection Agency: Canned and preserved seafood processing poit source category. (Effluent limitation guidelines). Fed. Register 39(124), 23133-23156.

Feniksova R.V., Tiunova N.A., Pirieva D.A. & Kuznetsov V.D.: Chitinase. USSR 413,189.

Gardner K.H. & Blackwell J.: The structure of native cellulose. Biopolymers 13, 1975-2001.

Gooday G.W.: Control and development of excised fruit bodies and stipes of *Coprinus cinereus*. Trans. Brit. Mycol. Soc. 62, 391-399.

Hackman R.H. & Goldberg M.: Light-scattering and infra-red spectrophotometric studies of chitin and chitin derivatives. Carbohydr. Res. 38, 35-45.

Hasilik A.: Inactivation of chitin synthetase in *Saccharomyces cerevisiae*. Arch. Microbiol. 101, 295-301.

Hayashi K., Inohara Y., Taira K., Doi Y. & Funatsu M.: Studies on lysozyme catalyzed transglycosylation. J. Fac. Agr. Kyushu Univ. 18, 107-137.

Hedges A. & Wolfe R.S.: Extracellular enzyme from Myxobacter AL-1 that inhibits both β-1,4--glucanase and chitosanase activities. J. Bacteriol. 120, 844-853.

Heinrich D., Surholt B. & Zebe E.: Autoradiographische Untersuchungen zum Einbau von ^{14}C--glukose and ^{3}H-N-acetylglukosamin bei der Wanderheuschrecke *Locusta migratoria*. Cytobiol. 9, 45-58.

Hori M., Eguchi J., Kakiki K. & Misato T.: Mode of action of polyoxins. 6. Effect of polyoxin B on chitin synthesis on polyoxin-sensitive and resistant strains of *Alternaria kikuchiana*. J. Antibiot. 27, 260-266.

Hori M., Kakiki K. & Misato T.: Mode of action of polyoxins. 5. Interactions between polyoxin and active center of chitin synthetase. Agr. Biol. Chem. 38, 699-705.

Hori M., Kakiki K. & Misato T.: Further sudies on the relation of polyoxin structure to chitin synthetase inhibition. Agr. Biol. Chem. 38, 691-698.

1974 CNTD

Iwata S. & Nakabayashi T.: Method for the elimination of coloring substances. Japan 110,588.

Jan Y.N.: Properties and cellular localization of chitin synthetase in *Phycomyces blakesleanus*. J. Biol. Chem 249, 1973-1979.

Kennedy J.F., Barker S.A. & Zamir A.: Active insolubilized antibiotics based on cellulose--metal chelates. Antimicr. Agents Chemot. 6, 777-782.

Kikuchi Y.: Polyelectrolyte complex of heparin with chitosan. Makromol. Chem. 175, 2209-2211.

Kikuchi Y. & Fukuda H;: Polyelectrolyte complex of sodium dextran sulfate with chitosan. Makromol. Chem. 175, 3593-3596.

Kina K., Tamura K. & Ishibashi N.: Use of 8-anilino-1-naphtalen-sulfonate as a fluorescent indicator in colloid titration. Japan Analyst 23, 1082-1083.

Kirkwood S.: Unusual polysaccharides, in Annual Review of Biochemistry, Vol. 43, Snell E.E., Ed., Annual Reviews Inc., Palo Alto.

Krishnan G.: Some less known aspects of the structure and composition of chitin and protein of the arthropod cuticle. J. Sci. Ind. Res. 33, 258-265.

Kunzru D. & Rei R.W.: Separation of aromatic amine isomers by high pressure liquid chromatography on cadmium-impregnated silica gel columns. J. Chromatogr. Sci. 12, 191-196.

Lee V.: Solution and shear properties of chitin and chitosan. Univ. Microfilms, Ann Arbor 74/29,446.

Lehmann P.F., White L.O. & Ride P.J.: Aspects of systemic *Aspergillus fumigatus* infection in untreated or immuno-suppressed mice: use of chitin assay to measure fungal growth. Proc. Soc. Gen. Microbiol. 2, 30.

Lundblad G., Hederstedt B., Lind J. & Steby M.: Chitinase in goat serum. Preliminary purification and characterization. Eur. J. Bioch. 46, 367-376.

Madhavan P. & Ramachandran N.K.G.: Utilization of prawn waste. Isolation of chitin and its conversion to chitosan. Fish. Technol. 11, 50-53.

Marchant W.N.: Modified cellulose adsorbent for removal of mercury from aqueous solutions. Envir. Sci. Technol. 8, 993-995.

Martineau J.P. & Charmantier-Daures M.: Effect of the intensive regeneration on the behavior of calcium in the crab *Pachygrapsus marmoratus* deprived of organ Y. Comp. Rend. D-279, 1281-1283.

Masri M.S. & Friedman M.: Effect of chemical modification of wool on metal ion binding. J. Appl. Polymer Sci. 18, 675-681.

Matsutani H., Kankichi K., Shinoda K., Nakagawa T. & Koike M.: Reverse osmosis membrane. Japan 91,079.

Miura T. & Takahashi R.M.: Toxicity of TH-6040 to freshwater crustacea and the use of a tolerance index as a method of expressing side effects to non-targets. Proc. 42nd Conf. Cal. Mosquito Control Ass.

Miura T. & Takahashi R.M.: Insect developmental inhibitors. Effects of candidate mosquito control agents on non-target aquatic organisms. Environ. Entomol. 3, 631-636.

Morozowich W.: The cellulose, a new support for high pressure anion exchagne liquid chromatography. J. Chromatogr. Sci. 12, 453-457.

Muzzarelli R.A.A.: Use of chitosan sulfate for the removal and the recovery of transition metal ions from natural and industrial waters and from solutions. Italy 3329 A/74.

Muzzarelli R.A.A.: Natural chelating polymers, Proc. CNRN Symp. Adv. Anal. Methodol. Rome 1974, 255-259.

Muzzarelli R.A.A.: Insecticides inhibiting chitin synthesis. Inquinamento 16(3), 29-30.

Muzzarelli R.A.A., Isolati A. & Ferrero A.: Chitosan membranes. Ion Exch. Membr. 1, 193-196.

Muzzarelli R.A.A., Nicoletti C. & Rocchetti R.: Interactions of metal ions with alginic acid, polygalacturonic acid, carboxymethylcellulose and cutin. Ion Exch. Membr. 2, 67-69.

Muzzarelli R.A.A. & Rocchetti R.: The determination of copper in sea-water by atomic absorption spectrometry with a graphite atomizer, after elution from chitosan. Anal. Chim. Acta 69, 35-42.

Muzzarelli R.A.A. & Rocchetti R.: The determination of vanadium in sea-water by hot graphite atomic absorption spectrometry on chitosan, after separation from salt. Anal. Chim. Acta 70, 283-289.

Muzzarelli R.A.A. & Rocchetti R.: Enhanced capacity of chitosan for transition metal ions in sulfate-sulfuric acid solutions. Talanta 21, 1137-1147.

Muzzarelli R.A.A. & Rocchetti R.: The use of chitosan columns for removal of mercury from waters. J. Chromatogr. 96, 115-121.

Nakatani H. & Hiromi K.: Kinetic studies of lysozyme-ethyleneglycolchitin interactions. J. Bioch. 76, 1343-1346.

Nanba H. & Kuroda H.: Fungicides. 10. Biosynthesis of β-glucan and chitinlike substances of cell wall of *Cochliobolus miyabeanus*. Chem. Pharm. Bull. 22, 610-616.

Nanba H. & Kuroda H.: Fungicides. 11. Biosynthesis of β-glucan and chitinlike substances of cell wall of *Cochliobolus miyabeanus*. Chem. Pharm. Bull. 22, 1895-1901.

Nud'ga L.A., Plisko E.A. & Danilov S.N.: The cyanoethylation of chitosan. Zhur. Obs. Khim. 45, 1145-1149.

Nud'ga L.A., Plisko E.A. & Danilov S.N.: Synthesis and properties of sulfoethylchitosan. Zhur. Prikl. Khim. 47, 872-875.

Odinokova L.Y. & Molodstov N.V.: Glycopeptides from chitin. Khim. Prir. Soedim. 10, 118-119.

1974 CNTD

Ou L.T. & Alexander M.: Effect of glass microbeads on the microbial degradation of chitin. Soil Sci 118, 164-167.

Post L.C.; De Jong B.J. & Vincent W.R.: 1-(2,6-disubstituted benzoyl)-3-phenylurea insecticides: inhibitors of chitin synthesis. Pest. Bioch. Physiol. 4, 473-483.

Preston R.D.F.R.S.: The Physical Biology of Plant Cell Walls. Chapman & Hall, London.

Rafestin M.E., Obrenovitch A., Oblin A. & Monsigny M.: Purification of N-acetyl-D-glucosamine binding proteins by affinity chromatography. FEBS Lett. 40, 62-66.

Rockstein M.: The Physiology of Insecta. Academic Press, New York.

Ruiz-Herrera J. & Bartnicki-Garcia S.: Synthesis of cell wall microfibrils in vitro by a soluble chitin synthetase from *Mucor rouxii*. Science 186, 357-359.

San-Blas G. & Carbonell L.M.: Chemical and ultrastructural studies on the cell walls of the yeast-like and mycelial forms of *Histoplasma farciminosum*. J. Bacteriol. 119, 602-611.

Shimahara K., Nagahata N. & Takiguchi Y.: Chitinase assay using 6-O-(hydroxypropyl)chitin. Seikei Daigaku Kogakubu Kogaku Okoku 18, 1371-1372.

Skujins J., Pukite A. & McLaren A.D.: Adsorption and activity of chitinase on kaolinite. Soil Biol. Bioch. 6, 179-182.

Smirnoff W.A.: Three years of aerial field experiments with *Bacillus thuringiensis* plus chitinase formulation against the spruce budworm. J. Invert. Pathol. 24, 344-348.

Spinelli J., Lehmen L. & Wieg D.: Composition, processing and utilization of red crab *Pleuroncodes planipes* as an aquacultural feed ingredient. J. Fish. Res. Board Can. 31, 1025-1029.

Tate L.G. & Wimer L.T.: Incorporation of ^{14}C from glucose into CO_2, chitin, lipid, protein and soluble carbohydrates during metamorphosis of the blowfly *Phormia regina*. Insect Bioch. 4, 85-98.

Train R.E., Agee J.L., Cywin A. & Forsht E.H.: Development document for effluent limitation guidelines and standards of performance, for the catfish, crab, shrimp and tuna segments of the canned and preserved seafood processing industry. EPA 440/1-74-020-a.

Turnipseed S.G., Heinrichs E.A., Silva R.F.P. & Todd J.W.: Response of soybean insects to foliar applications of a chitin synthesis inhibitor TH-6040. J. Econ. Entomol. 67, 760-762.

Ulane R.E. & Cabib E.: Activating system of chitin synthetase from *Saccharomyces cerevisiae*. Purification and properties of an inhibitor of the activating factor. J. Biol. Chem. 249, 3418-3422.

Voss-Foucart M.F., Jeuniaux Ch. & Gregoire Ch.: Resistance de la chitine de la nacre du Nautile (Mollusque cephalopode) à l'action de certain facteurs intervenant au cours de la fossilation. Comp. Bioch. Physiol. 48-B, 447-451.

Welinder B.S.: Crustacean cuticle. 1. Composition of the cuticle. Comp. Bioch. Physiol. 47-A, 779-787.

Winicur S. & Mitchell H.K.: Chitinase activity during *Drosophila* development. J. Insect Physiol. 20, 1795-1805.

Yano I.: Incorporation of glucose-1-^{14}C into chitin of the exoskeleton of a shore crab, with special reference to the period of endocuticle formation. Nippon Suisan Gakkaishi 40, 783--787.

1975

Allan G.G., Friedhoff J.F. Koppela M., Laine J.E. & Powell J.C.: Marine Polymers. 5. Modifications of paper with partially deacetylated chitin. ACS Symp. Ser. 10, 172-180.

Araki Y. & Ito E.: Pathway of chitosan formation in *Mucor rouxii*. Enzymic deacetylation of chitin. Eur. J. Bioch. 55, 71-78.

Austin P.R.: Purification of chitin. U.S. 3,879,377.

Austin P.R.: Solvents and purification of chitin. U.S. 3,892,731.

Averbach B.L.: The structure of chitin and chitosan. MITSG-75-17, NOAA 75102204.

Babu K., Satish Rao K. & Panduranga J.K.T.: Chitosan, a new auxiliary in leather manufacture. Leather Sci. 22, 244-246.

Bade M.L.: Time pattern of appearance and disappearance of active molting chitinase in *Manduca* cuticle. Endogenous activity. FEBS Lett. 51, 161-163.

Balassa L.L.: Promoting wound healing with chitin derivatives. U.S. 3,911,116.

Balassa L.L.: Process for facilitating wound healing with N-acetylated partially depolymerized chitin materials. U.S. 3, 914,413.

Banerjee S.K. & Rupley J.A: Turkey egg white lysozyme. Free energy, enthalpy and steady state kinetics of reaction with N-acetylglucosamine oligosaccharides. J. Biol. Chem. 250, 8267--8274.

Beddel C.R., Blake C.C. & Oatley S.J.: An x-ray study of the structure and binding properties of iodine-inactivated lysozyme. J. Mol. Biol.: 97, 643-654.

Benitez T., Villa T.G. & Garcia Acha I.: Chemical and structural differences in mycelial and regeneration walls of *Trichoderma viride*. Arch. Microbiol. 105, 277-282.

Berak L., Uher E. & Marhol M.: Sorbents for the purification of low- and medium-level radioactive waters. Atomic En. Rev. 13, 325-366.

Bijlo J.D.: A new insecticide: diflubenzuron. Phytiatr. Phytopharm. 24, 147-158.

Bihari-Varga M., Sepulchre C. & Moczar E.: Thermoanalytical studies on protein-polysaccharide complexes of connective tissue. J. Thermal Anal. 7, 675-683.

1975 CNTD

Bough W.A.: Coagulation with chitosan: an aid to recovery of byproducts from egg breaking wastes. Poultry Sci. 54, 1904-1912.

Bough W.A.: Reduction of suspended solids in vegetable canning waste effluents by coagulation with chitosan. J. Food Sci. 40, 297-301.

Bough W.A., Shewfelt A.L. & Salter W.L.: Use of chitosan for the reduction and recovery of solids in poultry processing waste effluents. Poultry Sci. 54, 992-1000.

Brimacombe J.S., MIller J.A. & Zakir U.: An approach to the synthesis of branched-chain amino sugar C-methylene sugars. Carbohydr. Res. 41, C3-C5.

Brine C.J. & Austin P.R.: Renatured chitin fibrils, films and filaments. ACS Symp. S. 18, 505-518.

Brown C.H.: Structural Materials in Animals. Pitman, London.

Butler L.G.: Cellulose enzyme immobilization by adsorption on hydrophobic derivatives of cellulose and other hydrophilic materials. Arch. Bioch. 171, 650.

Cabib E.: Molecular aspects of yeast morphogenesis. Ann. Rev. Microbiol. 29, 191-214.

Cabib E. & Bowers B.: Timing and function of chitin synthesis in yeast. J. Bacteriol. 124, 1586-1593.

Cabib E., Ulane R..& Bowers B.: A molecular model for morphogenesis: the primary septum of yeast. Curr. Topics Cell. Regul. USDHEW-NIH 8, 1-32.

Capozza R.C.: Enzymically decomposable biodegradable pharmaceutical carrier. Ger. 2,505,305.

Cherkasov I.A., Kravchenko N.A.: Complexing of lysozyme with chitosaccharides at pH 8.4. Izv. Akad. Nauk SSSR, Ser. Khim. 1975, 2619.

Cornelius C., Dandrifosse G. & Jeuniaux Ch.: Biosynthesis of chitinase by mammals of the order Carnivora. Bioch. System. Ecology 3, 121-122.

Dandrifosse G.: Purification of chitinase contained in pancreas or gastric mucosa of frog. Bioch. 57, 829-831.

Debruyn F.E.: Photographic material. Brit. 2,113,174.

Devries O.M.H. & Wessels J.G.H.: Chemical analysis of cell wall regeneration and reversion of protoplasts from Schizophyllum commune. Arch. Microbiol. 102, 209-218.

Dow J.M. & Rubery P.H.: Hyphal tip bursting in Mucor rouxii. Antagonistic effects of calcium ions and acids. J. Gen. Microbiol. 91, 425-428.

Duran A., Bowers B. & Cabib E.: Chitin synthetase zymogen is attached to the yeast plasma membrane. Proc. Nat. Acad. Sci. 72, 3952-3955.

Environment Protection Agency: Canned and preserved seafood processing point source category. (Effluent guidelines and standards and proposed guidelines and standards). Fed. Register 40(21), 4581-4619.

Environment Protection Agency: Seafood processing point source category. (Effluent guidelines and standards). Fed. Register 40(231), 55770-55801.

Ernst G., Emeric D.A. & Levine S.: Analysis of chitin in contaminated fuel. USAMERDC-2158.

Florence T.M. & Batley G.E.: Removal of trace metals from sea-water by a chelating resin. Talanta 22, 201-204.

Florkin M.: Biochemical evolution in animals (diversification in animals at the level of the central metabolic biosyntagm). Comp. Bioch. Evol. 29-B, 182-200.

Foster W.A.: Water-soluble polymers as flocculants in papermaking, in Water-soluble Polymers, Bikales N.M., Ed., Plenum Publishing Corp., New York.

Fukada e. & Sazaki S.: Piezoelectricity of α-chitin. J. Polymer Sci. 13, 1845-1847.

Gainey R., Malone M. & Rimpler M.: Chitin as causative mechanism for inflammatory disease. J. Theor. Biol. 52, 249-250.

Gardner K.H. & Blackwell J.: Refinement of the structure of β-chitin. Biopolymers 14, 1481-1595

Gooday.G.W. & Rousset-Hall A.: Properties of chitin synthetase from Coprinus cinereus. J. Gen. Microbiol. 89, 146-154.

Iizuka C. & Fumoto C.: Herbicides. Japan 10,926.

Hall H.S. & Hinkes T.M.: Air suspension encapsulation of moisture-sensitive particles using aqueous systems, in Microencapsulation: Process and Applications, Vandagaer J.E., Ed., Plenum Publishing Corp., New York.

Hata K.Y. & Sakai T.: Polyelectrolyte complexes from cellulose sulfates. Japan 123,179.

Herzog K.H., Grossmann H. & Lieflaender M.: Chemistry of a chitin protein from crayfish Astacus fluviatilis. Hoppe-Seyler Z. Physiol. Chem. 356, 1067-1071.

Hirano S., Kondo S. & Ohe Y.: Chitosan gel: a novel polysaccharide gel. Polymer 16, 622.

Hirano S. & Ohe Y.: Chitosan gels: a novel molecular aggregation of chitosan in acidic solutions on a facile acylation. Agr. Biol. Chem. 39, 1337-1338.

Hirano S. & Ohe Y.: A facile N-acylation of chitosan with carboxylic anhydrides in acidic solutions. Carbohydr. Res. 41, C1-C2.

Holland D.L. & Walker G.: Biochemical composition of the cypris larva of the barnacle Balanus balanoides. J. Cons., Cons. Int. Explor. Mer 36, 162-165.

Holler E., Rupley J.A. & Hess G.P.: Productive and unproductive lysozyme-chitosaccharide complexes. Kinteic investigation. Biochem. 14, 2377-2385.

Hsu S.C. & Lockwood J.L.: Powdered chitin agar as a selective medium for enumeration of Actinomycetes in water and soil. Appl. Microbiol. 29, 422-426.

Jakab L., Tenczer J. & Szilvasi R.: Effect of a glucosamine derivative on the in vitro synthesis of glycosaminoglycan by a granuloma tissue. Kiserl. Orvostud. 27, 482-485.

1975 CNTD

Jaspers H.T.A., Christianse K. & Van Steveninck J.: Improved method for the preparation of yeast enzymes in situ. Bioch. Biophys. Res. Comm. 65, 1434-1439.

Jeuniaux Ch.: Chitine et complexes chitino-protéiques de la cuticule, in Traité de Zoologie, Vol. 8(3), Grasse P.P., Ed., Paris.

Kancko T. & Colwell R.R.: Adsorption of *Vibrio parahemolyticus* onto chitin and copepods. Appl. Microbiol. 29, 269-274.

Katz F., Fishman L. & Levy H.: Method for isolation of lysozyme. Brit. 1,418,738.

Koaze Y., Kojima M. & Hara T.: Chitinase from molds. Japan 15,878.

Kokufuta E., Hirata M. & Iwai S.: Potentiometric titration behavior of glycol chitosan. Shikizai Kyokaishi 48, 493-497.

Kuramitsu S., Ikeda K., Hamaguchi K., Miwa S. & Nakashima K.: Binding of N-acetyl-chitotriose to human lysozyme. J. Bioch. 78, 327-333.

Lee H.Y. & Aronson J.M.: Composition of cellulin, the unique chitin-glucan granules of the fungus *Apodachlya* sp. Arch. Microbiol. 102, 203-208.

Lehmann P.F. & White L.O.: Chitin assay to demostrate renal localization and cortisone--enhanced growth of *Asperigillus fumigatus* mycelium in mice. Infection Immunity 12, 987-992.

Martin F.: Sur quelques chitinozoaires ordoviciens du Québec et de l'Ontario, Canada. Can. J. Earth Sci. 12, 1006-1018.

Masri M.S., Randall V.G. & Stanley W.L.: Chemical modification of insoluble polymers for certain end uses. Polymer Prepr. 16, 70.

Matsuda H. & Shinoda K.: Semipermeable membranes. Japan 154,182.

Michalenko G.O., Hohl H.R. & Rast D.: Chemistry and architecture of the mycelial wall of *Agaricus bisporus*. J. Gen. Microbiol. 92, 252-262.

Mileshkevich Y.G., Latosh M.V. & Reznikov V.M.: Aminolignin. USSR 464,597.

Mima S., Yoshikawa S. & Miya M.: Chitosan reinforced poly(vinylalcohol) membranes. Japan 130,870.

Mirelman D., Galun E., Sharon N. & Lotan R.: Inhibition of fungal growth by wheat germ agglutinins. Nature 256, 414-416.

Mody A.V.: Dyeing and printing cotton or polyester-cellulosic textiles. Ger. 2,438,454.

Monaghan R.L.: Discovery, distribution and utilization of chitosanases. Univ. Microfilms, 75-24,724, Ann Arbor.

Moore P.M. & Peberdy J.F.: Biosynthesis of chitin by particulate fractions from *Cunninghamella elegans*. Microbios 12, 29-39.

Morjani M.N., Achutha V. & Kha Sim D.I.: Parameters affecting the viscosity of chitosan from prawn waste. J. Food Sci. Technol. 12, 187-189.

Nakazaki N., Takeda T. & Yoshino T.: Fusion reactions involving activating agents of some amino-2-deoxy-β-D-glucopyranosides. Carbohydr. Res. 44, 215-226.

Navratil J.D., Murgia E. & Walton H.F.: Ligand-exchange chromatography of aminosugars. Anal. Chem. 47, 122-125.

Neville A.C.: Biology of the Arthropod Cuticle. Springer Verlag, Berlin.

Nud'ga L.A., Plisko E.A. & Danilov S.N.: Cyanoethylation of chitosan. Zhur. Obs. Khim. 45, 1145-1149.

Numazaki S.O. & Kiyoshi Kito H.: Chitosans. Japan 126,784.

Patton R.S. & Chandler P.T.: In vivo digestibility evaluation of chitinous materials. J. Dairy Sci. 58, 397-403.

Patton R.S., Chandler P.T. & Gonzales O.G.: Nutritive value of crab meal for young ruminating calves. J. Dairy Sci. 58, 404-409.

Peberdy J.F. & Moore P.M.: Chitin synthetase in *Mortierella vinacea*. Properties, cellular location and synthesis in growing cultures. J. Gen. Microbiol. 90, 228-236.

Peniston Q.P. & Johnson E.L.: Recovering chitosan and other byproducts from shellfish waste and the like. U.S. 3,862,122.

Peniston Q.P. & Johnson E.L.: Process for depolymerization of chitosan. U.S. 3,922,260.

Pervaiz S.M. & Haleem M.A.: Studies on the structure of β-chitin. Z. Naturforsch. 30-C, 571--574.

Ponomareva R.B., Kavunenko A.P., Kalacheva T.N., Tikhomirova-Sidrova N.S. & Samsonov G.V.: Modification of pancreatic ribonuclease activity in complexes with polyanions. Biokhimiya 40, 468-474.

Porter L.K.: Nitrogen transfer in ecosystem. Soil Bioch. 4, 1-30.

Potenzone R. & Hoffinger A.J.: Conformational analysis of glycosaminoglycans. 1. Charge disribution, torsional potentials and steric maps. Carbohydr. Res. 40, 323-336.

Prakasam V.R. & Azariah J.: An optimum pH for the demonstration of chitin in *Periplaneta americana* using Lugol's iodine. Acta Histoch. 53, 238-240.

Price J.S. & Stork R.: Production, purification and characterization of an extracellular chitosanase from *Streptomyces*. J. Bacteriol. 124, 1574-1585.

Privat J.P. & Monsigny M.: Luminescence studies of saccharide binding on wheat germ agglutinin (lectin). Eur. J. Bioch. 60, 555-567.

Ravindranath M.H.R. & Ravindranath M.H.: A simple procedure to detect chitin in delicate structures. Acta Histoch. 53, 203-205.

Ribrioux Y. & Dolbeau C.: Diflubenzuron control of processionary moths in pine. Phytiatr. Phytopharm. 24, 193-204.

1975 CNTD

Richmond D.V.: Effect of toxicants on the morphology and fine structure of fungi. Adv. Appl. Microbiol. 19, 289-319.

Rousset-Hall A. & Gooday G.W.: A kinetic study of a solubilized chitin synthetase preparation from *Coprinus cinereus*. J. Gen. Microbiol. 89, 146-154.

Ruiz-Herrera J., Sing V.O., Van der Woude W. & Bartnicki-Garcia S.: Microfibril assembly by granules of chitin synthetase. Proc. Nat. Acad. Sci. 72, 2706-2710.

Seichertova O., Beran K., Holan Z. & Pokorny V.: Chitin-glucan complex of *Saccharomyces cerevisiae*. 3. Electron microscopic study of the pre-budding stage. Folia Microbiol. 20, 371--378.

Shinoda K. & Nakajima A.: Complex formation of heparin or sulfate cellulose with glycol-chitosan. Bull. Inst. Chem. Res. Kyoto Univ. 53, 392-399.

Shinoda K. & Nakajima A.: Complex formation of hyaluronic acid or chondroitin sulfate with glycolchitosan. Bull. Inst. Chem. Res. Kyoto Univ. 53, 400-408.

Smirnoff W.A.: Extraction of chitinase. U.S. 3,862,007.

Smirnoff W.A.: Insecticidal composition. U.S. 3,911,110.

Sowa B.A.& Marks E.P.: In vitro system for quantitative measurement of chitin synthesis cockroach inhibition by TH-6040 and polyoxin D. Insect Bioch. 5, 855-859.

Stanely W.L. & Watters G.G.: Insolubilized enzymes. U.S. 3,909,358.

Stanley W.L., Watters G.G., Chan B. & Mercer J.M.: Lactase and other enzymes bound to chitin with glutaraldehyde. Biotechnol. Engin. 17, 315-326.

Stein G.S., Roberts R.M., Davis J.L., Head W.J., Stein J.L., Thrall C.L., Van Veen J. & Welch D.W.: Are glycoproteins and glycosaminoglycans components of the eukaryotic genome ? Nature 258, 639-641.

Stimizu Y.: Analysis of aminoacids by liquid chromatography. Shiga-Kenritsu Tanki Daigaku Gakujuutsu Zasshi 16, 34-35.

Subramanian V., Yoshinari T. & D'Anglejan B.: Studies on the formation of chitin-metal complexes. Manus. Rep. McGill Univ. Marine Sci. Cent. 1975, 27.

Surholt B.: Studies in vivo and in vitro on chitin synthesis during the larval-adult molting cycle of the migratory locust *Locusta migratoria*. J. Comp. Physiol. B-102, 135-147.

Surholt B.: Formation of glucosamine-6-phosphate in chitin synthesis during ecdysis of the migratory locust *Locusta migratoria*. Insect Bioch. 5, 585-594.

Tachiki J.O. & Miyagi T.: Regulation of aminosugar biosynthesis. Horumon to Rinsho. 23, 879--884.

Takanori S., Kurita K. & Iwakura Y.: Studies on chitin. 1. Solubility change by alkaline treatment and film casting. Makromol. Chem. 176, 1191-1195.

Temeriusz A.: Chemical composition of *Agaricus bisporus* cell walls. 1. Hydrolysis. Rocz. Chem. 49, 1803-1809.

Tometsko A.M. & Comstock J.: Evaluation of carbodiimide stoichiometry by resin probe analysis. Anal. Chem. 47, 2299-2301.

Tominaga Y. & Tsujisaka Y.: Purification and some enzymic properties of the chitosanase from *Bacillus* R-4 which lyses *Rhizopus* cell walls. Bioch. Biophys. Acta 410, 145-155.

Vichutinskii A.A. & Zaslavshii B.Y.: Reactions between lysozyme and some neutral saccharides, chitin oligosaccharides and their modified analogues. Conf. Int. Thermodyn. Chim. CR 4 TH5, 68-75.

Voss-Foucart M.F. & Gregoire Ch.: On biochemical and structural alterations of the nacre conchiolin of the *Nautilus* shell under conditions of protracted, moderate heating and pressure. Arch. Int. Physiol. Bioch. 83, 43-52.

Wang R., Sevier E.D., David G.S. & Reisfeld R.A.: An affinity adsorbent for the rapid purification of wheat germ agglutinin. J. Chromatogr. 114, 223-226.

Wargo P.M.: Lysis of the cell wall of *Armillaria mellea* by enzymes from forest trees. Physiol. Plant Pathol. 5, 99-105.

Welinder B.S.: The crustacean cuticle 2. Deposition of organic and inorganic material in the cuticle of *Astacus fluviatilis* in the period after molt. Comp. Bioch. Physiol. 51, 409-416.

Welinder B.S.: The crustacean cuticle. 3. Composition of individual layers in *Cancer pagurus* cuticle. Comp. Bioch. A-52, 659-663.

Wheatland A.B., Gledhill C. & O'Gorman J.V.: Developments in the treatment of metal-bearing eff' 'ents. Chem. Ind. 1975, 632-640.

Williams J.M.: Deamination of carbohydrate amines and related compounds, in Advances in Carbohydrate Chemistry and Biochemistry, Vol. 31, Tipson R.S. & Horton D., Eds., Academic Press, New York.

Wyss P.C. & Kiss J.: Sinthesis of heparin saccharides. 3. Synthesis of derivatives of D-glucosamine as starting materials for disaccharides. Helv. Chim. Acta 58, 1833-1847.

Yanase M.: Chemical composition of the exoskeleton of Antarctic krill. Tokaiku Suisan Kenkyu. Kenkyu Hokoku 83, 1-6.

Yoshida T. & Yamashita H.: Chitosan as a binder for pelletizing fertilizers and feeds. Japan 136,287.

Yoshida T. & Yamashita H.: Chitin or chitosan as antiaggregate agents for fertilizers. Japan 148,167.

1976

Allan G.G., Crosby G.D. & Sarkanen K.V.: Evaluation of chitosan as a strength additive for α-cellulose and unbleached sulfite papers.(Original results).

Ascher K.R.S. & Nemny N.E.: Toxicity of the chitin synthesis inhibitors, diflubenzuron and its dichloro-analog, to *Spodoptera littoralis* larvae. Pestic. Sci. 7, 1-9.

Asano T., Suzuki T. & Hayakawa N.: Centrifugal dewatering of municipal and industrial sludges. (Original results).

Ashcroft S.J.H. & Crossley J.R.: The effect of N-acylglucosamines on the biosynthesis and secretion in the rat. Bioch. J. 154, 701-707.

Baranova V.N., Plisko E.A. & Nud'ga L.A.: Modified chitosan used in paper production. Bum. Prom-St. 7, 9-10.

Benitez T., Villa T.G. & Bartnicki-Garcia S.: Effects of polyoxin-D on germination, morphological development and biosynthesis of cell wall of *Trichoderma viride*. Arch. Microbiol. 108, 183-188.

Bough W.A.: Coagulation with chitosan: aid to recovery of by-products from egg breaking wastes. Poultry Sci. 54, 1904-1912.

Bough W.A.: Chitosan, a polymer from sea food wastes, for use in treatment of food processing wastes and activated sludge. Process Bioch. 11, 13-16.

Bough W.A., Landes D.R., Miller J., Young C.T. & McWhorter T.R.: Utilization of chitosan for recovery of coagulated by products from food processing wastes and treatment systems. Proc. Sixth Natl. Symp. Food Process. Wastes, 1976.

Bourne B.J.: Industrial effluent treatment. Physico-chemical and biochemical options. 1. Removal of suspended matter and dissolved inorganic matter. Eff. Water Treat. 16, 455-464.

Bracker C.E., Ruiz-Herrera J. & Bartnicki-Garcia S.: Structure and transformation of chitin synthetase particles (chitosomes) during microfibril synthesis in vitro. (Original results)

Cherkasov I.A. & Kravchenko N.A.: Specifics of complexing lysozyme with chitosaccharides at pH 8.4. B. Acad. Sci. 24, 2511.

Chigaleichik A.G., Pirieva D.A. & Rylkin S.S.: Chitinase from *Serratia marcescens* BKM B-851. Prikl. Biokhim. Mikrobiol. 12, 581-586.

Chigaleichik A.G., Pirieva D.A. & Rylkin S.S.: Biosynthesis of chitinase by an *Achromobacter liquefaciens* culture. Mikrobiol. 45, 475-480.

Dahn U., Hagenmaier H., Hohne H., Konig W.A., Wolf G. & Zahner H.: Metabolic products of microorganisms. 154. Nikkomycin, a new inhibitor of fungal chitin synthesis. Arch. Micrbobiol. 107, 143-160.

Elsabee M.Z., Mattar S.M. & Habashi G.M.: Bonding between copper complexes with celluloses and related carbohydrates. J. Polymer Sc. C-14, 1773-1781.

Ernst G., Emeric D.A. & Levine S.: Chemical analysis for chitin as a measure of fungal infiltration of cellulose materials. Text. Res. J. 46, 616-619.

Fange R., Lundblad G. & Lind J.: Lysozyme and chitinase in blood and lymphomyeloid tissues of marine fish. Marine Biol. 36, 277-282.

Farkas V.: Regulation of biosynthesis of chitin during cell cycle of *Saccharomyces cerevisiae*. Folia Microbiol. 21, 193.

Florence T.M. & Batley G.E.: Trace metal species in sea water. 1. Removal of trace metals from sea water by a chelating resin. Talanta, 23, 179-186.

Gindrat D.: Components in unbleached commercial chitin stimulate *Pythium ultimum* in sugarbeet spermosphere. Phytopathol. 66, 312-316.

Grosscurt A.C.: Ovicidal effects of diflubenzuron on the housefly *Musca domestica*. Proc. Int. Symp. Crop Protection, Gent.

Hackman R.H. & Goldberg M.: Comparative chemistry of arthropod cuticular proteins. Comp. Bioch. Biophys. 55, 201-206.

Haleem M.A. & Parker K.D.: X-ray diffraction studies on the structure of α-chitin. Z. Naturforsch. 31-C, 383-388.

Hattis D. & Murray A.E.: Industrial prospects for chitin from seafood wastes. MIT Center of Policy Alternatives, Sea Grant Office MIT Cambridge, Report 27 August 1976.

Hepburn H.R. (Ed.): The Insect Integument. Elsevier, Amsterdam

Hettick B.P. & Stevenson J.R.: Metabolism of chitin precursors in crayfish. J. Cell. Biol. 70, A-74.

Hettick B.P. & Stevenson J.R.: Quantitative determination of chitin intermediate pools during incorporation of N-acetylglucosamine-^{14}C and glucose-^{3}H in crayfish. Am. Zool. 16, 237.

Hirano S., Ohe Y. & Ono H.: Selective N-acylation of chitosan. Carbohydr. Res. 47, 315-320.

Itaya M.: Study of the polyvinylalcohol fibers blended with chitosan. (Original results).

Kikuchi Y. & Fukuda H.: Polyelectrolyte complex of sodium dextransulfate with chitosan. Nippon Kagaku Kaishi 1976, 1505-1508.

Kikuchi Y. & Noda A.: Polyelectrolyte complex of heparin with chitosan. J. Appl. Polvmer 20, 2561-2563.

Kikuchi Y. & Toyota T.: Polyelectrolyte complexes of sodium dextransulfate with chitosan and their thrombous formation. Hiroshima Daigaku Kogakubu Kenkyu Hokoku 24, 7-9.

1976 CNTD

Kirinuki T., Ofuji H. & Suzuki N.: Chitinolytic and α-1,3-glucanase activity of Actinomycetes and their effects on preventing *Fusarium* in cucumbers. Kobe Daigaku Nogakubu Kenkyu Hokoku 12, 41-48.

Kohn P.M.: Shellfish wastes vie for CPI role. Chem. Engin. 1976, Sept. 13, 107-109.

Kratky Z., Vrsanka M. & Biely P.: Localization of synthesis of chitin during cell-cycle of *Saccharomyces cerevisiae*. Folia Microbiol. 21, 193.

Kreger D. & Kopecka M.: On the nature and formation of fibrillar nets produced by protoplast of *Saccharomyces cerevisiae* in liquid media: an electronmicroscopic, x-ray diffraction and chemical study. J. Gen. Microbiol. 92, 207-220.

Landes D.R. & Bough W.A.: Effects of chitosan, a coagulating agent for food processing wastes in the diet of rats on growth and liver and blood composition. Bull. Environ. Contam. Toxicol. 15, 555-563.

Linn C.C., Sicher R.C. & Aronson J.M.: Hyphal wall chemistry in *Apodachlya*. Arch. Microbiol. 108, 85-91.

Lukin O.V., Vigdorchik M.M. & Turchin K.F.: N-Acylation of glucosamine by 3-indolylacetic acid. Zhur. Org. Khim. 12, 565-567.

Lysenko O.: Chitinase of *Serratia marcescens* and its toxicity to insects. J. Invertebr. Pathol. 27, 385-386.

Marks E.P. & Sowa B.A.: Cuticle formation in vitro, in The Insect Integument, Hepburn H.R., Ed., Elsevier, Amsterdam.

Matsuda H., Kanki K. & Shinoda K.: Membrane for brine desalination. Japan 6879.

Matsuda H., Kanki K. & Nakagawa T.: Alkyl derivatives of chitin. Japan 68,689.

Michalenko G.O., Hohl H.R. & Rast D.: Chemistry and architecture of the mycelial wall of *Agaricus bisporus*. J. Gen. Microbiol. 92, 251-262.

Mo F. & Jensen L.H.: Structures of the α- and β-anomers of chitobiose. (Original results).

Moore P.M. & Peberdy J.F.: Particulate chitin synthase from *Aspergillus flavus Link*. Properties, location and levels of activity in mycelium and regenerating protoplast preparations. Can. J. Microbiol. 22, 915-921.

Morris O.N.: Two-year study of efficacy of *Bacillus thuringiensis* chitinase combination in spruce budworm (*Choristoneura fumiferana*) control. Can. J. Entomol. 108, 225-233.

Murayama K., Shindo N. & Koide H.: A modification of determination for glucosamine and galactosamine in glycoprotein with the amino acid analyzer. Application to total acid hydrolyzate of rat renal glomerular basement membrane. Anal. Bioch. 70, 537-541.

Muzzarelli R.A.A.: Biochemical Modifications of Chitin, in The Insect Integument, Hepburn H.R., Ed., Elsevier, Amsterdam.

Muzzarelli R.A.A., Barontini G. & Rocchetti R.: Immobilization of enzymes on chitosan columns. Acid phosphatase and α-chymotrypsin. Biotech. Bioengin., in press.

Neville A.C., Parry D.A.D. & Woodhead-Galloway J.: The chitin crystallite in arthropod cuticle. J. Cell. Sci. 21, 73-82.

National Oceanographic & Atmospheric Administration, N.M.F.S.: Ecuadorian shrimp industry prospers. Colombian shrimp catch and export reviewed. Marine Fish. Rev. 38(5), 37-38.

Ogata N.: Developemnt of special polymers. Kagaku Kojo 20, 28-31.

Philips-Duphar B.V.: Dimilin, a new insecticide interfering with chitin deposition. Technical Information.

Phillips D.R. & Loughton E.G.: Cuticle protein in *Locusta migratoria*. Comp. Bioch. Biophys. 55, 129-135.

Rudall K.M.: Molecular structure in arthropod cuticles, in The Insect Integument, Hepburn H.R., Ed., Elsevier, Amsterdam.

Schlotzhauer W.S., Chortyk O.T. & Austin P.R.: Pyrolysis of chitin, a potential tobacco extender. J. Agr. Food. Chem. 24, 177-180.

Selitrennikoff C.P., Allan D. & Sonneborn D.R.: Chitin biosynthesis during *Blastocladiella* zoospore germination: evidence that the hexosamine biosynthetic pathway is post-translationally activated during cell differentiation. Proc. Nat. Acad. Sci. 73, 534-538.

Sjogren R.D.: Preliminary evaluation of capsules containing a chitin synthesis inhibitor. Proc. Int. Symp. Controlled Release Pesticides, 217-221.

Spindler-Barth M.: Changes in the chemical composition of che common shore crab *Carcinus mae₁ ⃝* during the molting cycle. J. Comp. Physiol. 105, 197-205.

Stanely W.L., Watters G.G., Kelly S.H., Chan B.G., Garibaldi J.A. & Schade J.E.: Immobilization of glucose isomerase on chitin with glutaraldehyde and by simple adsorption. Biotech. Bioengin. 18, 439-443.

Stephens N.L., Bough W.A., Beuchat L.R. & Haton E.K.: Preparation and evaluation of two microbiological media from shrimp heads and hulls. Appl. Environ. Microbiol. 31, 1-6.

Stephens R.M. & Vincent J.F.V.: Infrared spectroscopy of locust extensible intersegmental cuticle and its components. J. Insect Physiol. 22, 601-605.

Takiguchi Y., Nagahata N. & Shimahara K.: A new method for chitinase assay using 6-O-hydroxy-propylchitin. J. Agr. Chem. Japan 50, 243-244.

Tiunova N.A., Pirieva D.A. & Feniksova R.V.: Production of chitinase by Actinomycetes during submerged cultivation. Mikrobiol. 45, 280-283.

1976 CNTD

Toei K. & Kohara T.: A conductometric method for colloid titration. Anal. Chim. Acta 83, 59-65.

Tokyo Tanabe Co., Ltd.: Alkylaminoglucopyranoside derivatives. Ger. 2,530,416.

Ulane R.E. & Cabib E.: The activating system of chitin synthetase from *Saccharomyces cerevisiae*. Purification and properties of the activating factor. J. Biol. Chem. 251, 3367-3374.

Van Laere A.J., Carlier A.R. & Van Assche J.A.: Effect of 5-fluorouracil and cyclo heximide on early development of *Phycomyces blakesleeanus* spores and activity of N-acetylglucosamine synthesizing enzymes. Arch. Microbiol. 108, 113-116.

Verloop A. & Ferrell C.D.: Benzoylphenyl ureas, a new group of larvicides interfering with chitin deposition. Proc. Symp. Pesticides 20th Century, New York, ACS, New York.

Welinder B.S., Roepstorff P. & Andersen S.O.: Crustacean cuticle. 4. Isolation and identification of cross-links from *Cancer pagurus* cuticle. Comp. Bioch. Physiol. 53-B, 529-533.

Whistler R.L. & BeMiller J.N., Eds.: Methods in Carbohydrate Chemistry, Vol. 7, Academic Press, New York.

Yamada M. & Miyazaki T.: Ultrasctructure and chemical analysis of cell wall of *Pythium debaryanum*. Jap. J. Microbiol. 20, 83-91.

Yoshinari T. & Subramanian V.: Adsorption of metals by chitin, in Environmental Biogeochemistry, Vol. 2, Metal Transfer and Ecological Mass Balance, Nriagu J.O., Ed., Ann Arbor Science Publ., Ann Arbor.

NOTE. The articles listed under 1976 are only those known to the author before September 30, 1976.

SUBJECT INDEX

Acetolysis, 174
N-Acetylglucosamine, 10, 167, 173,
    176, 255
N-Acetylmuramic acid, 167, 255
N-Acylation, 75, 132
Actinomycetes, 265
Acids, 108
Agglutination, 138
Aldehydes, 134, 226
Alkali chitin, 101, 124, 129
N-Alkyl derivatives, 126
Alkyl ethers, 125
Alginic acid, 137
Amino acids, 187
Analytical determinations, 150
Anhydrides, 134, 226
Anion-exchange, 145
Antibiotics, 259
Arthropods, 8, 178
Artificial kidney membrane, 257

Bacteria, 180, 252, 256
Batik dyeing, 235
Blood anticoagulants, 115, 260

Carbonyl groups, 131
Carboxyl groups, 113
Carboxymethylation, 120
Carboxymethylchitin, 78, 120, 259
Carotenoids, 24
Cements, 244
Chelating polymers, 147
Chelation, 139
    - chromatography, 193, 214
Chitin
    - alcoholized, 130
    - associations, 23
    - colloidal, 57, 110, 161
    - conformation, 45, 59, 110
    - fibers, 228
    - industrial production, 207
    - microfibrils, 22, 51
    - occurrence, 5
    - polymorphism, 45
    - preparation, 89

    - radioactive, 93
    - synthesis, 5, 10, 22, 38, 43, 44
    - viscose, 228
Chitinase, 11, 27, 44, 155, 176
    - occurrence, 177
    - production, 172
Chitin deacetylase, 30, 32
Chitin synthetase, 13, 17, 35, 53
Chitin xanthate, 76
Chitobiase, 12, 155, 177
Chitobiose, 12, 45
Chitosan, 62, 94
    - acyl, 132
    - fibers, 229
    - formylated, 116
    - industrial production, 207
    - membranes, 76
    - methylated, 125
    - papermaking, 220
    - polymolybdate, 107
    - potentiometric titrations, 105
    - salts, 103, 115, 235
    - sulfated, 115, 260
    - test, 151
Chitosanase, 164
Chitosomes, 23
Chromatography, 183, 193, 204
Chromium trioxide, 113
Coagulants, 247
Colloid titrations, 153
Coloring matters, 110
Conformation, 47, 50
Copper, 216
Cyanoethylchitosan, 120

DDT, 37, 43
Deacetylation, 94, 152, 175
Deacetylase, 33
Deamination, 112, 151
Depolymerization, 113, 117
Determination, 150
    - in cellulose, 152
    - in fuel, 152
    - in wheat, 132
Dextran sulfate, 137